The Paths of Plant Pathogens—Interactions with Host and Nonhosts and Insight of Mechanisms of Pathogenesis

The Paths of Plant Pathogens—Interactions with Host and Nonhosts and Insight of Mechanisms of Pathogenesis

Guest Editors

Yoana Kizheva
Petya Hristova

MDPI

Basel • Beijing • Wuhan • Barcelona • Belgrade • Novi Sad • Cluj • Manchester

Guest Editors
Yoana Kizheva
Department of General and Industrial Microbiology
Sofia University St. "Kliment Ohridski"
Sofia
Bulgaria

Petya Hristova
Department of General and Industrial Microbiology
Sofia University St. "Kliment Ohridski"
Sofia
Bulgaria

Editorial Office
MDPI AG
Grosspeteranlage 5
4052 Basel, Switzerland

This is a reprint of the Special Issue, published open access by the journal *Pathogens* (ISSN 2076-0817), freely accessible at: https://www.mdpi.com/journal/pathogens/special_issues/J8S3RI6SO8.

For citation purposes, cite each article independently as indicated on the article page online and as indicated below:

Lastname, A.A.; Lastname, B.B. Article Title. *Journal Name* **Year**, *Volume Number*, Page Range.

ISBN 978-3-7258-5589-6 (Hbk)
ISBN 978-3-7258-5590-2 (PDF)
https://doi.org/10.3390/books978-3-7258-5590-2

Contents

About the Editors

Yoana Kizheva

Yoana Kizheva, Associate Professor, PhD at Sofia university of St. Kliment Ohridski, Faculty of biology. Main areas of scientific expertise: phytopathogenic bacteria (biology, species diversity and distribution on sensitive and alternative hosts, interactions with plants and molecular aspects of pathogenicity), integrated plant disease management strategies, microorganisms as biocontrol agents, phage therapy and phage biocontrol, phage biology and phage–bacteria interactions. Member in numerous national and international scientific projects as well as European scientific networks dedicated mostly to phytopathogenic bacteria. Supervisor of 20+ bachelor's and master's theses in the field of phytopathogenic bacteria and bacteriophages.

Petya Hristova

Petya Hristova is a Professor of food microbiology in the Faculty of Biology, Sofia University of St. Kliment Ohridski. Her main research interests are cross-over pathogens as new biological hazards in plant foods, the identification of phytopathogenic bacteria, the pathogenicity and virulence of enterococci, antibiotic resistance and molecular mechanisms of transmission, biology and the diversity of lactic acid bacteria. She has published more than 70+ articles, one monograph and two textbooks for students. She has participated in numerous scientific and educational projects, and is the supervisor of more than 50+ graduate and doctoral students.

Preface

As Guest Editors we have the pleasure of presenting a curated collection of in-depth studies with high scientific quality, tracing the path of pathogenicity in different biological kingdoms caused by pathogenic microbes, viruses and parasitic plants. Plant diseases are considered a major problem leading to great yield and economical loses worldwide. Of serious interest during recent decades are the interaction between phyto- and animal cross-over pathogens with their susceptible and alternate hosts and their ability to jump between them. Moreover, the role of some parasitic plants and insects to act as vectors for their distribution is another concerning issue that warrants extensive study.

Gaining and summarizing knowledge about the molecular mechanisms of pathogenesis is essential for identifying and managing potential environmental threats. Additionally, it is proposed that animal pathogens inhabit plants as an obligatory stage of their life cycle. Thus, can the plant environment play a key role in modulating pathogenicity? Can the environmental factors stimulate the pathogens to switch to a pathogenic cycle even if they are not in their natural susceptible host and vice versa?

The main purpose of this Reprint is to collect recent data concerning the prevalence of cross-over pathogenic bacteria in plant tissue, the role of parasitic plants and insects in the distribution of phytopathogens, the key role of environmental factors in modulating plant defense systems, and, last but not least, the crucial role of microorganisms (bacteria and fungi) as biocontrol agents against severe pathogens. One of the main strengths of this selection is its detailed discussion of plant resistance at the molecular level, including pathogenicity, virulence factors, and regulatory systems. Moreover, increasing virulence and pathogenicity factor content has been tracked in certain cross-kingdom bacterial pathogens (enterococci), spanning biological kingdoms from plants to animals.

This Reprint, consisting of seven original research articles and two comprehensives reviews, could serve as a scientific guideline for researchers, clinicians, agronomists, phytopathologists, educators and any stakeholders who are highly involved in the field of plant diseases and the development of innovative approaches for their effective management. The accumulated and summarized knowledge on the molecular mechanisms of plant defense systems and the pathogenicity of different pathogens could serve as solid basis in this field.

We want to express our sincere gratitude to all authors who chose to publish their work with *Pathogens* and in this Special Issue. Our sincere thanks goes to the Editorial team for their continuous support, and to the reviewers, whose professionalism and insightful feedback have substantially elevated the quality of the publications. We sincerely hope that this Reprint will inspire new ideas and promote meaningful collaboration within the scientific community.

Yoana Kizheva and Petya Hristova
Guest Editors

pathogens

MDPI

Article

Characterization and Genomics of Pectinolytic Bacteria Isolated from Soft Rot Symptomatic Produce

Kyla Radke, Brandon Rivers, Mya Simpkins, Jacob Hardy and Jeffrey K. Schachterle *

Department of Microbiology and Molecular Biology, Brigham Young University, Provo, UT 84602, USA; kr539@byu.edu (K.R.)
* Correspondence: jeffrey_schachterle@byu.edu

Abstract: Bacterial soft rot causes major crop losses annually and can be caused by several species from multiple genera. These bacteria have a broad host range and often infect produce through contact with soil. The main genera causing bacterial soft rot are *Pectobacterium* and *Dickeya*, both of which have widespread geographical distribution. Because of many recent renaming and reclassifications of bacteria causing soft rot, identification and characterization of the causative agents can be challenging. In this work, we surveyed commercially available produce exhibiting typical soft rot symptoms, isolating pectinolytic bacteria and characterizing them genetically and phenotypically. We found that in our sampling, many samples were from the genus *Pectobacterium*; however, other genera were also capable of eliciting symptoms in potatoes, including an isolate from the genus *Chryseobacterium*. Genomic analyses revealed that many of the *Pectobacterium* isolates collected share prophages not found in other soft rot species, suggesting a potential role for these prophages in the evolution or fitness of these isolates. Our *Chryseobacterium* isolate was most similar to *C. scophthalmum*, a fish pathogen, suggesting that this isolate may be a crossover pathogen.

Keywords: soft rot; pectinolytic; *Pectobacterium*; *Pseudomonas*; *Chryseobacterium*

1. Introduction

Bacterial soft rot is a disease complex that causes severe crop loss worldwide. Multiple bacterial genera cause this disease, with the genera *Pectobacterium* and *Dickeya* being among the most common [1]. These pathogens leave necrotic spots on the soft tissues of many plants, such as stems, leaves, tubers, and flesh, of a wide variety of cultivated vegetables. Soft rot is usually observed as the gradual presence of water-soaked lesions where the tissue becomes depressed, soft, mushy, slimy, and discolored [2]. This results in significant yield losses, affecting both the pre-harvest period and the post-harvest period due to latent infection [3]. Soft rot can occur over a wide temperature range, with most pathogens infecting optimally in the range of 21–27 °C, and is most severe when oxygen is limited [4]. Crops in storage, such as potato tubers, may have restricted access to oxygen when stacked on top of each other with limited airflow [5]. *Pectobacterium* and *Dickeya* survive in the soil and on the surface of crops, infecting plants through natural openings and wound sites [6]. Currently, there are no effective treatments once soft rot has infected plant tissues.

Bacteria of the *Pectobacteriaceae* family release a variety of pectinases as exoenzymes and endoenzymes for pectin degradation [7]. These enzymes break down pectin, a polysaccharide present in the middle lamella of plant cell walls, allowing access to a nutrient-rich environment [8]. Pectinases are classified into hydrolases, esterases, and lyases based on their active sites. Hydrolases, also known as polygalacturonases (PGs), break down pectic substances by catalyzing the hydrolysis of α-1,4-glycosidic bonds between galacturonic monomers that form chains in the pectin molecule. Esterases, also known as pectin methylesterases (PMEs), remove the acetyl and methoxyl groups from pectin. Lyases catalyze the cleavage of glycosidic bonds through β elimination mechanisms [9]. These enzymes have been isolated and used for industrial applications [10]. When *Pectobacteriaceae* bacteria start to infect the plant and grow quickly in numbers, an overwhelming amount of pectinase enzymes are produced, which results in soft rot. Pectin binds plant cells together, thus causing plant structures to fall apart into mushy, discolored, and sunken pits when infected. Members of the *Pectobacterium* and *Dickeya* genera are rod-shaped, facultative anaerobes and are motile by means of peritrichous flagella [11].

Although *Pectobacterium* and *Dickeya* are perhaps the most established genera associated with soft rot [12], a variety of other species have been implicated, including *Pseudomonas, Bacillus,* and *Clostridium. Pseudomonas*, a large and diverse group of bacteria [13], has been identified through 16S rDNA sequencing as a causative agent of soft rot in potatoes [14]. The versatile, spore-forming genus *Bacillus* has also exhibited pathogenicity on apples, pears, and other produce [15]. However, it has also been proposed as a potential biological control method against the soft rot pathogen *Pectobacterium carotovorum* [6]. *Clostridium*, a spore-forming, primarily anaerobic genus, has been isolated from symptomatic sweet potatoes and other crops [16].

As classification techniques have become more sophisticated and accurate, methods of genome sequencing, such as whole-genome sequencing, have developed, leading to the necessity of the reclassification of previously identified species, including those causing soft rot [17]. For example, the once broad genera *Erwinia* became divided into subcategories *Pectobacterium* and *Dickeya* through 16s rRNA comparative analysis [17,18]. This is a critical differentiation in the field of soft rot and plant disease because *Pectobacterium* and *Dickeya* are distinct in their host range and pathogenicity, making proper classification crucial for studying and understanding the virulence mechanisms of pathogenic bacteria. Sophisticated classification efforts will almost certainly yield the identification of novel and previously unknown pathogenic bacteria contributing to soft rot [19]. However, as more novel bacteria are identified, the demand for increasingly advanced molecular techniques will rise to ensure accurate and rapid identification. Technology that is more expensive and less widely available will potentially be needed to make precise classifications between isolates of high similarity, complicating soft rot research endeavors in lower-resource areas and certain laboratories.

Additionally, there may be regulatory concerns implicated with advancements in diagnostic technology, especially in the field of agriculture. In August 2019, the U.S. Department of Agriculture ceased requiring permits for the interstate transfer of *Pectobacterium carotovorum* strains originating from within the continental United States (Federal Register #2019-13246). However, two months later, a novel species, *P. versatile*, was formed from isolates formerly considered *P. carotovorum* [20]. Although the ongoing refinement and revision of pathogen taxonomy can create improved classification and understanding of pathogen behavior, it can also foster confusion around regulatory requirements for newly formed groups. Rapidly changing taxonomy also creates differences in the reporting of diagnostic results from different laboratories, which limits the understanding of which pathogens are present in specific geographic areas.

In our study, we surveyed pectinolytic bacteria in commercial produce and characterized the isolates obtained via both genotypic and phenotypic methods. As expected, we identified *Pectobacterium* species; however, we also found isolates from several other genera, including isolates pathogenic to potatoes that represent potentially novel species. We further found an isolate of *Chryseobacterium scophthalmum*, closely related to fish pathogens [21,22], which may represent a crossover pathogen. While the *Pectobacterium* isolates were highly similar, phenotypic differences and distinct prophage profiles revealed differences between strains. Our survey advances our understanding of the pathogens currently causing soft rot in consumer-available produce and provides a foundation for further work into the mechanisms that make these pathogens successful.

2. Materials and Methods

2.1. Culture and Isolation Conditions

Unless otherwise specified, bacterial isolates were routinely cultured at 30 °C in LB medium (10 g/L tryptone, 5 g/L NaCl, 5 g/L yeast extract; for solid media: 15 g/L agar). Produce samples used in isolations were selected from commercially acquired produce and exhibited symptoms commonly associated with bacterial soft rot, primarily water-soaked tissues and sunken lesions. Samples used for isolation were all collected in 2024 and are specified in Supplemental Table S1. Because the produce was commercially obtained from consumer grocery stores, data regarding the location of cultivation or the specific varieties of plants are unknown.

Original isolations of bacteria from symptomatic plant tissues were conducted using crystal violet pectate medium (CVPM) [23]. Briefly, an inoculation loop was used to scrape water-soaked tissue from the plant sample and streaked for isolation on CVPM plates. Colonies forming sunken pits on CVPM were re-isolated using CVPM until pure cultures were obtained.

Genetic identification of the bacteria was conducted via Sanger sequencing of the 16S rDNA gene. Briefly, a fragment of the 16S rDNA gene was PCR-amplified using primers 27F and 1492R [24]. PCR fragments were purified using ethanol precipitation and digestion with exonuclease I and alkaline phosphatase (New England Biolabs, Ipswich, MA, USA). Sanger sequencing reactions and capillary electrophoresis of purified DNA fragments were conducted by the BYU DNA Sequencing Center (Provo, UT, USA) with either the 27F or 1492R primers. A BLAST [25] of the obtained sequences against the NCBI non-redundant database was used to determine the genus and, when possible, the predicted species of the isolate, taking the top BLAST hit as the putative identity of the isolate.

2.2. Phenotypic Characterizations

All pectinolytic isolates were tested using the following phenotypic tests: pathogenicity to potato and Chinese (Napa) cabbage, swimming motility, swarming motility, production of acyl homoserine lactones (AHLs), lactose fermentation, and protease production.

2.2.1. Pathogenicity Assays

For potato pathogenicity, 1 cm slices of russet potatoes were prepared and placed in sterile petri dishes. Potato inoculum was prepared by growing each isolate overnight in LB and normalizing the culture density to OD600 = 0.4. A 2 μL droplet (~2×10^6 CFU) was inoculated on the center of each potato slice. For Chinese (Napa) cabbage pathogenicity, cabbage leaves were placed in plastic totes cleaned with 70% ethanol. For cabbage inoculations, bacterial isolates were grown for 24 h on LB agar, and single colonies were picked with a pipette tip and inoculated by stabbing the tip 2 mm into the surface of the cabbage leaf. Inoculated produce was incubated at 30 °C and evaluated for symptom development every 24 h for 3 days. Symptoms included water soaking, tissue maceration, and discoloration. Potato slices were photographed and analyzed using ImageJ [26] to determine the surface area covered by disease 48 h post-inoculation.

2.2.2. Swimming and Swarming Motility

Swimming and swarming motility assays were conducted as described previously [27]. In short, soft agar plates containing 0.25% agar (swim) or 0.3% agar (swarm) were prepared. Swim plates were inoculated by stab inoculation, and swarm plates were surface inoculated with 3 μL of overnight culture. Plates were photographed 24 h post-inoculation, and the area covered by motile bacteria was quantified using ImageJ [26].

2.2.3. AHL Production

Acyl homoserine lactone production was assessed as described [28,29] using the *Agrobacterium fabrum* reporter strain NTL4 pZLR4. Briefly, soft agar plates containing the *A. fabrum* reporter as well as X-gal (5-bromo-4-chloro-3-indolyl-beta-D-galactopyranoside) were prepared. A 3 μL droplet of overnight culture of each isolate was dripped onto a plate containing the reporter strain and incubated overnight. A positive reaction was evidenced by the production of a blue color in the soft agar surrounding the test strain.

2.2.4. Lactose Fermentation

LB agar medium was prepared to contain 1% (*w*/*v*) lactose and phenol red dye as a pH indicator. At neutral pH, the media is a bright red color. When isolates were grown on this media, a color change from red to yellow was interpreted as a lowering of the pH in the media around colonies fermenting lactose.

2.2.5. Extracellular Protease Activity

LB agar containing 1% (*w*/*v*) skim milk powder was prepared, and isolates were cultured in this medium. A halo of clearing around colonies was taken as evidence of extracellular protease activity.

2.2.6. Biofilm Formation

Biofilm formation by each isolate was assayed using 96-well plates as described previously [30]. Briefly, isolates were grown overnight in LB broth. A 5 μL volume of overnight culture, standardized to an absorbance of 0.4, was added to 100 μL of LB in the well of a 96-well plate, and the plate was incubated at 30 °C for 48 h. Following incubation, media and planktonic cells were removed and plates were allowed to dry. Adherent cells were stained with 100 μL of 1% crystal violet, and cells were washed three times with deionized water. The dye was solubilized using 70% ethanol, and the absorbance of the solubilized dye was measured at 595 nm.

2.3. Genome Sequencing

For genome sequencing, select bacterial isolates were cultured in LB medium overnight at 30 °C. Genomic DNA was isolated from 5 mL of culture using the Quick DNA Miniprep Plus Kit (Zymo Research, Irvine, CA, USA) according to the manufacturer's instructions. Purified DNA sample concentration and quality were verified spectrophotometrically and by agarose gel electrophoresis. Samples were submitted to Plasmidsaurus (Eugene, OR, USA) for sequencing using long-read sequencing and assembly approach. All genome sequences generated are available on NCBI under BioProject number PRJNA1183984.

2.4. Computational Analyses

Reference genomes of all type strains from the following genera were obtained from the NCBI GenBank database: *Pectobacterium*, *Dickeya*, *Pseudomonas*, and *Chryseobacterium*. Where type strains were used, all GenBank genomes represent taxonomically named species. The average nucleotide identity between these type strains and our isolates with sequenced genomes was calculated using fastANI [31]. Further comparisons were made using additional genomes from the species *Pectobacterium carotovorum* and *Pectobacterium versatile*. Whole genome alignments were generated and visualized using the progressive-Mauve aligner [32] with default parameters.

Lysogenic bacteriophage within the sequenced *Pectobacterium* genomes were identified using Phastest [33]. Extracted sequences of identified lysogenic bacteriophages were further analyzed using fastANI [31]. The Newick tree was generated with ETEToolkit [34] for visualization.

3. Results

3.1. Isolation and Identification

We collected commercial produce samples exhibiting typical soft rot symptoms and plated them onto CVPM to isolate pectinolytic bacteria associated with these samples (Figure S1). From 47 produce samples, we collected 28 isolates of pectinolytic bacteria (Table S1). When apparently distinct isolates were obtained from a single sample, these isolates were designated 'a' or 'b' isolates. We used 16S rDNA sequencing to identify the genus and species of the pectinolytic isolates, when possible, and found isolates from nine different genera: *Achromobacter*, *Chryseobacterium*, *Escherichia*, *Lelliottia*, *Pantoea*, *Pectobacterium*, *Pseudomonas*, *Sphingobacterium*, and *Stenotrophomonas* (Tables 1 and S2).

Table 1. Pectinolytic isolates, their genetic identities, and phenotypic characteristics.

Isolate	Isolation Host	Genus/Species	Pathogenic to Potato	Pathogenic to Chinese (Napa) Cabbage)	Swim	Swarm	AHL Production	Lactose	Protease
M1	Onion	not determined	−	−	+	+	−	+	−
M2	Lettuce	*Pseudomonas* sp.	−	−	+	+	−	−	+
M3	Lettuce	*Pectobacterium carotovorum*	+	+	+	+	+	+	+
M4a	Lettuce	*Pectobacterium carotovorum*	+	−	+	+	+	+	+
M4b	Lettuce	not determined	−	−	+	+	−	−	+
M5	Lettuce	*Chryseobacterium indoltheticum*	+	−	+	+	−	+	+
M6	Lettuce	*Pseudomonas cedrina*	−	−	+	+	−	−	+
M7	Cilantro	*Pectobacterium carotovorum*	+	−	+	+	+	+	+
M8a	Cilantro	*Pectobacterium carotovorum*	+	−	+	+	+	+	+
M8b	Cilantro	*Pectobacterium carotovorum*	+	−	+	+	+	+	+
M9	Cilantro	*Lelliottia amnigena*	−	−	+	+	−	−	+
M10	Cilantro	*Pectobacterium carotovorum*	+	+	+	+	+	+	−
M11	Spinach	*Pseudomonas azotoformans*	−	−	+	+	−	−	+
M12a	Zucchini	*Pseudomonas allopoutida*	−	−	+	+	−	−	−
M12b	Zucchini	*Escherichia fergusonii*	−	−	+	+	−	−	−
M13	Onion	*Pseudomonas marginalis*	+	−	+	+	−	−	+
M14	Onion	*Pseudomonas petroselini*	−	−	+	+	−	+	+
M15	Potato	*Stenotrophomonas maltophila*	−	−	+	+	−	−	+
M16	Potato	not determined	−	−	+	+	−	+	+
M17	Spinach	*Achromobacter spanius*	−	−	+	+	−	−	+
M18	Spinach	*Pseudomonas cyclaminis*	+	−	+	+	−	−	+
M19	Spinach	*Stenotrophomonas rhizophila*	−	−	+	+	−	−	+
M20	Spinach	*Pseudomonas* sp.	+	−	+	+	−	−	+
M21	Celery	*Pseudomonas composti*	−	−	+	+	+	+	+
M22a	Celery	*Stenotrophomonas maltophila*	−	+	+	+	−	−	+
M22b	Celery	*Pectobacterium carotovorum*	+	−	+	+	+	+	+
M23	Carrot	*Pseudomonas grimontii*	−	−	+	+	−	−	+
M24	Carrot	*Pantoea* sp.	+	−	+	+	+	−	+

+ represents a positive reaction for each phenotype, and − represents a negative reaction for the phenotype.

3.2. Phenotypic Characterization

Because each of our isolates was collected from plant tissue exhibiting soft rot symptoms, we tested each of the isolates against potato and Chinese (napa) cabbage to determine whether these isolates were pathogenic to these hosts (Figures S2 and S3). We found that only 12 of the 28 isolates (43%) were pathogenic to potatoes, and only three (11%) were pathogenic to Chinese (napa) cabbage (Table 1). The *Pectobacterium* isolates, which were mainly isolated from different hosts, were all pathogenic to potatoes as they caused soft rot symptoms as they grew on the potatoes. We observed a high degree of variability in the size of lesions caused by our different isolates when used to inoculate potato slices (Figure 1A).

We assessed whether each of the 28 isolates exhibited various phenotypes associated with virulence, including swimming and swarming motility, biofilm formation, production of acyl-homoserine lactones, fermentation of lactose, and extracellular protease activity (Table 1). All of the isolates exhibited both swimming and swarming motility. All of the isolates appeared to form biofilms, with crystal violet staining higher than the negative control (Figure 1B). Several strains appeared to form strong biofilms with staining 10-fold higher than the negative control; however, strain M5 exhibited the lowest biofilm formation of all the isolates, at only about twice the crystal violet staining of the negative control. Nine of the isolates (32%) produced acyl homoserine lactones detectable with our *A. fabrum* reporter system [28,29]. Twelve of the isolates (43%) fermented lactose, and 24 of the isolates (86%) exhibited extracellular protease activity.

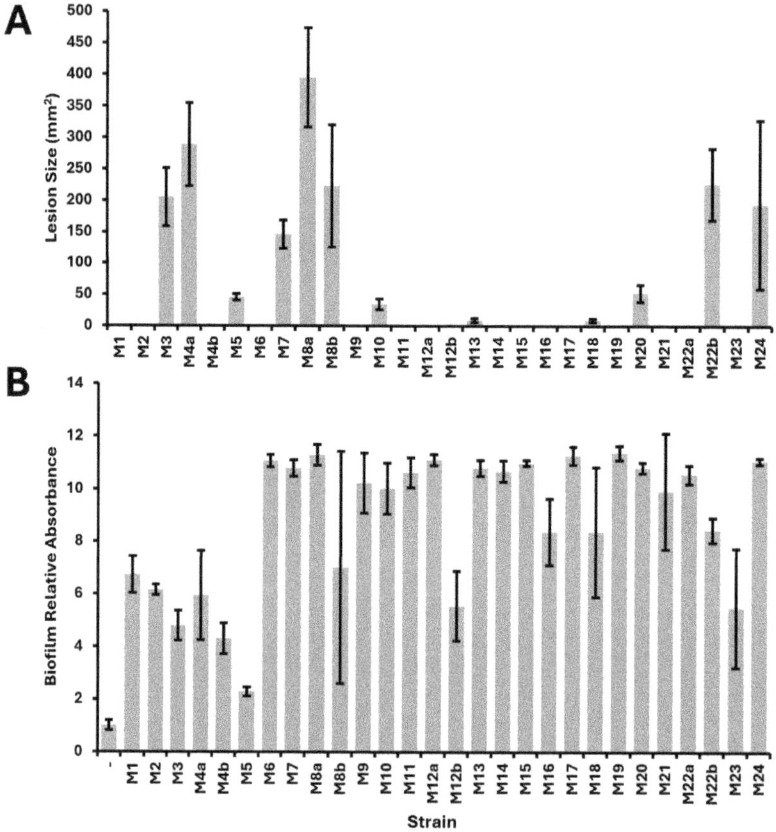

Figure 1. Virulence of isolates as measured by area of lesion on inoculated potato slices (**A**) and biofilm formation in 96-well plates following static growth for 48 h (**B**).

We used linear regression and computed F-statistics for pairwise correlation between the pathogenicity-to-potato trait and the other virulence-associated phenotypic traits we tested. The additional traits included motility, lactose fermentation, production of acyl homoserine lactones, and extracellular protease activity. We found a significant pairwise correlation ($p < 0.001$) between the production of acyl homoserine lactones and pathogenicity to potatoes. The correlations for all other virulence traits were not statistically significant.

3.3. Genome Sequencing and Assemblies

We selected a subset of 11 isolates obtained in our survey for whole genome sequencing, with a focus on isolates pathogenic to potatoes. The isolates sequenced and sequencing results are summarized in Table 2. The average sequencing depth was 76× across all samples. All of the *Pectobacterium* and the *Chryseobacterium* genomes were assembled to a single contig with no plasmids. The *Pectobacterium* isolates appeared to have the smallest genome size, ranging between 4.57 Mb and 5.14 Mb, while the *Pseudomonas* isolates appeared to have larger genome sizes, ranging between 6.27 Mb and 6.93 Mb. All of the *Pseudomonas* genomes sequenced had at least one plasmid present, suggesting that their genomes may be more complex.

Table 2. Sequencing data for isolates selected for genome sequencing.

Isolate	Species	Genome Size	Sequencing Coverage	Contigs	Putative Plasmids	Accession
M3	*Pectobacterium carotovorum*	4.80 Mb	103×	1	0	GCA_045038015.1
M4a	*Pectobacterium carotovorum*	4.79 Mb	46×	1	0	GCA_045038025.1
M5	*Chryseobacterium scophthalmum*	4.57 Mb	42×	1	0	GCA_045037995.1
M8a	*Pectobacterium carotovorum*	4.80 Mb	96×	1	0	GCA_045038005.1
M8b	*Pectobacterium carotovorum*	4.80 Mb	102×	1	0	GCA_045037985.1
M10	*Pectobacterium versatile*	4.88 Mb	101×	1	0	GCA_045037945.1
M12a	*Pseudomonas alloputida*	6.27 Mb	46×	4	3	GCA_045037975.1
M13	*Pseudomonas marginalis*	6.70 Mb	85×	4	3	GCA_045037965.1
M14	*Pseudomonas petroselini*	6.93 Mb	57×	2	1	GCA_045037875.1
M20	*Pseudomonas* spp. within *P. koreensis* group	6.76 Mb	62×	2	1	GCA_045037865.1
M22b	*Pectobacterium versatile*	5.14 Mb	100×	1	0	GCA_045037855.1

3.4. Genome Alignments

We compared the genomes sequenced in our study with genomes available in the RefSeq database by calculating average nucleotide identity (ANI). We first compared our *Pectobacterium* genomes to the 35 type strain genomes available from the genera *Pectobacterium* and *Dickeya*, which confirmed that isolates M3, M4a, M8a, and M8b are *P. carotovorum* and isolates M10 and M22b are *P. versatile* (Figure 2A, Table S3). We then calculated ANI across M3, M4a, M8a, M8b, and all 99 available *P. carotovorum* genomes (Figure 2B, Table S4), which indicated that our isolates formed their own highly related cluster, more similar to each other than any other *P. carotovorum* genome. We similarly calculated ANI for M10, M22b, and all 103 available *P. versatile* genomes (Figure 2C, Table S5) and found M10 and M22b to be closely related to *P. versatile* strains isolated from cabbage, potato, and surface water, and from geographically diverse areas across the United States and Europe.

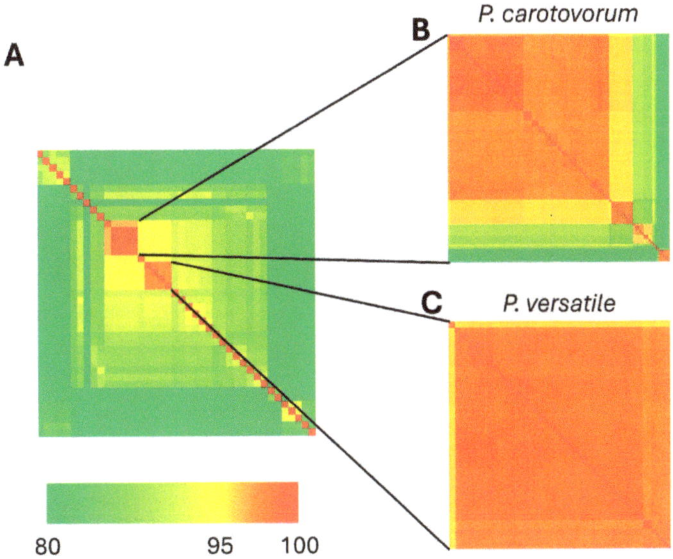

Figure 2. Average nucleotide identity of *Pectobacterium* isolates represented as a heatmap. (**A**) ANI across all type strains of *Pectobacterium* and *Dickeya* species in RefSeq database (n = 41 genomes). (**B**) ANI across all genomes entered in RefSeq database as *P. carotovorum* (n = 103 genomes). (**C**) ANI across all genomes entered in RefSeq database as *P. versatile* (n = 105 genomes).

For our four genomes from the genera *Pseudomonas*, we calculated ANI to compare these isolates to the 369 type strain genomes available (Figure 3, Table S6). This analysis confirmed that strain M12a is *Pseudomonas alloputida* (ANI = 96.71%), M13 is *Pseudomonas marginalis* (ANI = 99.22%), and M14 is *Pseudomonas petroselini* (ANI = 98.81%). The highest ANI found for strain M20 was for *Pseudomonas tensinigenes* (ANI = 93.27%), a member of the *P. koreensis* group.

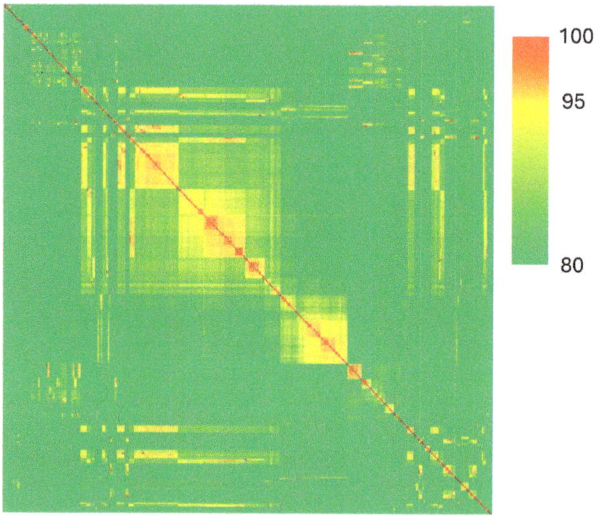

Figure 3. Average nucleotide identity of *Pseudomonas* isolates from our survey and type strains of all species in the RefSeq database (n = 375 genomes).

We also calculated ANI for our *Chryseobacterium* isolate, M5, and the 117 type strain genomes available for this genus (Figure 4, Table S7). We found that isolate M5 is similar to *Chryseobacterium scophthalmum* (ANI = 95.31%), which is a fish pathogen.

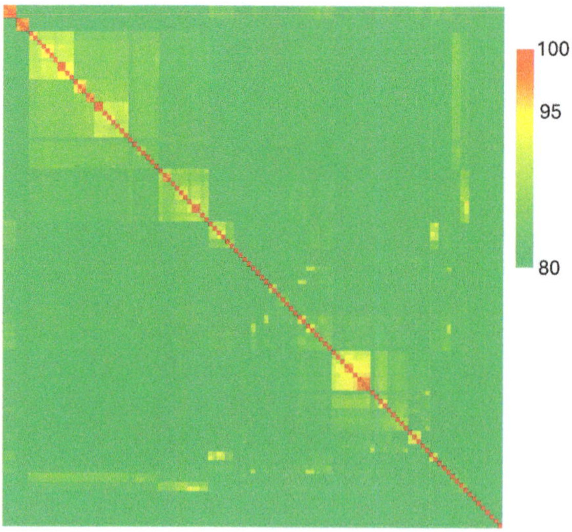

Figure 4. Average nucleotide identity of the *Chryseobacterium* isolate and all type strains from *Chryseobacterium* species in the RefSeq database (n = 118 genomes).

Because we found that average nucleotide identity had little power to resolve differences between our closely related *Pectobacterium carotovorum* isolates, we sought to better understand the relationships between these genomes and isolates. We first aligned the full-length genomes of these isolates and found that strains M3 and M8b had large rearrangements, but strains M4a and M8a shared the same structure (Figure 5).

To test whether the *Pectobacterium* strains also shared horizontally acquired genetic elements or if unique DNA fragments could serve to separate these strains, we computationally predicted prophage elements in each of the *Pectobacterium* genomes we sequenced (Table S8) and then compared the sequences of the predicted prophages (Figure 6). We found that all of the genomes contained at least one predicted prophage. We further found that although some isolates, such as M4a and M8a, shared their predicted prophages, other isolates, such as M3, carried almost entirely distinct prophages from their otherwise genomically close relatives. Top BLAST hits for each of the predicted prophages show that in addition to *P. carotovorum* and *P. versatile*, similar prophages are found in *P. brasiliense*, *P. parvum*, *P. odoriferum*, and *P. wasabiae* (Table S9).

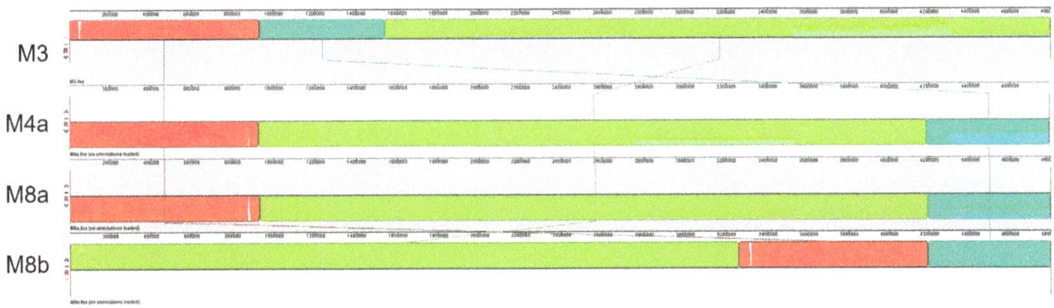

Figure 5. Whole genome alignments of *P. carotovorum* isolates M3, M4a, M8a, and M8b. Alignment and visualization generated by Mauve.

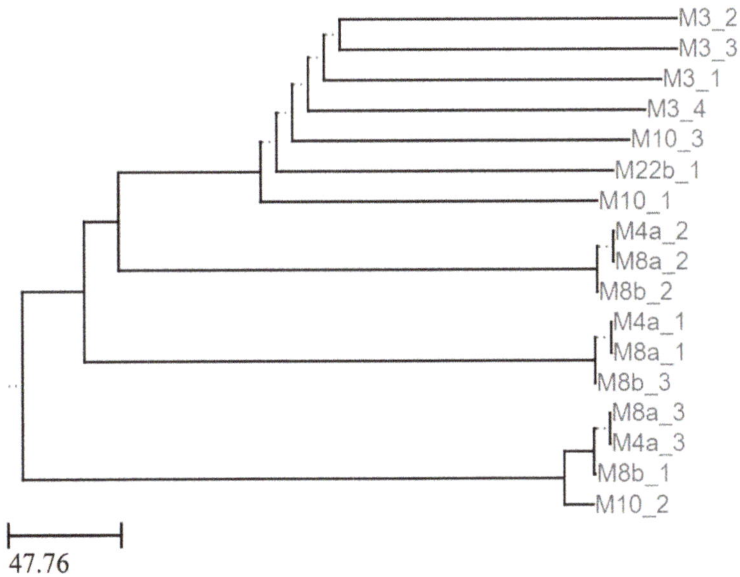

Figure 6. Prophage dendrogram. Newick tree generated by ANIclustermap using sequences of prophages predicted by Phastest in newly sequenced *Pectobacterium* genomes. Representation rendered by ETEToolkit.

4. Discussion

In this study, we isolated 28 pectinolytic bacteria from various produce samples exhibiting typical bacterial soft rot symptoms. These isolates came from nine different genera, but only isolates from three of these genera, *Chryseobacterium*, *Pectobacterium*, and *Pseudomonas*, were pathogenic to potatoes. Only isolates from *Pectobacterium* elicited disease symptoms on Chinese (Napa) cabbage. Although several species from the genera *Pectobacterium* and *Dickeya* are typically associated with bacterial soft rot, in our survey, we only obtained isolates from *P. carotovorum* and *P. versatile*. *P. versatile* represents a group of strains that was recently elevated to its own species and was formerly classified with *P. carotovorum* [20]. Our genomic observations that *P. versatile* and *P. carotovorum* have average nucleotide identity below 96% align with their division into two species [20]. Furthermore, our two *P. versatile* isolates, M10 and M22b, both infect Chinese (Napa) cabbage, but strain M3 was the only one of four *P. carotovorum* isolates to infect this host. Interestingly, strains M3, M10, and M22b each carry prophages in their genome distinct from the other *Pectobacterium* isolates and their predicted prophages. BLAST searches showed that nine of the predicted prophages are most similar to prophages found in *P. carotovorum* and *P. versatile*, while the remaining prophages are more similar to prophages from *P. brasiliense*, *P. odoriferum*, *P. parvum*, and *P. wasabiae*. Interestingly, one cluster of phages (M4a_2, M8a_2, and M8b_2) only had matches along 28% of the length of the prophage sequence. Another five phages had matches less than 80% of the length of the sequence. Future work is needed to address the underlying mechanisms of host specificity and the potential roles of horizontally acquired genetic elements, such as phage, in contributing to host range.

Phenotypic characterization of our isolates showed that many of them exhibited traits often associated with virulence in pathogens. Although the traits assessed are considered virulence factors for several pathogens, only acyl homoserine lactone production was significantly associated with pathogenicity in potatoes. All isolates exhibited swimming and swarming motility and most expressed extracellular protease activity. Because our isolates came from diverse host plants and represent various taxa, it is unclear whether these isolates use these phenotypic traits for virulence in other host plants but do not include potatoes as hosts. Thus, we hypothesize that the significant correlation between acyl homoserine lactone production and potato pathogenicity is because most of the potato pathogenic isolates are *Pectobacterium* isolates and all of the *Pectobacterium* isolates produce acyl homoserine lactones.

All of the *Pectobacterium* isolates, both *P. carotovorum* (M3, M4a, M8a, M8b) and *P. versatile* (M10, M22b), exhibited all measured phenotypic characteristics, with the exception of M10, which did not display protease activity. While strain M10 was pathogenic to potatoes, the lesion size caused by M10 was less than 50 mm^2, whereas the lesions caused by the other *Pectobacterium* isolates were closer to 200 mm^2. Additional work will reveal whether the reduced lesion size is directly or indirectly linked to the lack of extracellular protease activity in strain M10, as extracellular proteases are known to be important for full virulence in some plant-pathogenic bacteria but dispensable for others [35–37]. More broadly, there was no significant correlation between protease activity and pathogenicity to potatoes among all of our isolates.

Most of the *Pseudomonas* isolates we obtained were not pathogenic to potatoes, except for isolates M13, M18, and M20. Based on our average nucleotide identity analyses, strain M13 is a new isolate of *Pseudomonas marginalis*, a species that has previously been associated with soft rot disease [14,38]. For strain M20, the highest average nucleotide identity observed with another genome was with *Pseudomonas tensinigenes*, a member of the *P. koreensis* group [39]. However, the average nucleotide identity between these strains was only 93.27%, well below the typical species-level thresholds of 95–96%. We hypothesize that M20 represents a novel species that is capable of causing soft rot disease in potatoes and potentially other diverse hosts, as it was isolated from spinach. On potatoes, the lesion size caused by M20 was comparable to that of the *Pectobacterium* isolate M10. It is recognized that various species of *Pseudomonas* can cause soft rot, and the species most commonly associated therewith are *P. glycinae, P. cichorii, P. marginalis*, and *P. viridiflava* [38,40–42]. Of these species, we recovered one isolate of *P. marginalis* (M13) and a putatively novel species associated with soft rot (M20).

The genus *Chryseobacterium* is diverse, and its members occupy several environmental niches [43]. Many *Chryseobacteria* are not pathogenic to humans and animals, and of those that are, many tend to be weakly pathogenic, such as *C. indologenes*, formerly *Flavobacterium indologenes* [44]. *C. indologenes* has also been reported as a plant pathogen, causing root rot in Panax ginger [45]. On the other hand, *Chryseobacterium* isolates have been reported to exhibit plant-beneficial effects [46,47]. Some *Chryseobacterium* isolates also exhibit the ability to digest complex organic molecules, such as herbicides [48]. Our *Chryseobacterium* strain M5 contains predicted pectate lyase (ACI513_RS09935) and pectin esterase (ACI513_RS09930) genes located adjacent to each other in the genome, supporting a role for this strain in host cell wall degradation. Other strains of *Chryseobacterium* have previously been reported to express pectinolytic activity [10,49]. By our calculations, *Chryseobacterium* strain M5 was most similar to *Chryseobacterium scophthalmum* based on average nucleotide identity. *C. scophthalmum* is pathogenic to fish, being originally isolated from turbots in the Atlantic Ocean [21,22]. Because a fish pathogen is the closest relation to M5, this isolate may represent a crossover pathogen, evolving from an animal to a plant pathogen. It will be interesting to determine in future work whether this particular isolate also exhibits any animal pathogenic activity. However, because the average nucleotide identity between M5 and *C. scophthalmum* is 95.3%, we hypothesize that M5 may represent a novel species of *Chryseobacterium*. Further work to understand the evolution, host range, and virulence of this isolate and its relatives is warranted.

Among the other isolates obtained, both *Stenotrophomonas* isolates (M15 and M19), which were obtained from different hosts, displayed the same phenotypic characteristics and were not pathogenic to potatoes. Neither of the isolates that came from symptomatic potatoes (M15 and M16) were pathogenic to potatoes under our experimental conditions and time frame. Our isolation approach used crystal violet pectate medium to isolate from tissues. Because of the crystal violet component of the medium, it was anticipated that Gram-positive pectinolytic bacteria would not be isolated in this approach. Indeed, our isolations only yielded bacteria from Gram-negative taxonomic groups. We successfully isolated pectinolytic bacteria from more than half of our plant samples and may have found additional isolates had our methods also targeted Gram-positive pectinolytic bacteria; thus, some of our pectinolytic isolates may have been non-pathogenic because they are members of disease complexes in which multiple diverse bacteria work together to cause soft rot.

Altogether, our work demonstrates that there are abundant and diverse bacteria associated with and causing post-harvest soft rot disease in commercial produce, including a potential crossover pathogen. It is important to be aware of this diversity in future work to develop and improve disease control methods. Our survey work and the genomic resources we have developed lay an initial foundation for work on these pathogens to better understand their virulence mechanisms and host specificities. Furthermore, a lack of redundancy in the isolates we collected from our survey indicates that there is value in ongoing work to monitor the bacteria associated with soft rot disease.

Supplementary Materials: The following supporting information can be downloaded at: https://www.mdpi.com/article/10.3390/pathogens13121096/s1, Table S1: Produce samples collected and tested, Table S2: Top BLAST hits for 16S sequences, Table S3: Average nucleotide identity of *Pectobacterium* and *Dickeya* species, Table S4: Average nucleotide identity of *Pectobacterium carotovorum*, Table S5: Average nucleotide identity of *Pectobacterium versatile*, Table S6: Average nucleotide identity of *Pseudomonas* species, Table S7: Average nucleotide identity of *Chryseobacterium* species, Table S8: Sequences of predicted prophage elements in *Pectobacterium* genomes, Table S9: Top BLAST hits of prophages using all *Pectobacterium* genomes in NCBI; Figure S1: Representative CVP plate from isolation of pectinolytic bacteria, Figure S2: Potato slices inoculated with pectinolytic isolates, Figure S3: Chinese (napa) cabbage leaves inoculated with pectinolytic isolates.

Author Contributions: Conceptualization, K.R., B.R., M.S. and J.K.S.; Investigation: K.R., B.R., M.S., J.H. and J.K.S.; Writing—original draft preparation, K.R., B.R., M.S., J.H. and J.K.S.; Writing—review and editing, K.R., B.R., M.S., J.H. and J.K.S.; project administration and supervision: J.K.S. All authors have read and agreed to the published version of the manuscript.

Funding: This research received no external funding.

Institutional Review Board Statement: Not applicable.

Informed Consent Statement: Not applicable.

Data Availability Statement: All genome sequences generated as part of this study are available through NCBI under BioProject number PRJNA1183984.

Conflicts of Interest: The authors declare no conflicts of interest.

References

1. Charkowski, A.; Sharma, K.; Parker, M.L.; Secor, G.A.; Elphinstone, J. Bacterial diseases of potato. In *The Potato Crop: Its Agricultural, Nutritional and Social Contribution to Humankind*; Campos, H., Ortiz, O., Eds.; Springer: Cham, Switzerland, 2020; pp. 351–388.
2. Kamau, J.W.; Ngaira, J.; Kinyua, J.; Gachamba, S.; Ngundo, G.; Janse, J.; Macharia, I. Occurence of pectinolytic bacteria causing blackleg and soft rot of potato in Kenya. *J. Plant Pathol.* **2019**, *101*, 689–694. [CrossRef]
3. Czajkowski, R.; Perombelon, M.; Jafra, S.; Lojkowska, E.; Potrykus, M.; van der Wolf, J.; Sledz, W. Detection, identification and differentiation of Pectobacterium and Dickeya species causing potato blackleg and tuber soft rot: A review. *Ann. Appl. Biol.* **2015**, *166*, 18–38. [CrossRef] [PubMed]
4. Toth, I.K.; Bell, K.S.; Holeva, M.C.; Birch, P.R. Soft rot erwiniae: From genes to genomes. *Mol. Plant Pathol.* **2003**, *4*, 17–30. [CrossRef] [PubMed]
5. Pinhero, R.G.; Coffin, R.; Yada, R.Y. Post-harvest storage of potatoes. In *Advances in Potato Chemistry and Technology*; Elsevier: Amsterdam, The Netherlands, 2009; pp. 339–370.
6. Maung, C.E.H.; Choub, V.; Cho, J.Y.; Kim, K.Y. Control of the bacterial soft rot pathogen, by CE 100 in cucumber. *Microb. Pathog.* **2022**, *173*, 105807. [CrossRef]
7. Murata, H.; Chatterjee, A.; Liu, Y.; Chatterjee, A.K. Regulation of the production of extracellular pectinase, cellulase, and protease in the soft rot bacterium *Erwinia carotovora* subsp. carotovora: Evidence that aepH of *E. carotovora* subsp. carotovora 71 activates gene expression in *E. carotovora* subsp. carotovora, *E. carotovora* subsp. atroseptica, and *Escherichia coli*. *Appl. Environ. Microbiol.* **1994**, *60*, 3150–3159. [CrossRef]
8. Patidar, M.K.; Nighojkar, S.; Kumar, A.; Nighojkar, A. Pectinolytic enzymes-solid state fermentation, assay methods and applications in fruit juice industries: A review. *3 Biotech* **2018**, *8*, 199. [CrossRef]
9. Wu, P.; Yang, S.H.; Zhan, Z.C.; Zhang, G.M. Origins and features of pectate lyases and their applications in industry. *Appl. Microbiol. Biot.* **2020**, *104*, 7247–7260. [CrossRef]
10. Haile, S.; Ayele, A. Pectinase from Microorganisms and Its Industrial Applications. *Sci. World J.* **2022**, *2022*, 1881305. [CrossRef]
11. Toth, I.K.; Barny, M.-a.; Czajkowski, R.; Elphinstone, J.G.; Li, X.; Pédron, J.; Pirhonen, M.; Van Gijsegem, F. Pectobacterium and Dickeya: Taxonomy and evolution. In *Plant Diseases Caused by Dickeya and Pectobacterium Species*; Van Gijsegem, F., van der Wolf, J., Toth, I.K., Eds.; Springer Nature: Cham, Switzerland, 2021; pp. 13–37.
12. Charkowski, A.O. The Changing Face of Bacterial Soft-Rot Diseases. *Annu. Rev. Phytopathol.* **2018**, *56*, 269–288. [CrossRef]
13. Sawada, H.; Fujikawa, T.; Tsuji, M.; Satou, M. *Pseudomonas allii* sp. nov., a pathogen causing soft rot of onion in Japan. *Int. J. Syst. Evol. Micr.* **2021**, *71*, 004582. [CrossRef]
14. Li, J.; Chai, Z.; Yang, H.; Li, G.; Wang, D. First report of *Pseudomonas marginalis* pv. *marginalis* as a cause of soft rot of potato in China. *Australas. Plant Dis. Notes* **2007**, *2*, 71–73. [CrossRef]
15. Elbanna, K.; Elnaggar, S.; Bakeer, A. Characterization of *Bacillus altitudinis* as a New Causative Agent of Bacterial Soft Rot. *J. Phytopathol.* **2014**, *162*, 712–722. [CrossRef]
16. da Silva, W.L.; Yang, K.T.; Pettis, G.S.; Soares, N.R.; Giorno, R.; Clarks, C.A. Flooding-Associated Soft Rot of Sweetpotato Storage Roots Caused by Distinct Isolates. *Plant Dis.* **2019**, *103*, 3050–3056. [CrossRef]
17. Hauben, L.; Moore, E.R.; Vauterin, L.; Steenackers, M.; Mergaert, J.; Verdonck, L.; Swings, J. Phylogenetic position of phytopathogens within the Enterobacteriaceae. *Syst. Appl. Microbiol.* **1998**, *21*, 384–397. [CrossRef] [PubMed]
18. Samson, R.; Legendre, J.B.; Christen, R.; Saux, M.F.; Achouak, W.; Gardan, L. Transfer of *Pectobacterium chrysanthemi* (Burkholder et al. 1953) Brenner et al. 1973 and *Brenneria paradisiaca* to the genus *Dickeya* gen. nov. as *Dickeya chrysanthemi* comb. nov. and *Dickeya paradisiaca* comb. nov. and delineation of four novel species, *Dickeya dadantii* sp. nov., *Dickeya dianthicola* sp. nov., *Dickeya dieffenbachiae* sp. nov. and *Dickeya zeae* sp. nov. *Int. J. Syst. Evol. Microbiol.* **2005**, *55*, 1415–1427. [CrossRef]
19. Overbeek, R.; Begley, T.; Butler, R.M.; Choudhuri, J.V.; Chuang, H.Y.; Cohoon, M.; de Crecy-Lagard, V.; Diaz, N.; Disz, T.; Edwards, R.; et al. The subsystems approach to genome annotation and its use in the project to annotate 1000 genomes. *Nucleic Acids Res.* **2005**, *33*, 5691–5702. [CrossRef]

20. Portier, P.; Pedron, J.; Taghouti, G.; Fischer-Le Saux, M.; Caullireau, E.; Bertrand, C.; Laurent, A.; Chawki, K.; Oulgazi, S.; Moumni, M.; et al. Elevation of *Pectobacterium carotovorum* subsp. odoriferum to species level as *Pectobacterium odoriferum* sp. nov., proposal of *Pectobacterium brasiliense* sp. nov. and *Pectobacterium actinidiae* sp. nov., emended description of *Pectobacterium carotovorum* and description of *Pectobacterium versatile* sp. nov., isolated from streams and symptoms on diverse plants. *Int. J. Syst. Evol. Microbiol.* **2019**, *69*, 3207–3216. [CrossRef] [PubMed]
21. Mudarris, M.; Austin, B.; Segers, P.; Vancanneyt, M.; Hoste, B.; Bernardet, J.F. *Flavobacterium-scophthalmum* sp. nov., a Pathogen of Turbot (*Scophthalmus-maximus* L.). *Int. J. Syst. Bacteriol* **1994**, *44*, 447–453. [CrossRef]
22. Shahi, N.; Sharma, P.; Pandey, J.; Bisht, I.; Mallik, S.K. Characterization and pathogenicity study of *Chryseobacterium scophthalmum* recovered from gill lesions of diseased golden mahseer, *Tor putitora* (Hamilton, 1822) in India. *Aquaculture* **2018**, *485*, 81–92. [CrossRef]
23. Hélias, V.; Hamon, P.; Huchet, E.; Wolf, J.V.D.; Andrivon, D. Two new effective semiselective crystal violet pectate media for isolation of *Pectobacterium* and *Dickeya*. *Plant Pathol.* **2012**, *61*, 339–345. [CrossRef]
24. Hongoh, Y.; Yuzawa, H.; Ohkuma, M.; Kudo, T. Evaluation of primers and PCR conditions for the analysis of 16S rRNA genes from a natural environment. *FEMS Microbiol. Lett.* **2003**, *221*, 299–304. [CrossRef]
25. Johnson, M.; Zaretskaya, I.; Raytselis, Y.; Merezhuk, Y.; McGinnis, S.; Madden, T.L. NCBI BLAST: A better web interface. *Nucleic Acids Res.* **2008**, *36*, W5–W9. [CrossRef] [PubMed]
26. Abràmoff, M.D.; Magalhães, P.J.; Ram, S.J. Image processing with ImageJ. *Biophotonics Int.* **2004**, *11*, 36–42.
27. Schachterle, J.K.; Sundin, G.W. The Leucine-Responsive Regulatory Protein Lrp Participates in Virulence Regulation Downstream of Small RNA ArcZ in Erwinia amylovora. *mBio* **2019**, *10*, 10–1128. [CrossRef]
28. Luo, Z.Q.; Su, S.C.; Farrand, S.K. In situ activation of the quorum-sensing transcription factor TraR by cognate and noncognate acyl-homoserine lactone ligands: Kinetics and consequences. *J. Bacteriol.* **2003**, *185*, 5665–5672. [CrossRef] [PubMed]
29. Shaw, P.D.; Ping, G.; Daly, S.L.; Cha, C.; Cronan, J.E.; Rinehart, K.L.; Farrand, S.K. Detecting and characterizing N-acyl-homoserine lactone signal molecules by thin-layer chromatography. *Proc. Natl. Acad. Sci. USA* **1997**, *94*, 6036–6041. [CrossRef] [PubMed]
30. Santander, R.D.; Biosca, E.G. Erwinia amylovora psychrotrophic adaptations: Evidence of pathogenic potential and survival at temperate and low environmental temperatures. *PeerJ* **2017**, *5*, e3931. [CrossRef]
31. Jain, C.; Rodriguez-R, L.M.; Phillippy, A.M.; Konstantinidis, K.T.; Aluru, S. High throughput ANI analysis of 90K prokaryotic genomes reveals clear species boundaries. *Nat. Commun.* **2018**, *9*, 5114. [CrossRef]
32. Darling, A.E.; Mau, B.; Perna, N.T. progressiveMauve: Multiple genome alignment with gene gain, loss and rearrangement. *PLoS ONE* **2010**, *5*, e11147. [CrossRef]
33. Wishart, D.S.; Han, S.; Saha, S.; Oler, E.; Peters, H.; Grant, J.R.; Stothard, P.; Gautam, V. PHASTEST: Faster than PHASTER, better than PHAST. *Nucleic Acids Res.* **2023**, *51*, W443–W450. [CrossRef]
34. Huerta-Cepas, J.; Dopazo, J.; Gabaldon, T. ETE: A python Environment for Tree Exploration. *BMC Bioinform.* **2010**, *11*, 24. [CrossRef] [PubMed]
35. Dow, J.M.; Clarke, B.R.; Milligan, D.E.; Tang, J.L.; Daniels, M.J. Extracellular Proteases from Xanthomonas-Campestris Pv Campestris, the Black Rot Pathogen. *Appl. Environ. Microb.* **1990**, *56*, 2994–2998. [CrossRef] [PubMed]
36. Gouran, H.; Gillespie, H.; Nascimento, R.; Chakraborty, S.; Zaini, P.A.; Jacobson, A.; Phinney, B.S.; Dolan, D.; Durbin-Johnson, B.P.; Antonova, E.S.; et al. The Secreted Protease PrtA Controls Cell Growth, Biofilm Formation and Pathogenicity in *Xylella fastidiosa*. *Sci. Rep.* **2016**, *6*, 31098. [CrossRef] [PubMed]
37. Zou, H.S.; Song, X.; Zou, L.F.; Yuan, L.; Li, Y.R.; Guo, W.; Che, Y.Z.; Zhao, W.X.; Duan, Y.P.; Chen, G.Y. EcpA, an extracellular protease, is a specific virulence factor required by *Xanthomonas oryzae* pv. oryzicola but not by *X. oryzae* pv. oryzae in rice. *Microbiology* **2012**, *158*, 2372–2383. [CrossRef] [PubMed]
38. Janse, J.D.; Derks, J.H.J.; Spit, B.E.; Vandertuin, W.R. Classification of Fluorescent Soft Rot *Pseudomonas* Bacteria, Including *P. marginalis* Strains, Using Whole Cell Fatty-Acid Analysis. *Syst. Appl. Microbiol.* **1992**, *15*, 538–553. [CrossRef]
39. Girard, L.; Lood, C.; Höfte, M.; Vandamme, P.; Rokni-Zadeh, H.; van Noort, V.; Lavigne, R.; De Mot, R. The Ever-Expanding *Pseudomonas* Genus: Description of 43 New Species and Partition of the *Pseudomonas putida* Group. *Microorganisms* **2021**, *9*, 1766. [CrossRef]
40. Godfrey, S.A.C.; Marshall, J.W. Identification of cold-tolerant *Pseudomonas viridiflava* and *P. marginalis* causing severe carrot postharvest bacterial soft rot during refrigerated export from New Zealand. *Plant Pathol.* **2002**, *51*, 155–162. [CrossRef]
41. Wright, P.J.; Grant, D.G. Evaluation of Allium germplasm for susceptibility to foliage bacterial soft rot caused by *Pseudomonas marginalis* and *Pseudomonas viridiflava*. *N. Z. J. Crop Hort.* **1998**, *26*, 17–21. [CrossRef]
42. Liu, K.; Sun, W.S.; Li, X.L.; Shen, B.Y.; Zhang, T.J. Isolation, identification, and pathogenicity of *Pseudomonas glycinae* causing ginseng bacterial soft rot in China. *Microb. Pathog.* **2024**, *186*, 106497. [CrossRef]
43. Bernardet, J.F.; Hugo, C.J.; Bruun, B. Chryseobacterium. In *Bergey's Manual of Systematics of Archaea and Bacteria*; Whitman, W.B., Ed.; John Wiley & Sons, Inc.: Hoboken, NJ, USA, 2015; pp. 1–35.
44. Mukerji, R.; Kakarala, R.; Smith, S.J.; Kusz, H.G. *Chryseobacterium indologenes*: An emerging infection in the USA. *BMJ Case Rep.* **2016**, *2016*, bcr2016214486. [CrossRef]
45. Zou, H.; Ding, Y.F.; Shang, J.J.; Ma, C.L.; Li, J.H.; Yang, Y.; Cui, X.M.; Zhang, J.H.; Ji, G.H.; Wei, Y.L. Isolation, characterization, and genomic analysis of a novel bacteriophage MA9V-1 infecting *Chryseobacterium indologenes*: A pathogen of Panax notoginseng root rot. *Front. Microbiol.* **2023**, *14*, 1251211. [CrossRef] [PubMed]

46. Jung, H.; Lee, D.; Lee, S.; Kong, H.J.; Park, J.; Seo, Y.S. Comparative genomic analysis of *Chryseobacterium* species: Deep insights into plant- growth- promoting and halotolerant capacities. *Microb. Genom.* **2023**, *9*, 001108. [CrossRef] [PubMed]
47. Singh, A.V.; Chandra, R.; Goel, R. Phosphate solubilization by *Chryseobacterium* sp. and their combined effect with N and P fertilizers on plant growth promotion. *Arch. Agron. Soil Sci.* **2013**, *59*, 641–651. [CrossRef]
48. Zhao, H.H.; Xu, J.; Dong, F.S.; Liu, X.G.; Wu, Y.B.; Wu, X.H.; Zheng, Y.Q. Characterization of a novel oxyfluorfen-degrading bacterial strain *Chryseobacterium aquifrigidense* and its biochemical degradation pathway. *Appl. Microbiol. Biot.* **2016**, *100*, 6837–6845. [CrossRef]
49. Roy, K.; Dey, S.; Uddin, M.K.; Barua, R.; Hossain, M.T. Extracellular Pectinase from a Novel Bacterium *Chryseobacterium indologenes* Strain SD and Its Application in Fruit Juice Clarification. *Enzym. Res.* **2018**, *2018*, 3859752. [CrossRef]

pathogens

MDPI

Article

Pathogenic Potential and Antibiotic Susceptibility: A Comprehensive Study of Enterococci from Different Ecological Settings

Maria Pandova [1], Yoana Kizheva [1,*], Margarita Tsenova [1], Mariya Rusinova [2], Tsvetomira Borisova [2] and Petya Hristova [1]

[1] Department of General and Industrial Microbiology, Faculty of Biology, Sofia University, 1164 Sofia, Bulgaria; maria.pandova@abv.bg (M.P.); mhcenova@biofac.uni-sofia.bg (M.T.); pkabad@biofac.uni-sofia.bg (P.H.)

[2] Human Milk Bank Bulgaria, 1309 Sofia, Bulgaria; bmk@milkbank.bg (M.R.); cv_borisova@abv.bg (T.B.)

* Correspondence: joana_k@biofac.uni-sofia.bg or yokizheva@gmail.com

Abstract: The pathway and the lifestyle of known enterococcus species are too complicated. The aim of the present study is to trace the path of pathogenicity of enterococci isolated from seven habitats (*Cornu aspersum* intestine; Bulgarian yoghurt; goat and cow feta cheese—mature and young, respectively; Arabian street food—doner kebab; cow milk; and human breast milk) by comparing their pathogenic potential. In total, 72 enterococcal strains were isolated and identified by MALDI-TOF, sequencing, and PCR. Hemolytic and gelatinase activity were biochemically determined. PCR was carried out for detection of virulence factors (*cylB*, *esp*, *gls24*, *nucl*, *psaA*, *agg*, *gelE*, and *ace*) and antibiotic resistance (*erm*, *ermB*, *blaZ*, *vanA*, *aphA*, *mefA*, *gyrA*, cat_{pIP501}, and *aac6′-aph2″*). Phenotypic antibiotic resistance was assigned according to EUCAST. Eleven representatives of the genus *Enterococcus* were identified: *E. mundtii*, *E. casseliflavus*, *E. gilvus*, *E. pseudoavium*, *E. pallens*, *E. malodoratus*, *E. devriesei*, *E. gallinarum*, *E. durans*, *E. faecium*, and *E. faecalis*. Twenty-two strains expressed α-hemolysis. Thirteen strains had the *cylB* gene. Only two strains expressed α-hemolysis and possessed the *cylB* gene simultaneously. Positive amplification for *gelE* was found in 35% of the isolates, but phenotypic gelatinase activity was observed only in three strains. All isolates showed varying antibiotic resistance. Only *E. faecalis* BM15 showed multiple resistance (AMP-HLSR-RP). Correlation between genotypic and phenotypic macrolide resistance was revealed for two *E. faecalis* strains.

Keywords: enterococcus; pathogenic potential; virulence factors; antibiotic resistance; alimentary chain

1. Introduction

The members of the genus *Enterococcus* are bacteria that have a dual role in the environment: positive (as commensal and potential probiotic bacteria) and negative (opportunistic pathogens capable of infecting plants, animals, and humans) [1]. The pathway and the lifestyle of known enterococcus species in the natural environment are far too complicated and poorly studied. Most studies characterize enterococci isolated from particular ecological niches and do not track what features they develop when they jump from one biological kingdom to another.

Enterococci are ubiquitous Gram-positive bacteria that can be found in various ecological niches, such as environmental, clinical, and food. This genus of bacteria forms a part of natural biomes of soil, water, sewage, and arable land, as well as populations in the gastrointestinal tracts (GITs) of mammals, birds, fishes, invertebrates, and insects [2–6]. Similarly, enterococci have been isolated from fresh vegetables (olive, pepper, celery, cilantro, mustard greens, spinach, collards, parsley, dill, cabbage, and cantaloupe) and wild plants also [7–10]. Flowering plants and crops have also been known to be carriers of enterococci [11]. Mundt [10] states that relationships between enterococci and plants are based mainly on their epiphytic persistence. However, they have been considered temporary inhabitants as a result of wind and insect activity [10].

Enterococci are a diverse taxonomic group that includes 58 recognized species and 2 subspecies [12]. The most distributed members of the genus in the GITs of mammals have been reported to be *Enterococcus faecalis*, *Enterococcus faecium*, *Enterococcus durans*, and *Enterococcus hirae* [13]. Plant-associated epiphytic *Enterococcus* species most commonly belong to *E. faecalis*, *Enterococcus mundtii*, *Enterococcus casseliflavus*, *E. faecium*, and *Enterococcus sulfureus* [11,14,15]. A new taxonomic species, *Enterococcus plantarum* sp. nov., was identified during the study of the microflora of various plants from meadows [16]. However, most studies assume that enterococci, as part of the natural microbiota of the gastrointestinal tract of warm-blooded animals, can enter the environment through feces, contaminate soil and water, and then colonize plants. This pathway explains why enterococci predominate on plant surfaces and are resistant to a number of antibiotics [17–20], but at the same time, they can be identified as potential cross-over agents promoting the dissemination of antibiotic resistance [8,21]. Furthermore, it has been suggested that the infection strategies of some enterococci (*E. faecalis*) are similar in plants, mammals, and nematodes [1].

Moreover, evidence that some strains of *E. faecalis* can infect the roots and leaves of the plant *Arabidopsis thaliana*, causing local and systemic infection that leads to the death of the infected plant, has been reported [1]. *E. faecalis* has developed a significant biofilm-like pathogenic community that has colonized the root surface [1]. Enterococci have been previously reported as a component of the microbiome of pepper plants with symptoms of disease [8].

On the other hand, it can be assumed that enterococci are part of the plant microbiome and that they enter the intestinal tracts of animals and humans through the intake of plant food. Each gut microbiome selects the enterococcal species it needs to maintain eubiosis. Colonizing different microbiomes, from invertebrates to mammals, allows plant enterococci to acquire new genes, which they then spread into new environmental niches. This life cycle of passing through hosts from different biological kingdoms defines enterococci as important vectors for the horizontal transfer of antibiotic resistance and virulence genes, despite where they originate from [22].

Traditionally, enterococci have been considered to be normal commensal bacteria and may even be beneficial for a variety of gastrointestinal and systemic illnesses. Some enterococci species have the ability to stimulate the immune system and play an important role in the maintenance of intestinal homeostasis [23,24]. Similarly, enterococci have an active part in food technology as the starting culture in meat and cheese fermentation [25], as well as in food preservation [26–28].

However, enterococci can cause invasive infections if their relationship with the host is broken [29]. These bacteria exhibit remarkable adaptability in colonizing different hosts and show the ability to thrive as pathogens in diverse ecological niches [30]. However, some strains have acquired a wide range of virulence and antibiotic resistance genes, leading to an increase in their pathogenicity and posing a significant public health challenge [31,32]. Thus, enterococci, despite their commensal nature, have been identified as the most prevalent causes of urinary tract infections and nosocomial bacteremia. They also constitute the second most commonly reported cause of surgical wound infections and the third most often reported cause of bacteremia [33,34]. Moreover, enterococci have been reported as the main Gram-positive bacteria causing hospital-acquired infections during and after the COVID 19 pandemic [35–40].

The virulence factors that contribute to enterococcal pathogenesis include collagen-binding protein (Ace), aggregation substance (Agg), surface proteins (Esp), cytolysin (Cyl), gelatinase (Gel), general stress protein (Gls24), and immune evasion molecules [41]. Defined as effector molecules, virulence factors indicate a high potential of enterococci for host adherence, tissue invasion, immune evacuation, and nutrient acquisition. Ace is an adhesin, anchored to the cell wall, that helps enterococcal species to adhere to collagen. The agglutination substance (Agg) is a pheromone-inducible surface protein which helps in aggregation during the conjugation process. Cytolysin (Cyl) production is associated with the capacity of bacteria to access the bloodstream and trigger septicemia. Epidemiological research has found that the enterococcal surface protein, Esp, is typically linked with infectious strains, compared to commensal isolates, and is located on a large pathogenicity island [42]. Esp is also involved in initial adherence and biofilm formation and contributes to the pathogenesis of different infections. Gelatinase (Gel) is a zinc metallo-endopeptidase which takes part in pathogenesis by making nutrients available through degradation of host tissue and by taking part in biofilm formation [43].

A notable feature of enterococci is their intrinsic resistance to cephalosporin, cotrimoxazole, lincomycin, and low levels of penicillin and aminoglycosides. Enterococci can also acquire resistance genes from other microorganisms through horizontal gene transfer and thus become resistant to a variety of antibiotics such as chloramphenicol, tetracycline, streptogramin, macrolides, a high level of glycopeptide, aminoglycosides, and quinolones [44]. This acquired resistance along with their known remarkable ability to overcome and adapt to various environmental stress factors, give the enterococci the unique potential to realize complex lifestyles [45].

Therefore, the study of the diversity and distribution of pathogenicity-determining genes of enterococcal populations of different origins provides valuable insight into their adaptive strategies in different hosts and environments. It is also critical to understand the pathogenicity mechanisms that these multi-host pathogens possess. Moreover, the comparison of the virulence and resistance arsenal of the enterococcal populations, adapted to inhabit completely different niches, contributes to the global knowledge of enterococcal lifestyle and reveals the key role of the evolutionary pressure of the habitat on it. The present study considers enterococcal populations from different biological kingdoms/origins as a reservoir of genes for virulence and antibiotic resistance with respect to possible re-return into the environment and subsequent colonization of plants or other diverse ecological niches such as soil and water. The aim is to compare the pathogenic potential of enterococci isolated from herbivorous invertebrate animals, food products derived from herbivorous warm-blooded animals, and human breast milk.

2. Materials and Methods

2.1. Sample Collection and Isolation of the Bacteria

In total, twenty-seven samples from animal GIT (invertebrate herbivorous species *Cornu aspersum* at the hibernation stage of the life cycle) and food (Bulgarian yoghurt; goat and cow feta cheese—mature and young, respectively; Arabian street food—doner kebab; cow milk; and human breast milk) were used for *Enterococcus* species isolation. Breast milk samples were supplied by the Human Milk Bank, Bulgaria, *C. aspersum* samples were collected and processed according to Koleva et al. [46], and food samples were obtained randomly from artisanal markets. Approximately 1 g or 1 mL from each sample was homogenized in saline (at a ratio of 1:9) and all samples were directly cultivated on the selective medium Slanetz and Bartley agar (HiMedia Laboratories, Mumbai, India). The plates were cultivated at 37 °C for 24 h–48 h. The appearance of dark red-brownish colonies on the surface of the used media after the cultivation served as positive results for selection of enterococcal strains. Pure cultures from separate colonies were isolated as potential *Enterococcus* species after double purification. Three reference strains were also used in this study: *E. faecalis* NBIMCC 3915 and *E. faecium* NBIMCC 8754 as positive controls for the genus and species PCR identification, and *Bacillus cereus* NBIMCC 1085 as positive control for β-hemolytic activity in the hemolysis assay.

2.2. DNA Preparation

The bacterial cultures were cultivated in MRS broth (HiMedia, Mumbai, India) at 37 °C for 24 h prior to the genomic DNA extraction. The biomass was harvested by centrifugation at 10,000× *g* and was washed twice with 500 μL 1% NaCl. Total DNA was extracted by E.Z.N.A. Bacterial DNA Kit (Omega Biotek Inc., 400 Pinnacte Way, Suite 450, Norcross,

GA, USA). For improved lysis of the cells, 2 μL 1000 units/mg mutanolysin (Merck KGaA, Darmstadt, Germany) was added at the enzyme lysis step.

2.3. Species Identification

The isolates were identified by three different methods: PCR with genus- and species-specific primers [46,47], 16S rRNA sequencing [46], and MALDI-TOF (Matrix-Assisted Laser Desorption/Ionization Time-of-Flight) mass spectrometry [8]. Genus- and species-specific PCRs were performed in a total reaction volume of 25 μL containing 16.5 μL ultrapure H_2O, 0.5 μL (5 pmol/μL) of each primer, 6.5 μL VWR Red Taq polymerase master Mix (VWR International bvba/sprl, Haasrode Researchpark Zone 3, Geldenaaksebaan 464 B-3001, Haasrode Belgium), and 1 μL extracted DNA. The reactions conditions were as follows: initial denaturation at 95 °C for 5 min, followed by 25 cycles of denaturation at 94 °C for 45 s, annealing at 58 °C, 50 °C, 60 °C, and 55 °C, according to primer specificity [46,47] for 45 s, extension step at 72 °C for 45 s, and a final extension step at 72 °C for 7 min. PCR products were separated in 1.5% agarose gel electrophoresis at 100 V for 30 min, stained with ethidium bromide, and visualized under UV light. Molecular size marker 100 bp DNA ladder (SERVA FastLoad 100 bp DNA ladder, SERVA Electrophoresis GmbH, Carl-Benz-Str. 7, Heidelberg, Germany) was used. The universal primers 9F and 1542R were used to amplify the 16S rRNA gene [48]. Purified PCR products were sequenced in Macrogen Europe, Meibergdreef 57 1105 BA, Amsterdam, The Netherlands. The obtained sequences were subjected to comparative analyses using nucleotide BLAST (NCBI, accessed on June 2021).

2.4. Phenotypic Hemolytic Activity Assay

The evaluation of hemolytic activity was performed according to the method described by Carrillo et al. [49]. Pure bacterial cultures were cultivated overnight on brain heart infusion (BHI) agar (HiMedia Inc., Mumbai, India) to obtain log-phase cultures. Then, the cultures were surface spot inoculated on Columbia agar plates supplemented with 5% horse blood and incubated at a temperature of 37 °C for a duration of 24 to 48 h, after which the plates were examined for hemolysis. Clear zones around the colonies were interpreted as β-hemolysis (positive) and lack of zone was reported as gamma-hemolysis (negative). When greenish zones were observed, the strains were reported as α-hemolytic and taken as negative for the assessment of β-hemolytic activity [50].

2.5. Phenotypic Gelatinase Activity Assay

The evaluation of phenotypic gelatinase activity was carried out according to the procedure described by [41]. Pure bacterial cultures were cultivated overnight on BHI agar (HiMedia Inc., India) to obtain log-phase cultures. Then, the cultures were surface spot inoculated on agar plates containing 5 g/L peptone (Merck, Darmstadt, Germany), 30 g/L gelatin (Difco, Detroit, MI, USA), 3 g/L yeast extract (Gibco, Paisley, Scotland), and 15 g/L agar (Plant agar, Duchefa Biochemie, The Netherlands), with a pH of 7.0, and were incubated at 37 °C for 48 h. After the cultivation, the agar surface was flooded with a saturated solution of $(NH_3)_2SO_4$ (55 g/100 mL dH_2O). Gelatinase producers formed clear zones around the spots, and these results were interpreted as positive.

2.6. Antibiotic Susceptibility Testing

Susceptibility to antibiotic substances was performed using the Kirby–Bauer disc diffusion method [51]. For evaluation of antibiotic resistance of enterococcal isolates, fifteen antibiotics were tested: ampicillin 2 μg/disc (AMP), imipenem 10 μg/disc (IPM), ciprofloxacin 5 μg/disc (CP), levofloxacin 5 μg/disc (LE), norfloxacin 10 μg/disc (NX), gentamicin 30 μg/disc—test for high-level aminoglycoside resistance (GEN), streptomycin 300 μg/disc—test for high-level streptomycin resistance (HLS), teicoplanin 30 μg/disc (TEI), vancomycin 5 μg/disc (VA), quinupristin-dalfopristin 15 μg/disc (RP), eravacycline 20 μg/disc (ERV), tigecycline 15 μg/disc (TG), linezolid 10 μg/disc (LZ), nitrofurantoin 100 μg/disc (NIT), and trimethoprim 5 μg/disc (TR). The whole procedure of testing of the antibiotic susceptibility along with the interpretation of the obtained results was carried out according to European Committee on Antimicrobial Susceptibility Testing guidelines [52].

2.7. PCR Amplification of Virulence and Antibiotic Resistance Genes

PCR was carried out for the detection of eight virulence (*cylB*, *esp*, *gls24*, *nucl*, *psaA*, *agg*, *gelE*, and *ace*) and nine antibiotic resistance-related genes (*erm*, *ermB*, *blaZ*, *vanA*, *aphA*, *mefA*, *gyrA*, cat_{pIP501}, and *aac6′-aph2″*) commonly presented in clinical and environmental enterococci. PCR mixtures were prepared as described above (see Section 2.3). The reaction conditions were as follows: initial denaturation at 95 °C for 5 min, followed by 25 cycles of denaturation at 94 °C for 45 s, annealing temperature according to primer specificity (Table 1) for 45 s, extension step at 72 °C for 45 s, and a final extension step at 72 °C for 7 min. PCR products were visualized in a 1.5% agarose gel electrophoresis at 100 V for 30 min.

Table 1. Primer pairs used for detection of virulence and antibiotic resistance genes.

Primer	Sequence (5′ to 3′)	Tm (°C)	Product Size (bp)	Reference
	Primers for virulence-related genes			
cylB-F *cylB*-R	GGAGAATTAGTGTTTAGAGCG GCTTCATAACCATTGTTACTATAGAAAC	57	522	[53]
esp-F *esp*-R	CGATAAAGAGAGAGCGGAG GCAAACTCTACATCCACGTC	57	539	[53]
gls24-F *gls24*-R	GCATTAGATGAGATTGATGGTC GCGAGGTTCAGTTTCTTC	54	446	[53]
*psa*A-F *psa*A-R	CTATTTTGCAGCAAGTGATG CGCATAGTAACTATCACCATCTTG	54	540	[53]
agg-F *agg*-R	AAGAAAAAGAAGTAGACCAAC AAACGGCAAGACAAGTAAATA	54	1553	[54]
ace-F *ace*-R	AAAGTAGAATTAGATCACAC TCTATCACATTCGGTTGCG	51	320	[55]
*gel*E-F *gel*E-R	ACCCCGTATCATTGGTTT ACGCATTGCTTTTCCATC	51	419	[54]
nucl-F *nucl*-R	GTGTAAAAGAAGTTACTGAAAATGTTACTC GCGTTTTTTGTAGTAATGTTCCATCTACG	62	332	[53]
	Primers for antibiotic resistance-related genes			
aac6′-aph2″-F *aac6′-aph2″*-R	CTGATGAGATAGTCTATGGTATGGATC GCCACACTATCATAACCACTACCG	65	375	[53]
*aph*A-F *aph*A-R	GCCGATGTGGATTGCGAAAA GCTTGATCCCCAGTAAGTCA	55	292	[56]
*bla*Z-F *bla*Z-R	ACTTCAACACCTGCTGCTTTC TAGGTTCAGATTGGCCCTTAG	60	240	[57]
cat_{pIP501}-F cat_{pIP501}-R	GGATATGAAATTTATCCCTC CAATCATCTACCCTATGAAT	50	486	[58]
*gyr*A-F *gyr*A-R	ACTTGAAGATGTTTTAGGTGAT TTAGGAAATCTTGATGGCAA	55	559	[59]
erm-F *erm*-R	CATTTAACGACGAAACTGGC GGAACATCTGTGGTATGGCG	55	726	[59]

Table 1. *Cont.*

Primer	Sequence (5′ to 3′)	Tm (°C)	Product Size (bp)	Reference
*erm*B-F *erm*B-R	CATTTAACGACGAAACTGGC GGAACATCTGTGGTATGGCG	52	405	[59]
*mef*A-F *mef*A-R	ACTATCATTAATCACTAGTGC TTCTTCTGGTACTAAAAGTGG	52	346	[60]
vanA36-F vanA992-R	TTGCTCAGAGGAGCATGACG TCGGGAAGTGCAATACCTGC	65	957	[61]

2.8. Data Analysis

Welch's *t*-test was used to compare the number of resistance and virulence genes, as well as the number of phenotypic antibiotic resistance profiles of isolates from different origins and within different species. Results were considered significant when $p < 0.05$.

3. Results

3.1. Bacterial Isolation and Identification

In total, 72 presumptive enterococcal strains were isolated from various ecological niches. Seventeen strains were isolated from the GIT of *C. aspersum* at the hibernation stage of the life cycle, as described previously [46]. Thirty-nine strains were isolated from different food sources (27 from Bulgarian yogurt, 2 from matured goat feta cheese, 5 from young cow feta cheese, 1 from doner kebab, and 4 from cow milk). Sixteen strains were isolated from human breast milk. The latest strains were grouped as human enterococci with non-hospital origin. All isolates appeared as pink or dark red-brownish colonies when streaked on the selective Slanetz and Bartley medium (Figure 1a). Under the microscope, they were Gram-positive cocci or coccobacilli, grouped in clusters, chains, or pairs (Figure 1b).

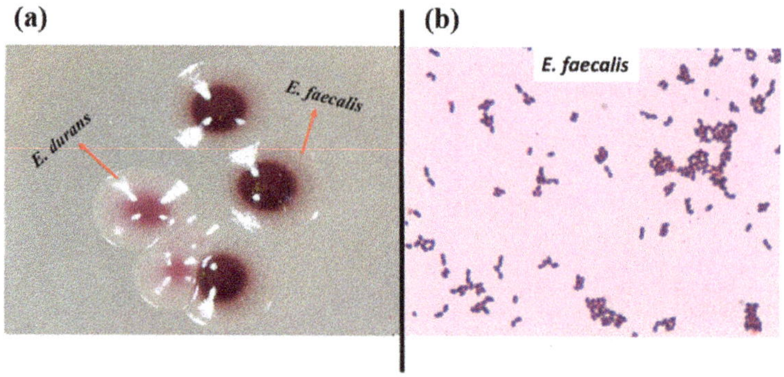

Figure 1. (**a**) Strains YFC1 (*E. faecalis*) and YFC2 (*E. durans*) on Slanetz and Bartley medium; (**b**) Gram staining of strain YFC3 (*E. faecalis* from young feta cheese).

Three different approaches were used for species identification: PCR with genus- and species-specific primers, 16S rRNA sequencing, and MALDI-TOF mass spectrometry. The comparative analyses of the obtained sequencing results showed similarity percentage above 98–99%, which is considered a very good species identification. All obtained results for the species identification with MALDI-TOF showed score values above 2.0, which represents reliable species-level identification. Detailed information for the species identification is given in Table 2.

Table 2. Enterococcal species identification.

No	Isolate	Origin	Species	Method of Identification	No.	Isolate	Origin	Species	Method of Identification
1	CA1	*C. aspersum*	*E. mundtii*	MALDI-TOF	37	BY8	Bulgarian yogurt	*Enterococcus* sp.	Sequencing
2	CA2	*C. aspersum*	*E. casseliflavus*	PCR, Sequencing	38	BY9	Bulgarian yogurt	*E. casseliflavus*	PCR, Sequencing
3	CA3	*C. aspersum*	*E. gilvus*	MALDI-TOF	39	BY10	Bulgarian yogurt	*E. faecalis*	PCR
4	CA4	*C. aspersum*	*E. mundtii*	MALDI-TOF	40	BY11	Bulgarian yogurt	*E. faecalis*	PCR
5	CA5	*C. aspersum*	*E. casseliflavus*	PCR, Sequencing	41	BY12	Bulgarian yogurt	*E. faecium*	PCR
6	CA6	*C. aspersum*	*E. mundtii*	Sequencing	42	BY13	Bulgarian yogurt	*E. faecium*	PCR
7	CA7	*C. aspersum*	*E. mundtii*	MALDI-TOF	43	BY14	Bulgarian yogurt	*E. faecium*	PCR
8	CA8	*C. aspersum*	*E. pseudoavium*	Sequencing	44	BY15	Bulgarian yogurt	*E. faecium*	PCR
9	CA9	*C. aspersum*	*E. pseudoavium*	Sequencing	45	BY16	Bulgarian yogurt	*E. faecium*	PCR
10	CA10	*C. aspersum*	*E. pallens*	Sequencing	46	BY17	Bulgarian yogurt	*E. gallinarum*	MALDI-TOF
11	CA11	*C. aspersum*	*E. malodoratus*	MALDI-TOF	47	BY18	Bulgarian yogurt	*E. casseliflavus*	PCR, Sequencing
12	CA12	*C. aspersum*	*E. casseliflavus*	PCR, Sequencing	48	BY19	Bulgarian yogurt	*E. casseliflavus*	PCR, Sequencing
13	CA13	*C. aspersum*	*E. devriesei*	Sequencing	49	BY20	Bulgarian yogurt	*E. casseliflavus*	PCR, Sequencing
14	CA14	*C. aspersum*	*E. gallinarum*	Sequencing	50	BY21	Bulgarian yogurt	*E. casseliflavus*	PCR, Sequencing
15	CA15	*C. aspersum*	*E. gallinarum*	Sequencing	51	BY22	Bulgarian yogurt	*E. faecalis*	PCR
16	CA16	*C. aspersum*	*E. devriesei*	MALDI-TOF	52	BY23	Bulgarian yogurt	*E. faecalis*	PCR
17	CA17	*C. aspersum*	*E. mundtii*	MALDI-TOF	53	BY24	Bulgarian yogurt	*E. faecalis*	PCR
18	CM1	Cow milk	*E. faecium*	PCR	54	BY25	Bulgarian yogurt	*E. faecalis*	PCR
19	CM2	Cow milk	*E. durans*	MALDI-TOF	55	BY26	Bulgarian yogurt	*E. faecalis*	PCR
20	CM3	Cow milk	*E. durans*	MALDI-TOF	56	BY27	Bulgarian yogurt	*E. faecalis*	PCR

Table 2. *Cont.*

No	Isolate	Origin	Species	Method of Identification	No.	Isolate	Origin	Species	Method of Identification
21	CM4	Cow milk	*E. faecalis*	MALDI-TOF	57	BM1	Breast milk	*E. faecalis*	PCR
22	YFC1	Young feta cheese	*E. faecalis*	PCR	58	BM2	Breast milk	*E. faecalis*	PCR
23	YFC2	Young feta cheese	*E. durans*	PCR	59	BM3	Breast milk	*E. faecalis*	PCR
24	YFC3	Young feta cheese	*E. faecalis*	PCR	60	BM4	Breast milk	*E. faecalis*	PCR
25	YFC4	Young feta cheese	*E. durans*	PCR	61	BM5	Breast milk	*E. faecalis*	PCR
26	YFC5	Young feta cheese	*E. durans*	PCR	62	BM6	Breast milk	*E. faecalis*	PCR
27	MFC1	Matured feta cheese	*E. faecium*	PCR	63	BM7	Breast milk	*E. faecalis*	PCR
28	MFC2	Matured feta cheese	*E. faecium*	PCR	64	BM8	Breast milk	*E. faecalis*	PCR
29	DK1	Doner kebab	*E. faecium*	PCR	65	BM9	Breast milk	*E. faecalis*	PCR
30	BY1	Bulgarian yogurt	*E. faecium*	MALDI-TOF	66	BM10	Breast milk	*E. faecalis*	PCR
31	BY2	Bulgarian yogurt	*E. faecalis*	PCR	67	BM11	Breast milk	*E. faecalis*	PCR
32	BY3	Bulgarian yogurt	*E. faecalis*	PCR	68	BM12	Breast milk	*E. faecalis*	PCR
33	BY4	Bulgarian yogurt	*E. faecalis*	PCR	69	BM13	Breast milk	*E. faecalis*	PCR
34	BY5	Bulgarian yogurt	*E. faecalis*	PCR	70	BM14	Breast milk	*E. faecalis*	PCR
35	BY6	Bulgarian yogurt	*E. faecalis*	PCR	71	BM15	Breast milk	*E. faecalis*	PCR
36	BY7	Bulgarian yogurt	*Enterococcus* sp.	Sequencing	72	BM16	Breast milk	*E. faecalis*	PCR

Eleven species were identified: *E. mundtii*, *E. casseliflavus*, *Enterococcus gilvus*, *Enterococcus pseudoavium*, *Enterococcus pallens*, *Enterococcus malodoratus*, *Enterococcus devriesei*, *Enterococcus gallinarum*, *E. durans*, *E. faecium*, and *E. faecalis*. Two isolates were identified at genus level as *Enterococcus* spp. (BY7 and BY8, isolated from Bulgarian yoghurt). The greatest species diversity was established in the GIT of *C. aspersum* as eight species were identified: 29% *E. mundtii*, 18% *E. casseliflavus*, 6% *E. gilvus*, 12% *E. pseudoavium*, 6% *E. pallens*, 6% *E. malodoratus*, 12% *E. devriesei*, and 12% *E. gallinarum* (Figure 2).

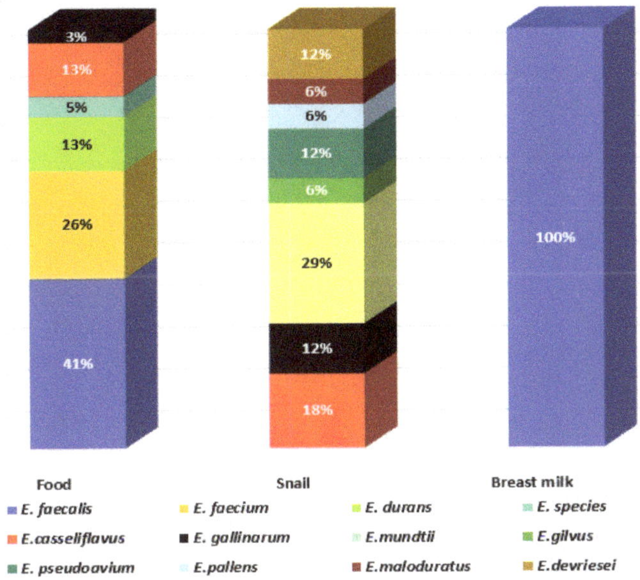

Figure 2. Species diversity in the analyzed habitats.

Two of the species (*E. mundtii* and *E. casseliflavus*), generally recognized as plant-associated enterococci [15], represent 47% of the enterococcal population of the GIT of the snail, which is a herbivore. *E. faecalis* and *E. faecium* were not detected in the GIT of *C. aspersum*. In contrast, these two species were predominantly identified in the food samples: 41% and 26%, respectively. The species *E. durans* (6%) was isolated from cow milk and young feta cheese from cow milk. All isolates from human breast milk were identified as *E. faecalis*.

3.2. Occurrence of cylB Gene and Production of Hemolysin

Hemolytic activity of the *Enterococcus* species is considered one of the basic virulence factors influencing their pathogenicity. Our results showed that there were no strains that showed phenotypic β-hemolytic activity on Columbia agar + 5% horse blood, but some strains expressed α-hemolysis (31% of all tested strains) (Figure 3a). Of the 17 strains isolated from *C. aspersum* GIT, 10 representatives of the species *E. casseliflavus* (n = 2), *E. gilvus* (n = 1), *E. gallinarum* (n = 2), *E. pseudoavium* (n = 1), *E. pallens* (n = 1), *E. malodoratus* (n = 1), and *E. devriesei* (n = 2) showed phenotypic α-hemolytic activity. The only species in this group not showing hemolytic activity was the plant-associated species *E. mundtii*. Of the 39 strains isolated from food samples (cow milk, Bulgarian yogurt, young feta cheese, and mature feta cheese), 11 were α-hemolytic. However, such activity was observed among *E. durans* (YFC2), *E. casseliflavus* (BY19), *E. gallinarum* (BY17), *Enterococcus* sp. (BY8), *E. faecalis* (n = 4), and *E. faecium* (n = 3). Surprisingly, only one strain isolated from human breast milk possessed α-hemolytic activity (*E. faecalis* BM5). Of great importance was the correlation between phenotypic hemolytic expression and the related genotypic determinants. The *cylB* gene is a member of the *cyl* operon, responsible for the synthesis of cytolysin and for β-hemolytic activity, respectively [62]. Thirteen out of all the tested strains had the *cylB* gene (9 from human breast milk and 4 from food samples), but none of them expressed β-hemolytic activity. Only two strains (*E. faecium* DK1 and *E. faecalis* BM5) expressed α-hemolysis and possessed the *cylB* gene simultaneously (Figure 3 b,c). None of the strains isolated from the GIT of the snail possessed the *cylB* gene.

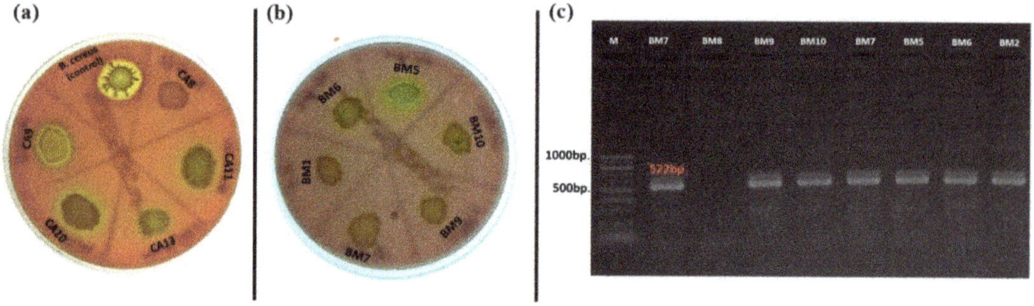

Figure 3. (**a**) Hemolysis test on Columbia agar plate supplemented with 5% horse blood; (**b**) α-hemolysis of *E. faecalis* isolated from breast milk; (**c**) Gel electrophoresis of PCR amplification products for the *cylB* gene of breast milk isolates.

3.3. Occurrence of *gelE* and Production of Gelatinase

The production of gelatinase and the occurrence of the related gene (*gelE*) were also investigated. Positive amplification for *gelE* was found in 35% of the tested isolates (Figure 4a). Of these, 11 had a food origin and 14 were isolated from breast milk. All of them belonged to the species *E. faecalis* and *E. faecium*. Simultaneous occurrence of phenotypic gelatinase activity and the related genotypic determinant (*gelE*) was observed only in three enterococcal strains (*E. faecalis* BM1, BM2, and BM11) isolated from human breast milk (Figure 4b). None of the snail isolates had the abovementioned gene.

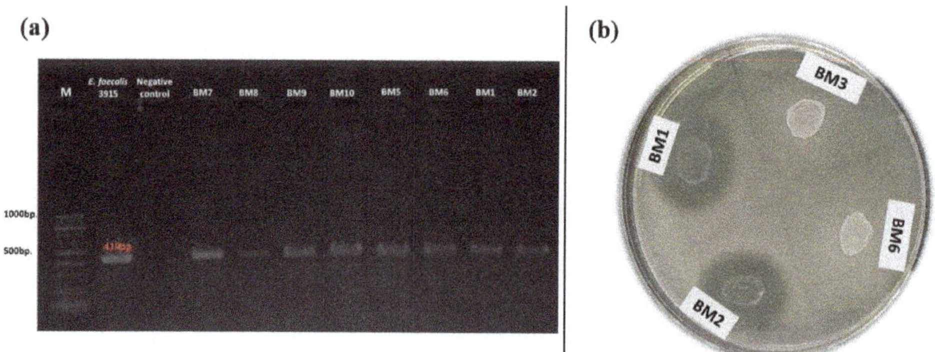

Figure 4. (**a**) Gel electrophoresis of PCR amplification products for the *gelE* gene of breast milk isolates; (**b**) Phenotypic gelatinase activity of breast milk isolates (BM1 and BM2) on gelatin agar.

3.4. Phenotypic Antibiotic Resistance

The obtained results from the phenotypic antibiotic resistance were interpreted according to EUCAST, 2019 [63]. All isolates were susceptible to fluoroquinolone antibiotics (ciprofloxacin, levofloxacin, and norfloxacin), teicoplanin, linezolid, nitrofurantoin, vancomycin, and imipenem. Resistance to ampicillin was observed in 21% (n = 15) of all tested strains (Table 3). Among them, 53% were *E. faecalis* isolated from human breast milk.

Table 3. Distribution of phenotypic antibiotic resistance among the enterococcal isolates.

ABR Phenotype	Number of Isolates	Species Identification	Origin of Isolation
One Antibiotic			
AMP	12	*E. faecium DK1*	Doner kebab
		E. gallinarum BY17	Bulgarian yogurt
		E. mundtii CA1 *E. malodoratus CA11* *E. devriesei CA13*	*C. aspersum*
		E. faecalis BM3 *E. faecalis BM4* *E. faecalis BM5* *E. faecalis BM6* *E. faecalis BM9* *E. faecalis BM12* *E. faecalis BM14*	Human breast milk
Two antibiotics			
AMP + ERV	1	*E. faecium CM1*	Cow milk
AMP + TG	1	*E. faecalis YFC1*	Young feta cheese
GEN + RP	1	*E. faecalis BM7*	Human breast milk
Three antibiotics			
AMP + HLS + RP	1	*E. faecalis* BM15	Human breast milk

High-level gentamicin resistance (HLGR), high-level streptomycin resistance (HLSR), and quinupristin-dalfopristin resistance (RP) were established for two isolates with human origin (Figure 5a).

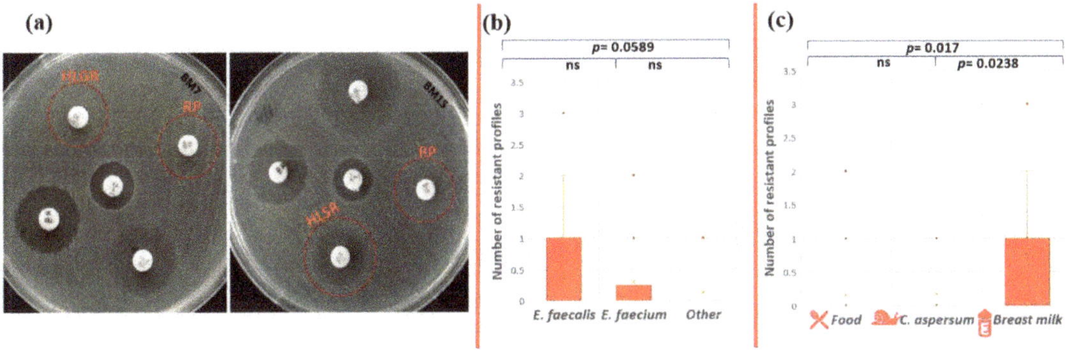

Figure 5. Phenotypic antibiotic resistance test. (**a**) Red circles indicate HLGR and RP resistance of *E. faecalis* BM7 and HLSR (colonies within the zone) and RP resistance of *E. faecalis* BM15; statistical analyses include comparison of the number of ABR profiles (**b**) between the different enterococcal species and (**c**) between the different strain origins. Significant difference was considered $p < 0.05$; ns corresponds to non-significant difference.

The only strain that showed multidrug phenotypic resistance profile to three antibiotics (AMP-HLSR-RP) was *E. faecalis* BM15 isolated from human breast milk (1.38% from all tested strains). Phenotypic resistance to two antibiotics was observed for strains *E. faecium* CM1 (AMP-ERV), *E. faecalis* YFC1 (AMP-TG), and *E. faecalis* BM7 (HLGR-RP). However, the statistical analysis showed that there is no significant difference in the number of resistance profiles between *E. faecalis* and *E. faecium* ($p = 0.674$); *E. faecalis* and other *Enterococcus* species

($p = 0.0589$); and *E. faecium* and other *Enterococcus* species ($p = 0.471$) (Figure 5b). The human isolates (breast milk) exhibited patterns of resistance to more antibiotics compared to the other two groups (food and snail isolates). A significant difference between antibiotic resistance phenotype profiles was established between strains from food and breast milk ($p = 0.017$), as well as strains from snail and breast milk ($p = 0.0238$), but not between isolates from snail and food ($p = 0.855$) (Figure 5c).

3.5. Screening for Antibiotic Resistance Genes

Overall, the abundance of antibiotic resistance genes in the analyzed strains was low. Only 16 (22%) of all isolates showed the presence of one or more antibiotic resistance genes (Figure 6a).

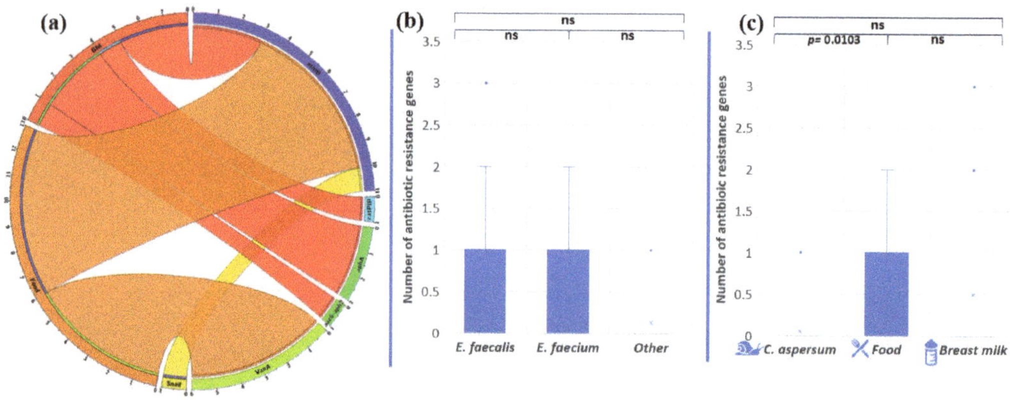

Figure 6. (**a**) Distribution of antibiotic resistance genes in enterococci from food, snail, and breast milk (labeled BM in the circus plot) (the image was generated with Circos Table Viewer v0.63-10). The outer ring of the circus plot represents the number of isolates that carry the analyzed genes. Connecting lines between the specific genes and the origin of the isolates are shown if the gene was detected in isolates from any of the three origins. Comparison of the number of antibiotic resistance genes between the different enterococcal species (**b**) and the different strains' origins (**c**). Significant difference is considered $p < 0.05$.

The gene *ermB*, associated with macrolide resistance, was most frequently found among the analyzed *Enterococcus* population (15, 2%), followed by *vanA* (8, 3%), *aphA3* (4, 2%), *aac6′-aph2″* (1, 4%), and cat_{pIP501} (1, 4%). We established that two *E. faecalis* strains (BM15 and BM7) had three genes encoding antibiotic resistance, which makes them unsusceptible to macrolide and aminoglycoside antibiotics. Strain BM7 showed the presence of *ermB*, *aphA3*, and *aac6′-aph2″*. Strain BM15 possesses the genes *ermB*, cat_{pIP501}, and *aphA3*. These data showed a correlation between genotypic and phenotypic antibiotic resistance to macrolides. However, the other gene responsible for macrolide resistance (*mefA*) was not detected. Unexpectedly, the gene *vanA*, associated with vancomycin resistance, was found in six of our strains: five *E. faecalis* strains from Bulgarian yogurt (BY2, BY3, BY4, BY5, BY6) and one *E. faecium* strain from mature feta cheese (MFC1), although no phenotypic appearance was observed.

The species comparison showed no significant differences in the number of antibiotic resistance genes ($p = 0.8897$ between *E. faecalis* and *E. faecium*; $p = 0.0665$ between *E. faecalis* and other *Enterococcus* species; and $p = 0.273$ between *E. faecium* and other *Enterococcus* species) (Figure 6b). On the other hand, the origin comparison showed significant differences between the snail and food distribution of antibiotic resistance genes ($p = 0.0103$). The above were not observed between food and breast milk isolates ($p = 0.569$) or between snail and breast milk isolates ($p = 0.135$) (Figure 6c).

Only one strain (BM7), having the HLGR gene *aac6′-aph2″*, showed the relevant phenotypic resistance to 30 μg/disc gentamicin. Fifteen of the isolates (20%) showed phenotypic resistance to ampicillin, but none of the strains had *blaZ* (codes β-lactamases) in its genome. The gene *gyrA* was also absent and, as expected, resistance to fluoroquinolones (ciprofloxacin, norfloxacin, and levofloxacin) was not observed.

3.6. Screening for Virulence-Associated Genes

The pathogenicity degree of the pathogenic microorganisms depends on genetically determined virulence factors. The presence of a total of eight virulence genes (*cylB*, *esp*, *gls24*, *nucl*, *psaA*, *agg*, *gelE*, and *ace*) among our enterococcal isolates was investigated (Table 4). The analyses of the distribution of the tested virulence-associated genes showed that the snail isolates did not possess any of the analyzed virulence genes. Stress protein regulator (*gls24-like*) was not found in the investigated enterococcal isolates. The most amplified gene among all the isolates was the gelatinase gene (*gelE*), followed by the Mn-transporter *psaA*: 31% and 28%, respectively. Both genes responsible for the synthesis of enterococcal surface protein (*esp*) and nuclease (*nucl*) were presented in 19.4% of all tested strains. Genes responsible for hemolytic activity (*cylB*) and collagen-binding protein (*ace*) were detected in 18% of the enterococcal population. Genetic determinants for aggregation substance (Agg) were found in 15.2% of the tested isolates. Four human breast milk isolates (BM5, BM6, BM9, and BM10) contain seven out of the eight screened virulence genes. The distribution of all tested virulence factors among food isolates were strain specific. The comparison of the distribution of the virulence factors in enterococcal strains isolated from the different ecological niches, showed significant differences (between snail and food isolates $p = 9.6 \times 10^{-5}$; food and breast milk $p = 5.43 \times 10^{-9}$; snail and breast milk $p = 7.12 \times 10^{-9}$) (Figure 7a). A similar tendency was observed between *E. faecalis* and *E. faecium* $p = 1.2 \times 10^{-5}$; *E. faecium* and other *Enterococcus* species $p = 0.00265$; *E. faecalis* and other *Enterococcus* species $p = 1.52 \times 10^{-7}$ (Figure 7b).

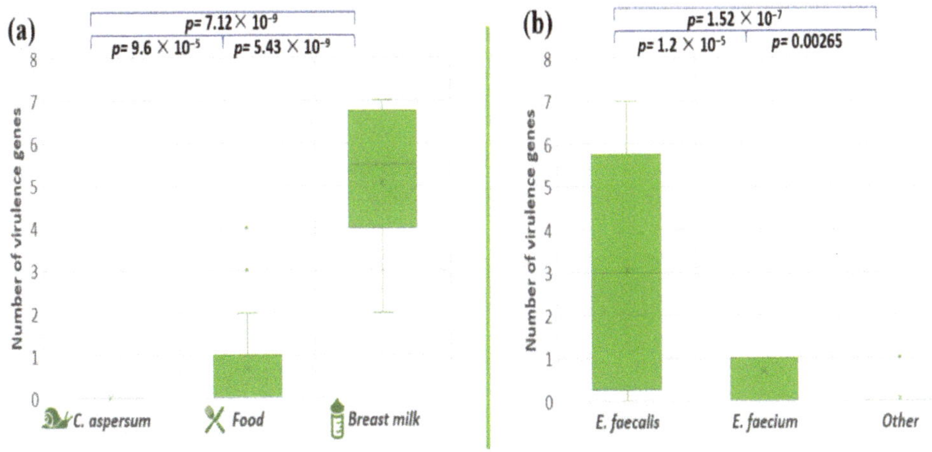

Figure 7. Comparison of the number of virulence genes between the different enterococcal species (**a**) and the different strains' origins (**b**). Significant difference is considered $p < 0.05$.

Table 4. Distribution of genes encoding virulence factors among the tested enterococcal population.

Strains	Virulence Genes								Strains	Virulence Genes							
	cyl B	*esp*	*gls 24*	*nucl*	*psa*	*agg*	*gel E*	*ace*		*cyl B*	*esp*	*gls 24*	*nucl*	*psa*	*agg*	*gel E*	*ace*
E. faecium CM1									*E. faecalis* BY25								
E. durans CM2									*E. faecalis* BY26								
E. durans CM3									*E. faecalis* BY27								
E. faecalis CM4									*E. mundtii* CA1								
E. faecalis YFC1									*E. casseliflavus* CA2								
E. durans YFC2									*E. gilvus* CA3								
E. faecalis YFC3									*E. mundtii* CA4								
E. durans YFC4									*E. casseliflavus* CA5								
E. durans YFC5									*E. mundtii* CA6								
E. faecium MFC1									*E. mundtii* CA7								
E. faecium MFC2									*E. pseudoavium* CA8								
E. faecium DK1									*E. pseudoavium* CA9								
E. faecium BY1									*E. pallens* CA10								
E. faecalis BY2									*E. malodurutus* CA11								
E. faecalis BY3									*E. casseliflavus* CA12								
E. faecalis BY4									*E. devriesei* CA13								
E.faecalis BY5									*E. gallinarum* CA14								
E.faecalis BY6									*E. gallinarum* CA15								
E. species BY7									*E. devriesei* CA16								
E. species BY8									*E. mundtii* CA17								
E. casseliflavus BY9									*E. faecalis* BM1								
E. faecalis BY10									*E. faecalis* BM2								
E. faecalis BY11									*E. faecalis* BM3								
E. faecium BY12									*E. faecalis* BM4								
E. faecium BY13									*E. faecalis* BM5								

Table 4. *Cont.*

Strains	Virulence Genes								Strains	Virulence Genes							
	cyl B	*esp*	*gls 24*	*nucl*	*psa*	*agg*	*gel E*	*ace*		*cyl B*	*esp*	*gls 24*	*nucl*	*psa*	*agg*	*gel E*	*ace*
E. faecium BY14	pink	pink	pink	pink	pink	pink	green	pink	*E. faecalis* BM6	green	green	pink	green	green	green	green	green
E. faecium BY15	pink	pink	pink	pink	pink	pink	green	pink	*E. faecalis* BM7	green	pink	pink	green	green	green	green	green
E. faecium BY16	pink	pink	pink	pink	pink	pink	green	pink	*E. faecalis* BM8	pink	pink	pink	green	green	green	green	pink
E. gallinarum BY17	pink	pink	pink	pink	pink	pink	pink	pink	*E. faecalis* BM9	green	green	pink	green	green	green	green	green
E. casseliflavus BY18	pink	pink	pink	pink	pink	pink	pink	pink	*E. faecalis* BM10	green	green	pink	green	green	green	green	green
E. casseliflavus BY19	pink	pink	pink	pink	pink	pink	pink	pink	*E. faecalis* BM11	pink	green	pink	green	green	green	green	green
E. casseliflavus BY20	pink	pink	pink	pink	pink	pink	pink	pink	*E. faecalis* BM12	pink	green	pink	green	green	pink	green	green
E. casseliflavus BY21	pink	pink	pink	pink	pink	pink	pink	pink	*E. faecalis* BM13	pink	pink	pink	green	green	pink	green	green
E. faecalis BY22	pink	pink	pink	pink	pink	pink	pink	pink	*E. faecalis* BM14	pink	green	pink	green	green	pink	pink	green
E. faecalis BY23	pink	pink	pink	pink	pink	pink	pink	pink	*E. faecalis* BM15	pink	green	pink	green	green	pink	pink	green
E. faecalis BY24	pink	pink	pink	pink	pink	pink	pink	pink	*E. faecalis* BM16	pink	green	pink	green	green	pink	green	pink

Pink color—negative result, no amplification product; green color—positive result—specific amplification product.

4. Discussion

The multi-host lifestyle and unique adaptability of enterococci lead to interconnected microbiomes between mammals, invertebrates, insects, and plants which facilitate the acquisition and spread of virulence and antibiotic resistance genes (ABR) [64]. Therefore, the enterococcal populations from different biological kingdoms/origins represent reservoirs of factors causing infections in humans and plants [1,65]. In this study, we compared the pathogenic potential of enterococci isolated from diverse habitats with respect to assess their possible virulent potential for subsequent colonization of plants after potential re-return into the environment.

Our main hypothesis was that, in passing through hosts from different kingdoms, enterococci successfully adapt to the current habitat by acquiring various virulence and ABR genes, which helps them in interspecies relationships. Therefore, it is mandatory to investigate in depth the pathogenic potential of enterococci originating from various ecological niches. This accumulated knowledge could be useful in evaluating the potential risk of undesired genetic burden in the environment after the eventual re-entering of the enterococci (with acquired virulence and ABR potential) into the environment.

As a primary source of food for many organisms, plants, along with soil and water, can act as reservoirs for enterococcal species, which can subsequently join the path of pathogenicity and be transmitted through the animal chain mainly by herbivorous animals. A good example is the *C. aspersum* species of snail, that is in touch with all these habitats and can itself be used as a food source for other animals, including humans. By studying the microbiome of the snail, the microbial presence in its food (plants) can be deducted.

In our study, in total, 72 enterococcal strains, representatives of 11 species and isolated from seven habitats, were characterized (Table 2). In this study, we found eight enterococci species in the snail intestinal tract, with *E. mundtii* and *E. casseliflavus* being the most prevalent. *E. casseliflavus* was also established in Bulgarian yoghurt (fermented cow milk), derived from herbivorous warm-blooded animal (cow), but not in our isolates from raw cow milk itself. However, the persistence of *E. casseliflavus* in raw bovine milk has been reported [66]. Surprisingly, in the GIT of the snail, none of the isolates belonged to the *E. faecium* or *E. faecalis* species but both species dominated in all other samples. The other six species found in the GIT of *C. aspersum* have been generally reported to have human and animal origins [15]. It can be suggested that these species of bacteria have moved into the plants from soil and water and from there into the GIT of the snail [21]. Our results showed that the potential plant-associated isolates in the GIT and food (*E. casseliflavus*) did not carry genes for virulence and antibiotic resistance. Only three isolates (*E. mundtii* CA1, *E. malodoratus* CA11, and *E. devriesei* CA13) were phenotypically resistant to ampicillin. However, a study found that enterococci isolated from raw and processed plant-derived foods have a quite different phenotypic and molecular profile of antibiotic resistance [21]. The authors of the study found that *E. faecium*, *E. faecalis*, and *E. casseliflavus* strains are resistant to erythromycin, streptomycin, tigecycline, fosfomycin, and rifampicin but not to ampicillin. In that study, correlation between phenotypic high aminoglycoside resistance (HLAR) and the related genetic determinants (*ant(6′)-Ia, aph(3′)-IIIa and aac(6′)-Ie-aph(2″)-la*) has been reported [21]. An interesting result was that the species *E. gallinarum* was found in two of our samples: snail GIT and Bulgarian yoghurt. We established that no virulence or ABR was found in the snail isolates (*E. gallinarum* CA14 and CA15), as opposed to the Bulgarian yoghurt isolate (*E. gallinarum* BY17), which was found to carry the gene for ampicillin resistance. Thus, we can conclude that enterococci from snails and their food source, namely plants, did not represent any threat to human health. The acquisition of ampicillin resistance may likely happen in some of the later stages of the alimentary chain.

An important reservoir for the dissemination of enterococcal populations is the products of the lactation of mammals. For example, the most commonly isolated species from goat and sheep raw milk and their products (cheese) have been reported to be *E. faecalis* and *E. faecium* [67]. In our investigation, two similar milk products were analyzed—from cows and humans. To our knowledge, the enterococcal population in human breast milk is poorly studied. Breast milk has complex nutrient composition and contains a variety of bacterial species which influence infant health and immunity [68]. Some authors have even suggested that the enterococcal abundance corelates with the infants' excessive weight gain [69]. The species *E. faecalis*, *E. faecium*, *E. hirae*, *E. casseliflavus*, and *E. durans* have been reported to be found in the milk of healthy women [69–71]. In our cow milk samples, we found three species—*E. faecalis*, *E. faecium*, and *E. durans*, which is in accordance with other authors' findings [67]. According to some authors, of all reported plant-associated enterococcal species, only *E. faecalis*, *E. faecium*, and *E. casseliflavus* are dominant and best adapted to mammals [69]. It has been suggested that this selection is due to the extreme genomic plasticity of these species, allowing for facile horizontal gene transfer [45]. However, in our study, only strains of *E. faecalis* were identified in human breast milk. Our results showed that the distribution of the virulence and ABR genes was greatest among the enterococcal population in this ecological niche. It has been reported that *E. faecalis* and *E. faecium* have the greatest potential for causing infections as these species are the primary isolates from infected patients [65]. Some authors have suggested that the virulence and antibiotic resistance capability of some enterococci is even strain specific, considering the ecological niche they inhabit [55,72]. The statistical analyses of our results showed that the distribution of ABR genes is dependent on the ecological level but not on the species belonging. Our observations showed that the only multidrug-resistant strain was found in human breast milk. However, multidrug-resistant enterococci have also been reported in dairy products [67].

The virulence genes were much more present in the strains and a better generalization can be made. Significant differences were observed between strains from different origins as well as between different species (Figure 7). The greatest number of virulence genes was detected in *E. faecalis* strains from breast milk. The distribution of tested virulence-associated genes among the strains from the other samples was found to be species and even strain specific, because *E. faecalis* and *E. faecium* were not found in the snail GIT. Comparing the number of virulence genes distributed among *E. faecalis* and *E. faecium* isolates from food and human samples, we can conclude that this number drastically increased in the latter. All breast milk isolates carry genes for virulence factors, and four of them (*E. faecalis* BM5, BM6, BM9, and BM10) contain seven out of the eight screened virulence genes. Our results differ from those reported by Santana et al. [73], who investigated the distribution of *ace*, *efaA*, *gelE*, *cylA*, *hyl*, and *esp* virulence genes among an enterococcal population isolated from raw human breast milk. In their investigation, only two genes were detected (*efaA* and *ace*). We also noticed that two of our strains of *E. durans* (CM2 and CM3) isolated from cow milk did not carry any genes for virulence factors, as opposed to two strains of *E. durans* (YFC4 and YFC5) isolated from young feta cheese, which had the *cylB* gene.

Hemolytic activity is another virulence trait with great importance, as it enhances the severity of the caused infections. The production of cytolysin is associated with induced septicemia and a fivefold increased risk of acutely terminal outcome in patients [74]. In this study, none of the isolates had β-hemolytic activity, although some of the strains amplified the *cylB* gene. However, the ability of *E. faecalis* to express β-hemolysis has been reported [70]. It is known that the operon for cytolysin production is composed of five genes. The genes $cylL_l$ and $cylL_s$ encode the two structural subunits, which are then modified intracellularly by the product of the *cylM* gene. Then, they are transported out of the cell by a transporter encoded by the *cylB* gene. Once they are out of the cell, the precursor components are then activated by the *cylA* product. The gene *cylI* is responsible for the immunity of the bacteria to cytolysin. The regulation of expression is carried out by the products of two other genes—*cylR1* and *cylR2* [75–78]. In the present study, we established that the *cylB* gene did not correlate with the phenotypic hemolytic activity of the strains, which could be explained by an incomplete *cyl* operon. An interesting finding was that the highest percentage of α-hemolytic strains (45%) was established among species isolated from the GIT of *C. aspersum* (*E. durans*, *E. casseliflavus*, *E. gilvus*, *E. pseudoavium*, *E. pallens*, *E. malodoratus*, *E. devriesei*, and *E. gallinarum*). On the other hand, only one

E. faecalis strain (BM4) isolated from human breast milk showed such activity. Moreover, α-hemolysis does not cause complete destruction of the red blood cells, which may limit the pathogenicity of the analyzed strains.

Gelatinase is an enzyme which is involved in the degradation of gelatin, collagen, casein, hemoglobin, etc. [79]. However, this feature of enterococcal isolates of non-hospital origin is poorly studied. For that reason, we examined the gelatinase phenotype and genotype in our collection. The expression of the *gelE* gene has been reported to be regulated by the products of different genes (*fsrA*, *fsrB*, and *fsrC*) in the *fsr* operon. Moreover, the expression of these genes has been described to be dependent on cell density [41]. Thus, the presence of *gelE* does not always produce a positive phenotype. Our results are in accordance with those reported from other authors [54,80,81]. Generally, our work demonstrates that *gelE* is present in 35% of our isolates, but only 4% were gelatinase producers (isolates from human breast milk). We can conclude that unexpressed *gelE* gene in most strains is due to one of the aforementioned reasons—lack of *fsr* operon or low cell density. Our observations indicate that the expression of the *gelE* gene and the manifestation of phenotypic gelatinase activity is a feature related to human isolates.

5. Conclusions

In this study, we tried to track the path of pathogenicity of potentially plant-associated enterococci in different levels of the alimentary chain. We established a step-by-step increase in the factors of virulence and ABR with maximal persistence in the human product—breast milk. This creates a serious problem and ambiguity—what will happen with these acquired pathogenic potential when these strains re-enter the environment and colonize the plant again? This study's findings can be considered as a solid basis for future investigations.

Author Contributions: Conceptualization: M.P., Y.K. and P.H.; methodology: P.H.; software: M.P.; validation: P.H. and Y.K.; formal analysis: M.P.; investigation: M.P., M.T., Y.K. and P.H.; resources: T.B. and M.R.; data curation: P.H. and Y.K.; writing—original draft preparation: M.P.; writing—review and editing: Y.K. and P.H.; visualization: M.P. and Y.K.; supervision: P.H. All authors have read and agreed to the published version of the manuscript.

Funding: This research received no external funding.

Institutional Review Board Statement: Not applicable.

Informed Consent Statement: Not applicable. In Bulgaria, breast milk deposited at the Human Milk Bank, Bulgaria, is registered as food and is regulated by the Bulgarian Food Safety Authority.

Data Availability Statement: The data presented in this research are available in the manuscript.

Acknowledgments: The authors want to thank the Human Milk Bank, Bulgaria, for donating the breast milk samples.

Conflicts of Interest: The authors declare no conflicts of interest.

References

1. Jha, A.K.; Bais, H.P.; Vivanco, J.M. Enterococcus faecalis mammalian virulence-related factors exhibit potent pathogenicity in the Arabidopsis thaliana plant model. *Infect Immun.* **2005**, *73*, 464–475. [CrossRef] [PubMed]
2. Micallef, S.A.; Rosenberg Goldstein, R.E.; George, A.; Ewing, L.; Tall, B.D.; Boyer, M.S.; Joseph, S.W.; Sapkota, A.R. Diversity, distribution and antibiotic resistance of *Enterococcus* spp. recovered from tomatoes, leaves, water and soil on U.S. Mid-Atlantic farms. *Food Microbiol.* **2013**, *36*, 465–474. [CrossRef] [PubMed]
3. Sadowy, E.; Luczkiewicz, A. Drug-resistant and hospital-associated Enterococcus faecium from wastewater, riverine estuary and anthropogenically impacted marine catchment basin. *BMC Microbiol.* **2014**, *14*, 66. [CrossRef] [PubMed]
4. Nowakiewicz, A.; Ziółkowska, G.; Trościańczyk, A.; Zieba, P.; Gnat, S. Determination of resistance and virulence genes in *Enterococcus faecalis* and *E. faecium* strains isolated from poultry and their genotypic characterization by ADSRRS-fingerprinting. *Poult. Sci.* **2017**, *96*, 986–996. [CrossRef] [PubMed]
5. Abriouel, H.; Omar, N.B.; Molinos, A.C.; López, R.L.; Grande, M.J.; Martínez-Viedma, P.; Ortega, E.; Cañamero, M.M.; Galvez, A. Comparative analysis of genetic diversity and incidence of virulence factors and antibiotic resistance among enterococcal populations from raw fruit and vegetable foods, water and soil, and clinical samples. *Int. J. Food Microbiol.* **2008**, *123*, 38–49. [CrossRef] [PubMed]

6. Dubin, K.; Pamer, E.G. Enterococci and Their Interactions with the Intestinal Microbiome. *Microbiol Spectr.* **2014**, *5*, 5–6. [CrossRef]
7. Franz, C.M.A.P.; Schillinger, U.; Holzapfel, W.H. Production and characterization of enterocin 900, a bacteriocin produced by Enterococcus faecium BFE 900 from black olives. *Int. J. Food Microbiol.* **1996**, *29*, 255–270. [CrossRef]
8. Kizheva, Y.; Georgiev, G.; Donchev, D.; Dimitrova, M.; Pandova, M.; Rasheva, I.; Hristova, P. Cross-Over Pathogenic Bacteria Detected in Infected Tomatoes (*Solanum lycopersicum* L.) and Peppers (Capsicum annuum L.) in Bulgaria. *Pathogens* **2022**, *11*, 1507. [CrossRef]
9. Johnston, L.M.; Jaykus, L.A. Antimicrobial resistance of Enterococcus species isolated from produce. *Appl. Environ. Microbiol.* **2004**, *70*, 3133–3137. [CrossRef]
10. Mundt, J.O. Occurrence of Enterococci on Plants in aWild Environment. *Appl. Microbiol.* **1963**, *11*, 141–144. [CrossRef]
11. Müller, T.; Ulrich, A.; Ott, E.M.; Müller, M. Identification of plant-associated enterococci. *J. Appl. Microbiol.* **2001**, *91*, 268–278. [CrossRef] [PubMed]
12. Miller, W.R.; Murray, B.E.; Rice, L.B.; Arias, C.A. Resistance in Vancomycin-Resistant Enterococci. *Infect. Dis. Clin. N. Am.* **2020**, *34*, 751–771. [CrossRef] [PubMed]
13. Godfree, A.F.; Kay, D.; Wyer, M.D. Faecal streptococci as indicators of faecal contamination in water. *Soc. Appl. Bacteriol. Symp. Ser.* **1997**, *26*, 110S–119S. [CrossRef] [PubMed]
14. Ott, E.M.; Müller, T.; Müller, M.; Franz, C.M.A.P.; Ulrich, A.; Gabel, M.; Seyfarth, W. Population dynamics and antagonistic potential of enterococci colonizing the phyllosphere of grasses. *J. Appl. Microbiol.* **2001**, *91*, 54–66. [CrossRef] [PubMed]
15. Byappanahalli, M.N.; Nevers, M.B.; Korajkic, A.; Staley, Z.R.; Harwood, V.J. Enterococci in the Environment. *Microbiol. Mol. Biol.* **2012**, *76*, 685. [CrossRef] [PubMed]
16. Švec, P.; Vandamme, P.; Bryndová, H.; Holochová, P.; Kosina, M.; Mašlaňová, I.; Sedláček, I. Enterococcus plantarum sp. nov., isolated from plants. *Int. J. Syst. Evol. Microbiol.* **2012**, *62 Pt 7*, 1499–1505. [CrossRef] [PubMed]
17. Giraffa, G. Enterococci from foods. *FEMS Microbiol. Rev.* **2002**, *26*, 163–171. [CrossRef]
18. Hammerum, A.M. Enterococci of animal origin and their significance for public health. *Clin. Microbiol. Infect.* **2012**, *18*, 619–625. [CrossRef]
19. Leff, J.W.; Fierer, N. Bacterial communities associated with the surfaces of fresh fruits and vegetables. *PLoS ONE* **2013**, *8*, e59310. [CrossRef]
20. Mogren, L.; Windstam, S.; Boqvist, S.; Vågsholm, I.; Söderqvist, K.; Rosberg, A.K.; Lindén, J.; Mulaosmanovic, E.; Karlsson, M.; Uhlig, E.; et al. The hurdle approach-A holistic concept for controlling food safety risks associated with pathogenic bacterial contamination of leafy green vegetables. A review. *Front. Microbiol.* **2018**, *9*, 1965. [CrossRef]
21. Chajęcka-Wierzchowska, W.; Zarzecka, U.; Zadernowska, A. Enterococci isolated from plant-derived food—Analysis of antibiotic resistance and the occurrence of resistance genes. *Food Sci. Technol.* **2021**, *139*, 110549. [CrossRef]
22. Jahan, M.; Zhanel, G.G.; Sparling, R.; Holley, R.A. Horizontal transfer of antibiotic resistance from Enterococcus faecium of fermented meat origin to clinical isolates of E. faecium and Enterococcus faecalis. *Int. J. Food Microbiol.* **2015**, *199*, 78–85. [CrossRef] [PubMed]
23. Saillant, V.; Lipuma, D.; Ostyn, E.; Joubert, L.; Boussac, A.; Guerin, H.; Brandelet, G.; Arnoux, P.; Lechardeur, D. A novel enterococcus faecalis heme transport regulator (Fhtr) senses host heme to control its intracellular homeostasis. *MBio* **2021**, *12*, e03392-e20. [CrossRef] [PubMed]
24. Laissue, J.A.; Chappuis, B.B.; Müller, C.; Reubi, J.C.; Gebbers, J.O. The intestinal immune system and its relation to disease. *Dig. Dis.* **1993**, *11*, 298–312. [CrossRef] [PubMed]
25. García-Díez, J.; Saraiva, C. Use of starter cultures in foods from animal origin to improve their safety. *Int. J. Environ. Res. Public Health* **2021**, *18*, 2544. [CrossRef] [PubMed]
26. Gelsomino, R.; Vancanneyt, M.; Condon, S.; Swings, J.; Cogan, T.M. Enterococcal diversity in the environment of an Irish Cheddar-type cheesemaking factory. *Int. J. Food Microbiol.* **2001**, *71*, 177–188. [CrossRef] [PubMed]
27. Franz, C.M.A.P.; Huch, M.; Abriouel, H.; Holzapfel, W.; Gálvez, A. Enterococci as probiotics and their implications in food safety. *Int. J. Food Microbiol.* **2011**, *151*, 125–140. [CrossRef]
28. Giraffa, G. Functionality of enterococci in dairy products. *Int. J. Food Microbiol.* **2003**, *88*, 215–222. [CrossRef]
29. Lengfelder, I.; Sava, I.G.; Hansen, J.J.; Kleigrewe, K.; Herzog, J.; Neuhaus, K.; Ho_man, T.; Sartor, R.B.; Haller, D. Complex bacterial consortia reprogram the colitogenic activity of Enterococcus faecalis in a gnotobiotic mouse model of chronic, immune-mediated colitis. *Front. Immunol.* **2019**, *10*, 1420. [CrossRef]
30. Staley, C.; Dunny, G.M.; Sadowsky, M.J. Environmental and Animal-Associated Enterococci. *Adv. Appl. Microbiol.* **2014**, *87*, 147–186. [CrossRef]
31. Shankar, P.; Chung, R.; Frank, D.A. Association of food insecurity with children's behavioral, emotional, and academic outcomes: A systematic review. *J. Dev. Pediatr.* **2017**, *38*, 135–150. [CrossRef] [PubMed]
32. Arias, C.A.; Murray, B.E. The rise of the Enterococcus: Beyond vancomycin resistance. *Nat. Rev. Microbiol.* **2012**, *10*, 266–278. [CrossRef] [PubMed]
33. Klein, G.; Pack, A.; Reuter, G. Antibiotic resistance patterns of enterococci and occurrence of vancomycin-resistant enterococci in raw minced beef and pork in Germany. *Appl. Environ. Microbiol.* **1998**, *64*, 1825–1830. [CrossRef] [PubMed]
34. Tendolkar, P.M.; Baghdayan, A.S.; Shankar, N. Pathogenic enterococci: New developments in the 21st century. *Cell. Mol. Life Sci.* **2003**, *60*, 2622–2636. [CrossRef] [PubMed]

35. Giacobbe, D.R.; Battaglini, D.; Ball, L.; Brunetti, I.; Bruzzone, B.; Codda, G.; Crea, F.; De Maria, A.; Dentone, C.; Di Biagio, A.; et al. Bloodstream infections in critically ill patients with COVID-19. *Eur. J. Clin. Investig.* **2020**, *50*, e13319. [CrossRef] [PubMed]
36. Kokkoris, S.; Papachatzakis, I.; Gavrielatou, E.; Ntaidou, T.; Ischaki, E.; Malachias, S.; Vrettou, C.; Nichlos, C.; Kanavou, A.; Zervakis, D.; et al. ICU-acquired bloodstream infections in critically ill patients with COVID-19. *J. Hosp. Infect.* **2021**, *107*, 95–97. [CrossRef] [PubMed]
37. Patrier, J.; Villageois-Tran, K.; Szychowiak, P.; Ruckly, S.; Gschwind, R.; Wicky, P.H.; Gueye, S.; Armand-Lefevre, L.; Marzouk, M.; Sonneville, R.; et al. Oropharyngeal and intestinal concentrations of opportunistic pathogens are independently associated with death of SARS-CoV-2 critically ill adults. *J. Crit. Care.* **2022**, *26*, 300. [CrossRef]
38. Protonotariou, E.; Mantzana, P.; Meletis, G.; Tychala, A.; Kassomenaki, A.; Vasilaki, O.; Kagkalou, G.; Gkeka, I.; Archonti, M.; Kati, S.; et al. Microbiological characteristics of bacteremias among COVID-19 hospitalized patients in a tertiary referral hospital in Northern Greece during the second epidemic wave. *FEMS Microbes* **2022**, *2*, xtab021. [CrossRef]
39. Tao, S.; Chen, H.; Li, N.; Fang, Y.; Xu, Y.; Liang, W. Association of CRISPR-Cas System with the Antibiotic Resistance and Virulence Genes in Nosocomial Isolates of Enterococcus. *Infect. Drug Resist.* **2022**, *15*, 6939–6949. [CrossRef]
40. Toc, D.A.; Butiuc-Keul, A.L.; Iordache, D.; Botan, A.; Mihaila, R.M.; Costache, C.A.; Colosi, I.A.; Chiorean, C.; Neagoe, D.S.; Gheorghiu, L.; et al. Descriptive Analysis of Circulating Antimicrobial Resistance Genes in Vancomycin-Resistant Enterococcus (VRE) during the COVID-19 Pandemic. *Biomedicines* **2022**, *10*, 1122. [CrossRef]
41. Lopes, M.; Simões, A.; Tenreiro, R.; Joaquim, J.; Marques, F.; Crespo, M. Activity and expression of a virulence factor, gelatinase, in dairy enterococci. *Int. J. Food Microbiol.* **2006**, *112*, 208–214. [CrossRef] [PubMed]
42. Leavis, H.; Top, J.; Shankar, N.; Borgen, K.; Bonten, M.; Van Embden, J.; Willems, R.J.L. A novel putative enterococcal pathogenicity island linked to the esp virulence gene of Enterococcus faecium and associated with epidemicity. *J. Bacteriol.* **2004**, *186*, 672–682. [CrossRef] [PubMed]
43. Kiruthiga, A.; Padmavathy, K.; Shabana, P.; Naveenkumar, V.; Gnanadesikan, S.; Malaiyan, J. Improved detection of esp, hyl, asa1, gelE, cylA virulence genes among clinical isolates of Enterococci. *BMC Res. Notes.* **2020**, *13*, 170. [CrossRef] [PubMed]
44. Frazzon, A.P.; Gama, B.A.; Hermes, V.; Bierhals, C.G.; Pereira, R.I.; Guedes, A.G.; d'Azevedo, P.A.; Frazzon, J. Prevalence of antimicrobial resistance and molecular characterization of tetracycline resistance mediated by tet(M) and tet(L) genes in *Enterococcus* spp. isolated from food in Southern Brazil. *World J. Microbiol. Biotechnol.* **2010**, *26*, 365–370. [CrossRef]
45. Cattoir, V. The multifaceted lifestyle of enterococci: Genetic diversity, ecology and risks for public health. *Curr. Opin. Microbiol.* **2022**, *65*, 73–80. [CrossRef] [PubMed]
46. Koleva, Z.; Dedov, I.; Kizheva, J.; Lipovanska, R.; Moncheva, P.; Hristova, P. Lactic acid microflora of the gut of snail *Cornu aspersum*. *Biotechnol. Biotechnol. Equip.* **2014**, *28*, 627–634. [CrossRef]
47. Jackson, C.R.; Fedorka-Cray, P.J.; Barrett, J.B. Use of a genus- and species-specific multiplex PCR for identification of enterococci. *J. Clin. Microbiol.* **2004**, *42*, 3558–3565. [CrossRef]
48. Yoon, J.H.; Lee, S.T.; Park, Y.H. Inter- and intraspecific phylogenetic analysis of the genus Nocardioides and related taxa based on 16S rRNA gene sequences. *Int. J. Syst. Bacteriol.* **1998**, *48*, 187–194. [CrossRef]
49. Carrillo, P.G.; Mardaraz, C.; Pitta-Alvarez, S.I.; Giulietti, A.M. Isolation and selection of biosurfactant-producing bacteria. *World J. Microbiol. Biotechnol.* **1996**, *12*, 82–84. [CrossRef]
50. Mangia, N.P.; Saliba, L.; Deiana, P. Functional and safety characterization of autochthonous Lactobacillus paracasei FS103 isolated from sheep cheese and its survival in sheep and cow fermented milks during cold storage. *Ann. Microbiol.* **2019**, *69*, 161–170. [CrossRef]
51. Bauer, A.W.; Kirby, W.M.M.; Sherris, J.C.; Turck, M. Antibiotic susceptibility testing by a standard single disk method. *Am. J. Clin. Pathol.* **1966**, *36*, 493–496. [CrossRef]
52. EUCAST Version 13.1, 2023. (n.d.). Available online: http://www.eucast.org/expert_rules_and_intrinsic_resistance (accessed on 10 August 2023).
53. McBride, S.M.; Fischetti, V.A.; LeBlanc, D.J.; Moellering, R.C.; Gilmore, M.S. Genetic diversity among Enterococcus faecalis. *PLoS ONE* **2007**, *2*, e582. [CrossRef] [PubMed]
54. Eaton, T.J.; Gasson, M.J. Molecular Screening of *Enterococcus virulence* Determinants and Potential for Genetic Exchange between Food and Medical Isolates. *Appl. Environ. Microbiol.* **2001**, *67*, 1628–1635. [CrossRef] [PubMed]
55. Mannu, L.; Paba, A.; Daga, E.; Comunian, R.; Zanetti, S.; Duprè, I.; Sechi, L.A. Comparison of the incidence of virulence determinants and antibiotic resistance between Enterococcus faecium strains of dairy, animal and clinical origin. *Int. J. Food Microbiol.* **2003**, *88*, 291–304. [CrossRef] [PubMed]
56. van Asselt, G.J.; Vliegenthart, J.S.; Petit, P.L.C.; van de Klundert, J.A.M.; Mouton, R.P. High-level aminoglycoside resistance among enterococci and group A streptococci. *J. Antimicrob. Chemother.* **1992**, *30*, 651–659. [CrossRef]
57. Garofalo, C.; Vignaroli, C.; Zandri, G.; Aquilanti, L.; Bordoni, D.; Osimani, A.; Clementi, F.; Biavasco, F. Direct detection of antibiotic resistance genes in specimens of chicken and pork meat. *Int. J. Food Microbiol.* **2007**, *113*, 75–83. [CrossRef] [PubMed]
58. Aarestrup, F.M.; Agerso, Y.; Gerner-Smidt, P.; Madsen, M.; Jensen, L.B. Comparison of antimicrobial resistance phenotypes and resistance genes in Enterococcus faecalis and Enterococcus faecium from humans in the community, broilers, and pigs in Denmark. *Diagn. Microbiol. Infect. Dis.* **2000**, *37*, 127–137. [CrossRef] [PubMed]
59. Jensen, L.B.; Frimodt-Møller, N.; Aarestrup, F.M. Presence of erm gene classes in gram-positive bacteria of animal and human origin in Denmark. *FEMS Microbiol. Lett.* **1999**, *170*, 151–158. [CrossRef]

60. Jia, W.; Li, G.; Wang, W. Prevalence and Antimicrobial Resistance of Enterococcus Species: A Hospital-Based Study in China. *Int. J. Environ.* **2014**, *11*, 3424–3442. [CrossRef]
61. Liu, C.; Zhang, Z.Y.; Dong, K.; Yuan, J.P.; Guo, X.K. Antibiotic resistance of probiotic strains of lactic acid bacteria isolated from marketed foods and drugs. *Biomed. Environ. Sci.* **2009**, *22*, 401–412. [CrossRef]
62. Ferchichi, M.; Sebei, K.; Boukerb, A.M.; Karray-Bouraoui, N.; Chevalier, S.; Feuilloley, M.G.J.; Connil, N.; Zommiti, M. *Enterococcus* spp.: Is It a Bad Choice for a Good Use—A Conundrum to Solve? *Microorganisms* **2021**, *9*, 2222. [CrossRef] [PubMed]
63. Gunnar Kahlmeter and the EUCAST Steering Committee 2019, Redefining Susceptibility Testing Categories S, I and R. Available online: https://www.eucast.org/newsiandr (accessed on 10 August 2023).
64. Hiltunen, T.; Virta, M.; Laine, A.L. Antibiotic resistance in the wild: An eco-evolutionary perspective. *Philos. Trans. R Soc. Lond B Biol. Sci.* **2017**, *372*, 20160039. [CrossRef] [PubMed]
65. de Kraker, M.E.; Jarlier, V.; Monen, J.C.; Heuer, O.E.; van de Sande, N.; Grundmann, H. The changing epidemiology of bacteraemias in Europe: Trends from the European Antimicrobial Resistance Surveillance System. *Clin. Microbiol. Infect.* **2013**, *19*, 860–868. [CrossRef] [PubMed]
66. McAuley, C.M.; Britz, M.L.; Gobius, K.S.; Craven, H.M. Prevalence, seasonality, and growth of enterococci in raw and pasteurized milk in Victoria, Australia. *J. Dairy Sci.* **2015**, *98*, 8348–8358. [CrossRef]
67. Gołaś-Prądzyńska, M.; Łuszczyńska, M.; Rola, J.G. Dairy Products: A Potential Source of Multidrug-Resistant Enterococcus faecalis and Enterococcus faecium Strains. *Foods* **2022**, *11*, 4116. [CrossRef]
68. Martín, R.; Langa, S.; Reviriego, C.; Jiménez, E.; Marín, M.L.; Xaus, J.; Fernández, L.; Rodríguez, J.M. Human milk is a source of lactic acid bacteria for the infant gut. *J. Pediatr.* **2003**, *143*, 754–758. [CrossRef]
69. Laursen, M.F.; Larsson, M.W.; Lind, M.V.; Larnkjær, A.; Mølgaard, C.; Michaelsen, K.F.; Bahl, M.I.; Licht, T.R. Intestinal Enterococcus abundance correlates inversely with excessive weight gain and increased plasma leptin in breastfed infants. *FEMS Microbiol. Ecol.* **2021**, *96*, fiaa066. [CrossRef]
70. Wajda, Ł.; Ostrowski, A.; Błasiak, E.; Godowska, P. Enterococcus faecium Isolates Present in Human Breast Milk Might Be Carriers of Multi-Antibiotic Resistance Genes. *Bacteria* **2022**, *1*, 66–87. [CrossRef]
71. Huang, M.S.; Cheng, C.C.; Tseng, S.Y.; Lin, Y.L.; Lo, H.M.; Chen, P.W. Most commensally bacterial strains in human milk of healthy mothers display multiple antibiotic resistance. *MicrobiologyOpen* **2019**, *8*, e00618. [CrossRef]
72. Franz, C.M.; Muscholl-Silberhorn, A.B.; Yousif, N.M.; Vancanneyt, M.; Swings, J.; Holzapfel, W.H. Incidence of virulence factors and antibiotic resistance among Enterococci isolated from food. *Appl. Environ. Microbiol.* **2001**, *67*, 4385–4389. [CrossRef]
73. Santana, L.A.M.; Andrade, N.N.N.; da Silva, L.S.C.; Oliveira, C.N.T.; de Brito, B.B.; de Melo, F.F.; Souza, C.L.; Marques, L.M.; Oliveira, M.V. Identification and characterization of resistance and pathogenicity of *Enterococcus* spp. in samples of donor breast milk. *World J. Clin. Pediatr.* **2020**, *9*, 53–62. [CrossRef] [PubMed]
74. Huycke, M.M.; Spiegel, C.A.; Gilmore, M.S. Bacteremia caused by hemolytic, high-level gentamicin-resistant Enterococcus faecalis. *Antimicrob. Agents Chemother.* **1991**, *35*, 1626–1634. [CrossRef] [PubMed]
75. Gilmore, M.S.; Segarra, R.A.; Booth, M.C.; Bogie, C.P.; Hall, L.R.; Clewell, D.B. Genetic structure of the Enterococcus faecalis plasmid pAD1-encoded cytolytic toxin system and its relationship to lantibiotic determinants. *J. Bacteriol.* **1994**, *176*, 7335–7344. [CrossRef] [PubMed]
76. Haas, W.; Shepard, B.D.; Gilmore, M.S. Two-component regulator of Enterococcus faecalis cytolysin responds to quorum-sensing autoinduction. *Nature* **2002**, *415*, 84–87. [CrossRef] [PubMed]
77. Haas, W.; Gilmore, M.S. Molecular nature of a novel bacterial toxin: The cytolysin of *Enterococcus faecalis*. *Med. Microbiol. Immunol.* **1999**, *187*, 183–190. [CrossRef]
78. Ike, Y.; Clewell, D.B.; Segarra, R.A.; Gilmore, M.S. Genetic analysis of the pAD1 hemolysin/bacteriocin determinant in Enterococcus faecalis: Tn917 insertional mutagenesis and cloning. *J. Bacteriol.* **1990**, *172*, 155–163. [CrossRef]
79. Vergis, E.N.; Shankar, N.; Chow, J.W.; Hayden, M.K.; Snydman, D.R.; Zervos, M.J.; Linden, P.K.; Wagener, M.M.; Muder, R.R. Association between the presence of enterococcal virulence factors gelatinase, hemolysin, and enterococcal surface protein and mortality among patients with bacteremia due to Enterococcus faecalis. *Clin. Infect. Dis.* **2002**, *35*, 570–575. [CrossRef]
80. Semedo, T.; Santos, M.A.; Lopes, M.F.; Figueiredo Marques, J.J.; Barreto Crespo, M.T.; Tenreiro, R. Virulence factors in food, clinical and reference Enterococci: A common trait in the genus? *Syst. Appl. Microbiol.* **2003**, *26*, 13–22. [CrossRef]
81. Creti, R.; Imperi, M.; Bertuccini, L.; Fabretti, F.; Orefici, G.; Di Rosa, R.; Baldassarri, L. Survey for virulence determinants among Enterococcus faecalis isolated from different sources. *J. Med. Microbiol.* **2004**, *53 Pt 1*, 13–20. [CrossRef]

pathogens

MDPI

Article

Impact of *Sclerotinia sclerotiorum* Infection on Lettuce (*Lactuca sativa* L.) Survival and Phenolics Content—A Case Study in a Horticulture Farm in Poland

Violetta Katarzyna Macioszek [1,*,†], Paulina Marciniak [2,3,†] and Andrzej Kiejstut Kononowicz [4]

1 Laboratory of Plant Physiology, Department of Biology and Plant Ecology, Faculty of Biology, University of Bialystok, 15-245 Bialystok, Poland
2 Wiesław and Izabela Królikiewicz Horticulture Market Farm, 97-306 Majków Średni, Poland
3 Faculty of Biology and Environmental Protection, University of Lodz, 90-237 Lodz, Poland
4 Department of Plant Ecophysiology, Faculty of Biology and Environmental Protection, University of Lodz, 90-237 Lodz, Poland; andrzej.kononowicz@biol.uni.lodz.pl
* Correspondence: v.macioszek@uwb.edu.pl
† These authors contributed equally to this work.

Abstract: *Sclerotinia sclerotiorum* is a cause of a prevalent and destructive disease that attacks many horticultural food crops, such as lettuce. This soil-borne necrotrophic fungal pathogen causes significant economic losses in worldwide lettuce production annually. Furthermore, current methods utilized for management and combatting the disease, such as biocontrol, are insufficient. In this study, three cultivars of lettuce (one Crispy and two Leafy cultivars of red and green lettuce) were grown in central Poland (Lodz Voivodeship), a widely known Polish horticultural region. In the summer and early autumn, lettuce cultivars were grown in control and *S. sclerotiorum*-infected fields. The lettuce cultivars (*Templin*, *Lollo Rossa*, and *Lollo Bionda*) differed phenotypically and in terms of the survival of the fungal infection. The Crispy iceberg *Templin* was the most susceptible to *S. sclerotiorum* infection compared to the other cultivars during both vegetation seasons. The total content of phenolic compounds, flavonoids, and anthocyanins varied among cultivars and fluctuated during infection. Moreover, phenolic content was affected by vegetation season with alterable environmental factors such as air temperature, humidity, soil temperature, and pH. The most increased levels of phenolics, both flavonoids and anthocyanins in infected plants, were observed in the Leafy red *Lollo Rossa* cultivar in both crops. However, the highest survival/resistance to the fungus was noticed for *Lollo Rossa* in the summer crop and *Lollo Bionda* in the autumn crop.

Keywords: anthocyanins; epidemiology; flavonoids; vegetation season; lettuce; phenolics; soil-borne fungus; temperature

1. Introduction

Lettuce (*Lactuca sativa* L.) belongs to the largest family of dicotyledons, i.e., the *Asteraceae* family, and it is an annual self-pollinating plant [1,2]. Lettuce comprises a straight stem, the height of which ranges from 30 to 100 cm. Leaves are arranged in a spiral pattern and may form a dense rosette or head. They can be oblong, round, triangular, or transversely elliptical. Lettuce forms a deep root with horizontal lateral roots for maximum water and nutrient uptake [1]. There are six main types of lettuce: Crispy, Butterhead, Romain, Leafy, Stem, and Latin. Moreover, each type has a great variety, including its color and shape [3].

Lettuce is a leafy vegetable that is widely consumed across the globe, particularly in Europe, Asia, and North and Central America [1,4], due to its traditional and healthy qualities [5]. In salads or sandwiches, lettuce is typically consumed fresh and raw [6], as whole heads or cut leaves of lettuce are available on the market [7]. Fresh lettuce is rich in bioactive compounds such as phenolics and fiber [8–10], carotenoids, and tocopherol [9], as well as vitamins C, B1, B2, PP, and K [11], and minerals such as iron, calcium, and magnesium [10]. Lettuce leaf tissues contain phenolic compounds responsible for their health-protecting and antioxidant properties. Two main phenolic compound groups in lettuce are caffeic acid derivatives and flavonols (flavonoid class). Red lettuce cultivars contain high levels of another flavonoid class, anthocyanins [12,13]. Phenolic content depends on a variety and even cultivars during normal development under physiological conditions. Many studies have shown that the concentration of phenolics alters in plant tissues due to variable environmental factors such as light intensity or color [14]. Furthermore, the reduction in phenolic content in some cultivars is affected by processing and storage [15]. It is thought that lettuce polyphenols may have a more substantial antioxidant effect than vitamins C and E [16].

Despite the relatively short growing season, lettuce can be grown outdoors in the field and caterpillar tunnels and indoors in a greenhouse hydroponically [7]. Therefore, this vegetable is available to consumers all year. Lettuce is cultivated outdoors in temperate climate zones from early spring to early autumn. The optimal temperature range for lettuce growth is between 12 and 24 °C, as it is a cool-season crop [17]. Cultivating lettuces in the field employs various soil classes, with the optimal soil pH value ranging from 6.0 to 6.8 [18]. However, growing outdoors or indoors, lettuce crops are exposed to many pathogens, especially pathogenic fungi such as downy mildew (*Bremia lactucae*) [19], powdery mildew (*Erysiphe cichoracearum*), gray mold (*Botrytis cinerea*) [20–22], and white mold (*Sclerotinia sclerotiorum*) [22–24]. Among fungus-induced lettuce diseases, *S. sclerotiorum* is considered the most destructive one, and it causes stem rot, often called lettuce drop or white mold. *S. sclerotiorum* belongs to the division *Ascomycota*, family *Sclerotiniaceae*, and this necrotrophic fungus is a soil-borne polyphage that infects almost 600 plant species worldwide [25]. Its most important and preferred hosts are dicotyledons such as sunflower, oilseed rape, soybean, bean, pea, lentil, chickpea [24], potato, lettuce, carrot, cabbage, celery, pepper, poppy seeds [26], and from monocotyledons onions and tulips are at risk [25]. Dicotyledonous plants, such as lettuce, are more susceptible to the fungus because it secretes pathogenicity determinants such as oxalic acid (OA) and necrosis- and ethylene-inducing peptide 1-like proteins (NLPs) [27]. Oxalic acid suppresses host defense responses because it does not metabolize in dicotyledons but detoxifies in monocotyledons [27–29]. NLPs elicit cell death by binding to specific membrane lipids-glycosyl inositol phosphoryl ceramides (GIPCs) characteristic for dicotyledons but absent in monocotyledons [27]. One of the factors determining the fitness and difficulties in the management of *S. sclerotiorum* infection of lettuce is that up to 90% of the life cycle of this fungus takes place in the soil in the form of sclerotia [28]. Sclerotia serve as survival structures and can survive in the soil for up to eight years. Depending on the host plant, sclerotia can vary in size and shape [30]. However, contrary to *S. minor*, the disease caused by *S. sclerotiorum* is not correlated with the number of sclerotia in the soil [23]. The fungus can spread through contact from an infected plant to another healthy one [22,24]. In lettuce, *S. sclerotiorum* infection becomes visible when the rotting and decaying of the tissues begins at the stem, near the soil. Emerging lesions then spread down the plant until the roots break down. The watery spots also develop on the leaves and spread to the stem [23]. Sclerotia appear on the leaf surface close to mycelium and damage host tissues [22].

Complete lettuce resistance to *S. sclerotiorum* is lacking and to this date no genetic source of resistance is known [31,32]. The lower level of lettuce susceptibility to the fungus is rather based on developmental traits, such as, among others, stem mechanical strength, low leaf area or rapid bolting [31], and variable fungal factors such as ascospore and mycelium density [33]. Physiological traits, including phenolic compounds content, are rarely investigated in *S. sclerotiorum*-infected lettuce [34].

The aim of this study, performed under field conditions in a market farm in central Poland, was to estimate the influence of *S. sclerotiorum* on three lettuce cultivars during two vegetation seasons (summer and early autumn) in 2016. The impact of the *S. sclerotiorum*-

induced disease on the lettuce cultivars' survival and changes in their phenolic content, such as total phenolic compounds, flavonoids, and anthocyanins, were investigated. To our knowledge, this is the first study to show spontaneous *S. sclerotiorum* infection without additional artificial plant inoculation in functioning farm-producing food crops.

2. Materials and Methods

2.1. Lettuce Cultivars, Planting, and Cultivation

In all experiments, two varieties of lettuce (*Lactuca sativa* L.) were used Crispy and Leafy one Crispy iceberg lettuce *Templin* and two cultivars of Leafy lettuce: green *Lollo Bionda* and red *Lollo Rossa*. Iceberg lettuce *Templin* is fragile but has a perfect head shape, and it can be harvested easily from spring to autumn due to its raised and compact base. Leafy lettuces *Lollo Bionda* and *Lollo Rossa* are recommended for cultivation in the ground throughout a whole growing season, and they can be sold as heads or as ready-to-eat leaf mixtures.

The lettuces were grown in the Wiesław and Izabela Królikiewicz Horticulture Market Farm in Majków Średni, in central Poland (Grabica Commune, Piotrków Trybunalski District, Łódź Province; 51°25′57.5″ N, 19°36′04.5″ E). Commercially available lettuce cultivated for production as human food has to be purchased from certified companies and grown according to agricultural rules and standardized practices. Therefore, lettuce seeds of *Templin* have been purchased from the Rijk Zwaan (Blonie, Poland), and *Lollo Bionda* and *Lollo Rossa* from the Nunhems Company (Warsaw, Poland); germination and production of young seedlings have been performed by the Schwanteland Jungpflanzen GmbH (Oberkrämer, Germany). Six-week-old seedlings have been planted in two crops - in summer and early autumn, from July 2016 until October 2016. In each crop, plants were grown in two experimental fields, one infected field containing *S. sclerotiorum*-contaminated soil and another, a control field located in the uninfected field in close proximity to the infected field. The individual lettuce cultivars were grown in beds of 8 m long and 1 m wide. Additionally, each bed was divided into 5 sectors, and in each sector, 20 lettuce seedlings were planted in 4 plots, with 5 seedlings per plot. Therefore, in each crop, 100 plants per bed of every cultivar were grown in the control field and 100 plants in the infected field (Figure 1). The beds were formed one day after the application of fertilizers. For each field (control and infected), 5 kg of ammonium nitrate, 5 kg of potassium sulfate, and 5 kg of triple super phosphate were applied. No additional applications of fertilizers were used during any of the vegetation seasons.

All experiments performed on the market farm started in the spring of 2016. However, the spring vegetation season lasted 81 days due to unfavorable weather conditions. Moreover, this crop was established with two cultivars-iceberg *Diamentinas* and Leafy *Lollo Rossa*. Thus, it could not be compared with two other seasons in 2016. Therefore, the spring vegetation season was not considered in this study (Appendix A, Figure A1).

2.1.1. The Summer (1st) Crop, 2016

The summer crop was established by planting three cultivars of lettuce in the infected field: Crispy lettuce *Templin*, Leafy green lettuce *Lollo Bionda*, and Leafy red lettuce *Lollo Rossa* (Figure 1a). Control plants were planted in the uninfected field right next to the *S. sclerotiorum*-contaminated field. Plant irrigation with a net of sprinklers was carried out as a standard practice throughout the entire cultivation period, i.e., from 7 July to 18 August 2016 (for 43 days).

2.1.2. The Autumn (2nd) Crop, 2016

This crop was established with the same parameters and lettuce cultivars as the summer one (Figure 1b), and it was carried out from 3 September to 22 October 2016 (for 50 days).

Figure 1. Establishment of the crops in the 2016 vegetation seasons. Seven-week-old seedlings: (**a**) the summer (1st) crop, from the left *Templin*, *Lollo Rossa*, and *Lollo Bionda* in the infected field; (**b**) the autumn (2nd) crop, from the left *Templin*, *Lollo Bionda*, and *Lollo Rossa* in the control field.

2.2. Air and Soil Parameters

From the first day of cultivation during both vegetation seasons, daily measurements of soil temperature and pH were carried out in the control and infected plots at 3 p.m. CEST, using a Flo 4in1 Soil Survey Instrument (Toya, Wroclaw, Poland). However, data on air temperature and humidity have been collected from the Polish website interia.pl, which publishes meteorological parameters obtained from the meteorological stations for every specific area/place in Poland every hour of the day (e.g., https://pogoda.interia.pl/prognoza-szczegolowa-majkow-sredni,cId,20064 (accessed on 30 October 2023)).

2.3. S. sclerotiorum Soil Contamination and Re-Isolation of the Fungus

The lettuce plants infected with *S. sclerotiorum* on the farm were first observed in early autumn of 2015 in a small part of the lettuce field surrounded by radish and cabbage plots, and the fungus was identified then [Marciniak and Macioszek, unpublished data]. Most lettuce plants were heavily infected and damaged at that time because the disease spread and decayed the whole plants. The contamination of the soil was caused by the spillage and plowing of the soil containing *S. sclerotiorum* derived from the tunnel lettuce cultivation carried out on the farm in early spring 2015. In 2016, lettuce was cultivated in the infected plots during the whole growing season without any additional or artificial inoculation of plants with the fungus.

The fungus was re-isolated from the infected lettuce heads of cultivar *Diamentinas*, harvested in early autumn 2015. The lettuce showed typical features for the late *S. sclerotinia* infection, with lower leaves wilted and covered with white mycelium and sclerotia and brownish rot tissues of the stem, with the blight separating the lettuce head off the stem. The samples of fungal mycelium were grown on PDA plates (Difco, Leeuwarden, The Netherlands) under laboratory conditions: temperature of 20 °C, approximately 65% humidity, and photoperiod 16 h day/8 h night under a fluorescent light 120 μmol m^{-2} s^{-1} for 14–21 days. After this time, a microscopic estimation of the fungal morphology was performed to evaluate whether the obtained fungus culture met Koch's postulates. After 21 days, the whole PDA plates were covered with fungal culture comprising white mycelium (the same as the collected one) and many black hard sclerotia of 2–3 mm diameter, characteristic of *S. sclerotiorum*. Subsequently, Crispy lettuce heads of cultivar *Diamentinas* were infected with samples of the re-grown fungus under laboratory conditions. The same infection symptoms were observed as in the case of infected *Diamentinas* plants harvested for re-isolation of the fungus.

2.4. Evaluation of Lettuce Infection

The disease symptoms on plants in the infected and control plots were observed each week of cultivation. A plant was considered dead when almost all its leaves were wilted and brownish (Table 1). Individual plant survival was estimated every week of cultivation in each crop, and the percentage of surviving still-growing plants was calculated for each cultivar. In the case of disease progression, estimation of infection progression in lettuce cultivars at the 7th week was performed based on a 6-point scale (0–5) prepared for each lettuce cultivar grown in the early autumn crop (Table 1).

Table 1. The scale of disease progression in the tested lettuce cultivars by *S. sclerotiorum*.

Stage of Infection	Presence of the Fungus	Symptom on Plant
0	No fungus	No symptoms
1	Small white mycelium	Small necroses on the stem
2	White mycelium and black sclerotia on the stem	Wilted leaf base, necroses on leaves and stem
3	White mycelium and sclerotia on leaves and stems	Necroses on leaves and stems, marginal leaves wilted and brown
4	Mycelium spread and many black sclerotia	Wilted, rotten, and discolored leaves, rotten stem
5	Spreading fungus on the whole plant	Decoyed, dead plant

2.5. Analyses of Phenolics Contents

2.5.1. Sample Collection and Preparation of Plant Extracts

Samples were harvested from the control and infected plants of each cultivar. However, two samples per leaf were harvested from the leaf blade and leaf base. In each crop, 12–50 samples from each treatment (control or infected) per each cultivar were collected.

About 100 mg samples were collected from each leaf and kept frozen at −80 °C until used. The samples were extracted in 1 mL of 80% methanol (HPLC grade, POCH, Gliwice, Poland) and centrifuged for 15 min at 14,000 rpm and temp. 4 °C. The supernatants were placed in new tubes and kept on ice. A total of 1 mL of methanol was added to the remaining pellets and vortexed. The samples were then centrifuged again. Thereafter, supernatants were added to the previously prepared ones and gently mixed. The obtained plant extracts were used to determine the total phenolic compounds (TPCs), flavonoid, and anthocyanin contents using a Power Wave plate spectrophotometer (BioTek Instruments, Winooski, VT, USA).

2.5.2. Total Phenolic Compounds (TPCs)

Total phenolic compounds were determined using the Folin-Ciocalteau method, as described by Macioszek et al. [35]. Briefly, 100 μL of the plant extract was added to 3.85 mL of distilled water, and then 100 μL of the Folin-Ciocalteau reagent (Sigma-Aldrich, Poznan, Poland) was added. A blank (with 80% methanol instead of plant extract) was prepared in parallel. After 3 min of incubation, 1 mL of 10% sodium carbonate was added to each sample, thoroughly mixed, and incubated in the dark at room temperature for 30 min. Next, the samples were analyzed spectrophotometrically at a wavelength of 725 nm. The total phenolic compounds content was calculated using a gallic acid (Sigma-Aldrich, Poznan, Poland) calibration curve and expressed as mg g^{-1} FW.

2.5.3. Flavonoids

The content of flavonoids in the plant extracts was determined as described by Macioszek et al. [35]. Briefly, 600 μL of plant extract, 1800 μL of 80% methanol, 120 μL of 10% aluminum chloride hexahydrate, and 120 μL of 1 M sodium acetate were mixed. After 30 min of incubation in the dark, the samples were transferred to a 96-well plate and analyzed spectrophotometrically at a wavelength of 415 nm. The content of flavonoids was calculated based on the absorbance of the standard quercetin solution (Sigma-Aldrich, Poznan, Poland) and expressed as mg g^{-1} FW.

2.5.4. Anthocyanins

The content of anthocyanins was determined using an adapted method described by Lee et al. [36]. Briefly, 200 μL of plant extract was added to 800 μL of 0.025 M potassium chloride, pH 1, and mixed thoroughly. Another 200 μL of plant extract was added to 800 μL of 0.4 M sodium acetate, pH 4.5, and mixed thoroughly. The samples were incubated at room temperature for 30 min and analyzed spectrophotometrically at 520 and 700 nm wavelengths. The content of anthocyanins was calculated based on a cyanidin-3-glucoside equivalent and expressed in μg g^{-1} FW [36].

2.6. Statistical Analysis

All the statistical analyses, including analysis of variance (ANOVA) and a post hoc Duncan's test ($p < 0.05$), were performed using Microsoft Office Excel 2021 (Microsoft Corporation, Redmond, WA, USA) and STATISTICA 13.3 (Tibco Software Inc., StatSoft, Krakow, Poland).

The charts were prepared using Microsoft Office Excel 2021. All the figures were composed using Adobe Photoshop CS3 version 10.0 (Adobe System Incorporated, San Jose, CA, USA).

3. Results

3.1. Changes in the Air and Soil Parameters

During two vegetation seasons in 2016, air temperature and humidity, as well as soil temperature and pH, were measured daily. In the summer crop (the first crop, 7 July–18 August), the range of air day temperatures was between 15 and 28 °C (Figure 2a). The temperature below the optimum (18 °C) for growing lettuce in the field was observed only for two days during the entire summer. The air temperature in the fourth week (when the first significant decrease in the survival of lettuces was observed) was high between 21 and 28 °C, but it only ranged between 19 and 22 °C at the end of the summer crop (Figure 2a, Table S1).

In the summer, the soil temperature for 21 days was lower, about 1–2 °C, than the air temperature, and it was higher than the air temperature only for 13 days. Air and soil temperatures were equal for the remaining seven days of the summer crop. The average air temperature in the summer was 22.0 °C, and the average soil temperature was 21.3 °C. Humidity was highly alterable during this period and ranged between 28 and 79%, with an average humidity of 52.9% (Figure 2a, Table S1).

In the autumn crop (the second crop, 3 September–22 October), the air temperature ranged between 5 and 29 °C (Figure 2b, Table S2). The lowest soil temperature was 6 °C, and the highest was 27 °C. Air and soil temperatures were lower in October (from the fifth week of this crop) than in September. The average air temperature in September was 21.7 °C and only 10.0 °C in October. The autumnal humidity ranged from 26 to 97%, with a significantly higher average value of 72% in October compared to only 41.6% in September. However, the average humidity of 55% in the autumn was only about 3% higher than in the summer crop (Figure 2b, Table S2). The temperature and humidity values remained moderately equal in the summer crop compared to variable ones in the autumn crop (Figure 2).

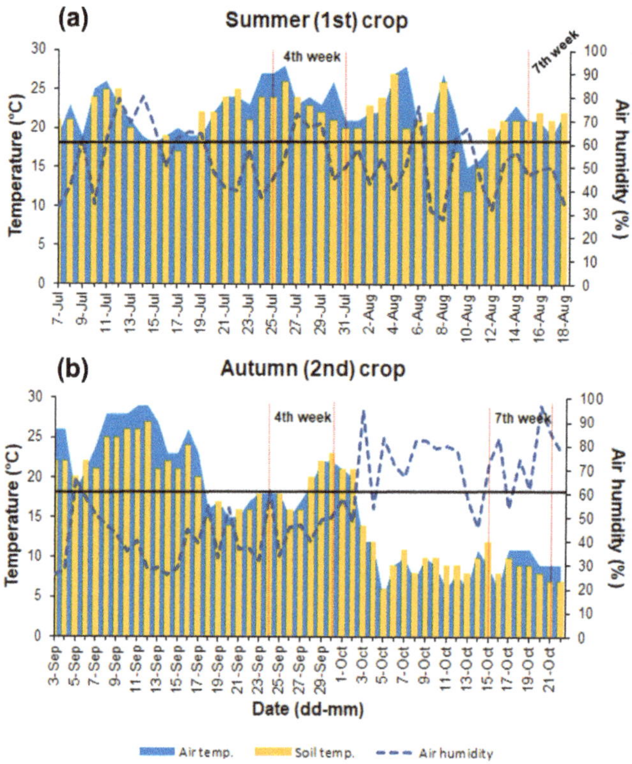

Figure 2. Changes in the soil and air temperature, and relative air humidity during vegetation seasons of lettuce in 2016. (**a**) The summer (first) crop; (**b**) early autumn (second) crop. The horizontal black line on each chart indicates a temperature of 18 °C, which is an optimal temperature for lettuce cultivation under field conditions in Poland. Detailed data are included in Supplementary Materials (Tables S1 and S2).

In the first summer crop, the pH of the soil in the control field for the first three weeks of cultivation was lower (pH 5.5–6.5) than that of the infected soil (pH 6.0–7.0). Three weeks after cultivation, the pH values equalized to a similar level in both fields. The average pH value of soil in the control field was 6.4 and 6.8 in the infected one during the summer crop. The soil pH remained constant (pH 7) during cultivation in the autumn crop, both in the case of control and infected fields (Figure 3).

Analysis of variance (one-way ANOVA) revealed that the soil and air temperatures, as well as soil pH, differed significantly between crops ($p < 0.001$). However, the humidity was not significantly variable between crops.

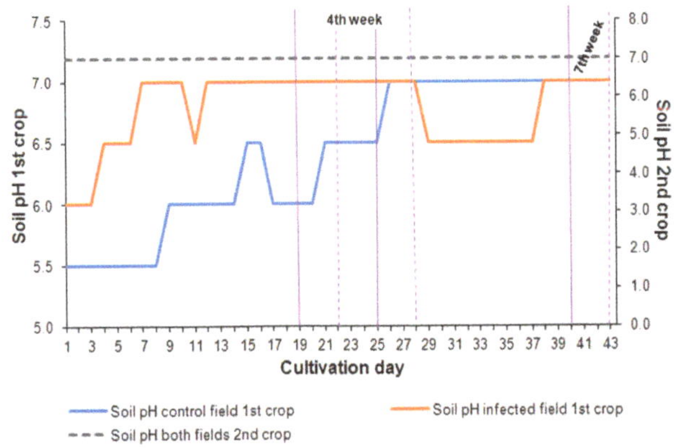

Figure 3. Average pH of the soil in the control and infected fields based on measurements taken daily during each vegetation season. Vertical solid violet lines indicate the fourth and seventh weeks during the summer crop, and vertical dashed violet lines indicate the corresponding weeks during the autumn crop. Detailed data are included in Supplementary Materials (Tables S1 and S2).

3.2. Re-Isolation of the Pathogen

The pathogenic fungus was re-isolated from the infected lettuces (Figure 4a,b) and harvested in early autumn 2015. After 14–21 days of growth on the PDA plates under laboratory conditions, the fungus developed a fluffy white mycelium, accompanied by numerous black, hard sclerotia of 2–3 mm diameter (Figure 4c,d). The morphology of mycelium and sclerotia on plates was similar to that observed in infected lettuce heads, and based on this, the fungus was recognized as *S. sclerotiorum* [25]. Infection of healthy lettuce heads (cultivar *Diamentinas*) with the re-isolated fungus caused the same symptoms as in the case of infected lettuce from the autumn crop in 2015. Mature plant tissues were covered with white mycelium and sclerotia, and wilting leaves and stem rot were visible after 20 days of infection.

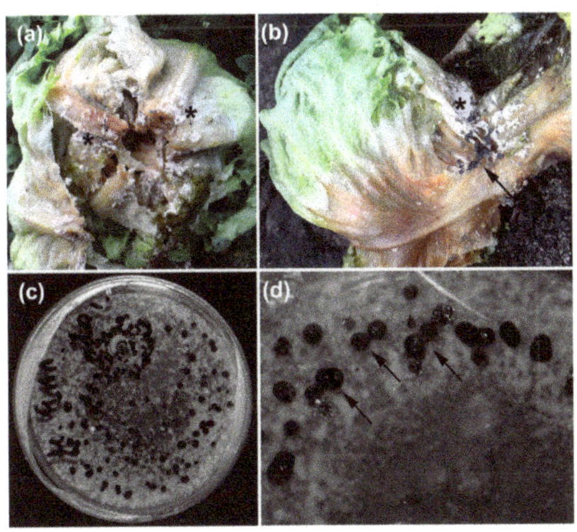

Figure 4. *Sclerotinia sclerotiorum* growth. (**a**,**b**) The fungus on heads of iceberg lettuce *Diamentinas* grown in the field in October 2015; (**c**,**d**) the re-isolated pathogen growing in vitro. Arrows indicate sclerotia and asterisks indicate white fluffy mycelium.

3.3. Evaluation of Disease Progression and Lettuce Survival

In the control field, Crispy iceberg *Templin* plants formed compact heads on short stems, whereas both Leafy cultivars *Lollo Rossa* and *Lollo Bionda* grew, forming dense rosettes on short stems (Figure 5a,g,m). The infection of lettuce plants with *S. sclerotiorum* was observed exclusively in the infected field during both crops; there were no *S. sclerotiorum*-infected lettuces in the control field. The loss of plants in the control field was related to the abnormal, weak development of individual seedlings and the incidents of lettuce heads (*Templin*) being eaten by hares (Tables 2 and 3).

Figure 5. Progression of the disease symptoms in lettuce cultivars observed during the autumn crop. (**a–f**) Crispy iceberg lettuce *Templin*; (**g–l**) Leafy lettuce *Lollo Rossa*; (**m–r**) Leafy lettuce *Lollo Bionda*. Detailed description in the text.

Table 2. Percentage of the survival of individual lettuce plants in the control and infected fields during the summer crop (July–August).

Cultivation Week	I	II	III	IV	V	VI	VII
Cultivar	Plants in control field (%)						
Crispy iceberg *Templin*	98	92 [1]	90 [1]	82 [1]	80 [1]	80 [1]	80 [1]
Leafy red *Lollo Rossa*	100	100	99	99	99	99	99
Leafy green *Lollo Bionda*	99	99	93	92	92	92	92
	Plants in infected field (%)						
Crispy iceberg *Templin*	100	97	95	80	70	21	10
Leafy red *Lollo Rossa*	97	96	90	89	89	84	83
Leafy green *Lollo Bionda*	100	94	93	93	79	70	70

[1] Due to a few lettuce heads eaten by hares.

Table 3. Percentage of the survival of individual lettuce plants in the control and infected fields during the autumn crop (September–October).

Cultivation Week	I	II	III	IV	V	VI	VII
Cultivar	Plants in control field (%)						
Crispy iceberg *Templin*	100	100	100	98	96	96	92
Leafy red *Lollo Rossa*	100	100	100	100	99	99	97
Leafy green *Lollo Bionda*	100	100	100	99	99	99	97
	Plants in infected field (%)						
Crispy iceberg *Templin*	100	97	94	70	52	48	30
Leafy red *Lollo Rossa*	98	88	81	80	76	65	51
Leafy green *Lollo Bionda*	100	99	98	88	85	80	60

The infection process in all lettuce cultivars began with the appearance of white mycelium on stems, leaf petioles, and bases close to the soil (stage 1, Table 1; Figure 5b,h,n). In the next second stage of disease development, wilting and browning of a few leaves and stems with watery spots were observed (Figure 5c,i,o). In the case of iceberg *Templin*, dark spots on leaf blades developed, and the spreading of mycelium from the stem to the leaves was visible (Figure 5c). The third stage characterized (Table 1) the first dark brown, decaying, death leaves that appeared in the plants of Crispy *Templin* and Leafy red *Lollo Rosa* (Figure 5d,j). Brown leaf petioles and discoloration of leaf blades could be observed in Leafy green *Lollo Bionda* plants at this stage of disease (Figure 5p). As the disease progressed, the disintegration of heads and rosettes could be found, and most of the leaves in *Templin* and *Lollo Rossa* plants were black and decoyed (Figure 5e,k). However, still-green leaves could be observed in *Lollo Bionda* plants (Figure 5q). The final fifth stage of the disease was characterized by discoloration and almost total decomposition of the infected plants of all three cultivars (Figure 5f,l,r). It has to be emphasized that such a sequence of disease symptoms similar to the ones described above appeared in all three cultivars during both crops.

In both vegetation seasons, the survival of individual plants of three planted cultivars was determined each week of cultivation in the control and infected fields. In the summer crop, the number of growing lettuce plants ranged between 80 and 100 per cultivar in the control field (Table 2). The lowest number of Crispy iceberg *Templin* lettuces was observed in the control field every week due to several lettuce heads eaten by hares. In the case of Leafy cultivars grown in the control field, at most, eight lettuce plants per cultivar were lost at the end of the summer crop. Overall, the cultivars grown in the control field showed significantly higher survival than lettuces planted in the one infected with *S. sclerotiorum* (Table 2). A gradual, significant decrease in survival of Crispy iceberg *Templin* was observed in the infected field every week. Beginning from the fourth week, when 80% of plants were still growing, the survival of this cultivar decreased to only 10% at the end of the summer crop (Table 2). The survival of Leafy green *Lollo Bionda* decreased only slightly up to the fourth week in both control and infected fields. However, there was a decline in the number of *Lollo Bionda* plants during the fifth week, with 79% of the plants surviving in the infected field compared to 92% still growing healthy plants in the control field. The downward trend in the survival of the *Lollo Bionda* cultivar continued until the seventh week of cultivation in the infected field, as only 70% of the lettuces ultimately survived. The leafy red lettuce *Lollo Rossa* showed the highest survival among lettuces planted in the infected field. Until the last week of cultivation, 83% of the plants of this cultivar still grew (Table 2).

In the autumn crop, 97% of Leafy red and green lettuce plants were still growing in the control field at the seventh final week of cultivation (Table 3). In the case of Crispy iceberg *Templin*, 92% of the healthy plants were noticed in the control field. A gradual decrease in the number of surviving plants in the infected field was observed for all the cultivars every week. The highest survival of 60% was observed for the Leafy green lettuce *Lollo Bionda*, and the lowest of 30% was observed for Crispy *Templin* at the end of the autumn crop. In the case of red lettuce *Lollo Rossa*, 51% of the plants survived the potential *S. sclerotiorum* infection in this crop (Table 3).

Analysis of the main effects of ANOVA revealed that the survival of all cultivars was highly dependent on both the crop and treatment ($p < 0.05$; Table S3). Furthermore, the survival of all lettuce cultivars in the infected field was correlated significantly to the soil and air temperatures and humidity ($p < 0.05$). However, only the survival of the *Lollo Rossa* was correlated to the soil pH in the infected field ($p < 0.05$) (Table S4). Interestingly, the survival of *Lollo Rossa* and *Lollo Bionda* was correlated to air and soil temperatures in the control field, and only *Lollo Rossa* survival was correlated significantly with humidity and soil pH (Table S5).

3.4. *Analysis of Phenolics Content*

In three lettuce cultivars, the contents of phenolic compounds, i.e., total phenolic compounds, flavonoids, and anthocyanins, were analyzed in control and infected samples. The infected samples were obtained from plants grown in the field with *S. sclerotiorum*-contaminated soil without additional plant inoculation. The control samples were collected from plants grown in the control field at corresponding sites as infected plants. The samples were obtained from the leaf base and leaf blade during the seventh week of cultivation in both crops.

3.4.1. Total Content of Phenolic Compounds (TPCs)

Analysis of the samples of three lettuce cultivars during the first summer and second autumn crops revealed that the TPC content in the leaf base and leaf blade significantly differed (Figure 6).

In the first summer crop, the content of TPCs in infected plants was the lowest in both green lettuce cultivars *Templin* and *Lollo Bionda* compared to the control plants. The levels of TPCs in control and infected leaf blade samples from both cultivars did not differ, although the content of TPCs was higher in *Lollo Bionda* than in *Templin* (53.9–59.9 and 22.9–24.9 mg g^{-1} FW, respectively). A significant increase in TPC content was observed in infected samples of the red *Lollo Rossa* cultivar in the leaf base and blade compared to the corresponding control samples (Figure 6).

In the second autumn crop, the levels of TPCs significantly increased in the infected green lettuce cultivars *Templin* and *Lollo Bionda* in both the leaf base and leaf blade compared to the control plants. The highest levels of TPCs were observed in control and infected plants of the *Lollo Rossa* cultivar, compared to other cultivars. However, the highest levels of TPCs were observed in infected samples from the leaf blade of *Lollo Rossa* (55.05 ± 5.05 mg g^{-1} FW) in this crop. The content of TPCs in the leaf base of *Lollo Rossa* was at a similar level in the control and infected samples (Figure 6).

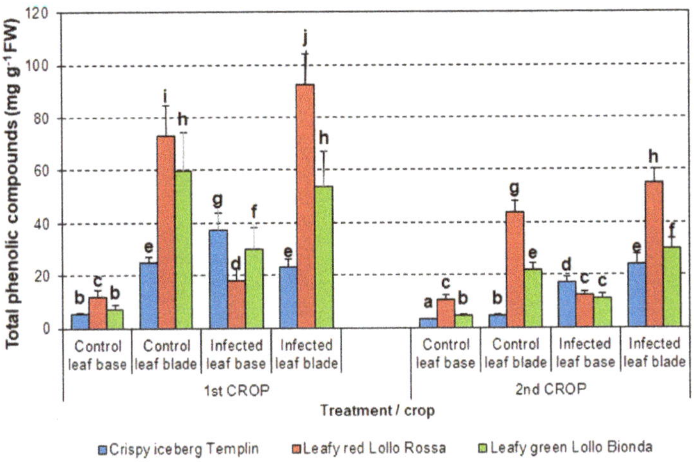

Figure 6. Total phenolic compounds content in lettuce cultivars in two vegetation seasons in 2016. The means ± SE were obtained from independent samples harvested from individual plants per treatment (n = 9–25). Different letters indicate a significant difference between the means according to Duncan's test ($p < 0.05$).

The TPC content was lower in all investigated lettuce cultivars from the second autumn crop than in plants from the first summer crop ($F = 75.87$; $p < 0.001$). Moreover, factorial analysis of variance revealed that TPC content was also significantly influenced by lettuce cultivar ($F = 47.22$, $p < 0.001$) and plant treatment (control/infected and leaf base/leaf blade, $F = 85.29$, $p < 0.001$).

3.4.2. Flavonoid Content

As for TPC content, the flavonoid content in samples collected from the leaf blade was higher than in the leaf base. This trend was the case in both control and infected samples of all three cultivars during both crops, but a few exceptions were found (Figure 7). In the first summer crop, the flavonoid content of Crispy iceberg *Templin* was the lowest compared to other cultivars, despite treatment. Moreover, the levels of flavonoids in infected *Templin* plants were higher than those of control plants in samples obtained from the leaf base only. Surprisingly, the infected leaf blades of the *Templin* cultivar had significantly lower flavonoid content than the control samples. Flavonoid content increased significantly in the infected leaf base and leaf blade of Leafy green *Lollo Bionda* and red *Lollo Rossa* compared to the control samples. However, the levels of flavonoids did not differ in the infected leaf base and infected leaf blade of *Lollo Bionda* (Figure 7).

In the second autumn crop, the flavonoid content was higher in all infected samples of investigated cultivars obtained from the leaf blade compared to the control and leaf base samples. Only the *Lollo Rossa* samples obtained from the infected leaf base contained lower levels of flavonoids than the control. The flavonoid content in *Lollo Bionda* control and infected leaf blades was similar (Figure 7).

Similarly, as in the case of TPCs, factorial analysis of variance revealed that the flavonoid content was influenced significantly by crop ($F = 47.9$, $p < 0.001$), lettuce cultivar ($F = 92.98$, $p < 0.001$), and treatment ($F = 147.6$, $p < 0.001$). Flavonoid contents in the first summer crop were higher than in the second autumn crop for all cultivars.

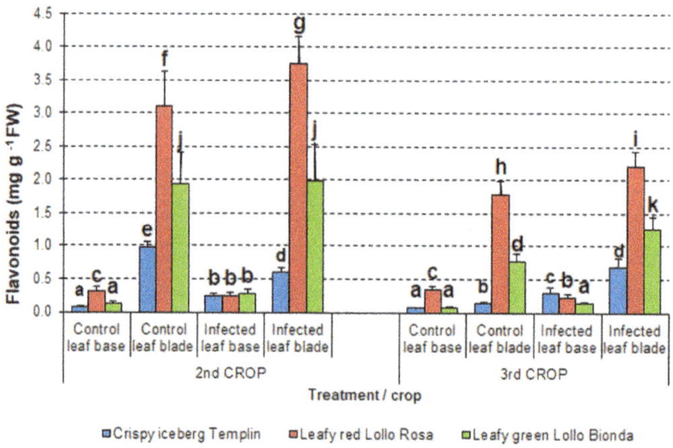

Figure 7. Changes in flavonoid content in lettuce cultivars in two vegetation seasons in 2016. The means ± SE were obtained from independent samples harvested from individual plants per treatment (n = 9–25). Different letters indicate a significant difference between the means according to Duncan's test ($p < 0.05$).

3.4.3. Anthocyanin Content

Analysis of anthocyanins in the first summer crop revealed that their contents did not change in the Crispy *Templin* and Leafy *Lollo Bionda* cultivars regardless of treatment (Figure 8). However, several times more anthocyanins were noticed in the leaf blades of the Leafy red *Lollo Rossa* cultivar than in both green cultivars. Furthermore, the anthocyanin levels in infected *Lollo Rossa* leaf blades significantly increased compared to the control. Similar trends in anthocyanin contents for all cultivars were observed in the second autumn crop (Figure 8).

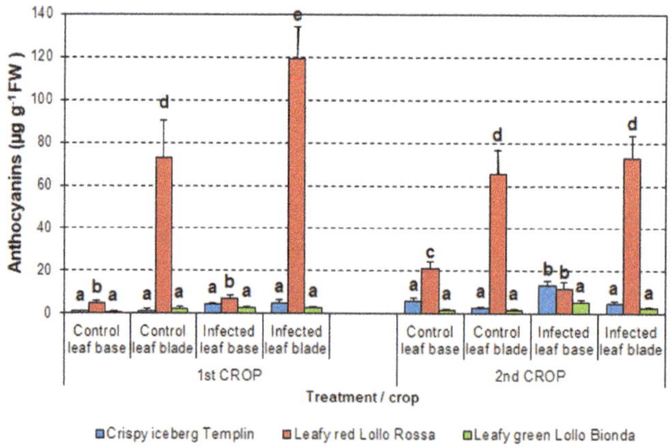

Figure 8. Differences in anthocyanin content in lettuce cultivars in two vegetation seasons in 2016. The means ± SE were obtained from independent samples harvested from individual plants per treatment (n = 3–24). Different letters indicate a significant difference between the means according to Duncan's test ($p < 0.05$).

Factorial analysis of variance indicated that the anthocyanin content was influenced significantly by lettuce cultivar ($F = 97.33$, $p < 0.001$) and treatment ($F = 22.41$, $p < 0.001$). Nevertheless, anthocyanin contents did not significantly differ between both crops.

4. Discussion

Lettuce is one of the most popular vegetables around the world. This leafy vegetable is investigated extensively in terms of improvement of its cultivation strategies to obtain high yields both in the field and indoors using various substrates or hydroponic culture [37–39]. Modifications to lettuce growth conditions to improve its qualities, resulting in a higher content of valuable nutrients and minerals, are also under study [40–42]. However, the extensive cultivation of crops as monocultures and the negative impact of global climate changes highly influence the quality and yield of crops, including lettuce. Moreover, these factors contribute to the increased activity of pathogens and pests, which threaten agriculture production, food security, and farmer incomes [42–45]. Simulation models predict that environmental conditions will impact especially incidents of soil-borne or air-borne fungus-induced diseases of crops and their severity due to the warming climate [46–48]. Forecasting models and research indicated temperature and humidity as factors influencing the development of *S. sclerotiorum* and the formation of ascospores in such crops as dry beans, canola, and lettuce. Higher average temperature (range of 15–25 °C) and humidity (above 90%) may intensify the development of the fungus on lettuce, extending the range of its occurrence [49–51]. In this study, we investigated the emergence of *S. sclerotiorum*-induced disease in three cultivars of lettuce grown on the market farm during two vegetation seasons, the summer and early autumn. Despite the stable temperature and humidity during the summer crop (over 18 °C, the optimal temperature for lettuce cultivation), their high variability in the autumn crop was observed. The average high air and soil temperatures in the first four weeks of the autumn crop (around 21 °C) were contrary to the much lower temperature in the last three weeks (around 10 °C). However, the severe disease symptoms (Figure 5) were observed first around the fourth week of each crop, exclusively on lettuces grown in the field with *S. sclerotiorum*-contaminated soil (Tables 2 and 3).

The research on Polish isolates of *S. sclerotiorum* from canola has indicated that temperatures between 15 and 20 °C significantly accelerated the fungus development and sclerotia formation [52]. Sclerotia are produced most often during unfavorable environmental conditions or when nutrients are lacking. Several factors, such as temperature, humidity, and pH, can influence the formation and development of sclerotia [25]. Although it was not determined quantitatively, more sclerotia were observed at the end of the autumn crop, when the air and soil temperatures ranged between 6 and 14 °C (Figure 2). Moreover, *S. sclerotiorum* can grow and form sclerotia in a wide range of soil pH from 2.5 to 9 [26]. However, sclerotia formation in the culture at neutral or alkaline pH can be arrested [25]. Despite the fluctuating pH in the first summer crop and neutral pH in the autumn crop, the fungus developed extensively, causing severe disease symptoms and forming numerous sclerotia (Figure 3).

Furthermore, the development of lettuce drop depends on other factors, not only environmental conditions. It has been reported that host plants' resistance to *Sclerotinia* spp. depends on their morphological and physiological attributions, such as lifting leaves off the soil or lignifying stems [53]. In our study, this phenomenon could also be noticed as Crispy iceberg *Templin* with watery leaves and a formed head had the lowest survival of 10–30% during both seasons, compared to Leafy lettuce cultivars forming rosettes such as *Lollo Bionda* and *Lollo Rossa* (Tables 2 and 3). Moreover, various levels of commercial lettuce varieties/cultivars' resistance to *S. sclerotiorum* have been tested under field and growth room conditions [33], suggesting other physiological disease-restricting factors.

Individual lettuce cultivar features influencing the pathogen development and disease progression may be related to secondary metabolites such as phenolic compounds regarding their composition and content, similar to many other pathosystems [54,55]. Under normal growth conditions, lettuce leaf tissue contains a certain quantity of a wide range of phenolic compounds depending on the cultivar [56,57]. Environmental conditions affect the concentration of phenolic compounds in lettuce [12,13,15]. However, when exposed to stress conditions, the activity of the phenylpropanoid pathway intensifies, leading to the extensive biosynthesis and storage of various phenolic compounds such as flavonols and anthocyanins, protecting plants and reducing oxidative stress [15,58,59]. The lowest contents of total phenolic compounds and flavonoids were noticed in green lettuce cultivars Crispy iceberg *Templin* and Leafy *Lollo Bionda* in both vegetation seasons. In the case of red lettuce *Lollo Rosa*, the samples from infected plants had the highest content of phenolics (Figures 6 and 7). Thus, activation of the phenylpropanoid pathway is closely related to enhanced resistance to *S. sclerotiorum* also in other plant species, such as a pea [60].

Purple-bluish anthocyanins are water-soluble pigment flavonoids, and they play a relevant role in lettuce defense against pathogenic fungi, showing strong antioxidant and antimicrobial properties [61,62]. Anthocyanin content is influenced by temperature and depends on a cultivar with higher content in red lettuce cultivars than green ones [62–64]. Accordingly, in our study, green lettuce cultivars *Templin* and *Lollo Bionda* contained low levels of anthocyanins compared to red lettuce *Lollo Rossa*. Moreover, we have shown first that anthocyanin content increased several times in *S. sclerotiorum*-infected lettuce *Lollo Rossa* compared to the control, while it remained at the same level or increased only slightly in infected green lettuce cultivars (Figure 8). Enhanced anthocyanin biosynthesis and accumulation have been reported in the wild-type Indian mustard infected with *S. sclerotiorum*, showing relatively small disease symptoms contrary to the spreading lesions in anthocyanin-devoid mutant plants [65]. One possible explanation of the antifungal effect of phenolics is the induction of cell death by phenolic acid (ferulic acid) described for the maize fungal pathogen *Cochliobolus heterostrophus* [66].

Lettuce infection with *S. sclerotiorum* and severe symptoms of the lettuce drop incidents in many market farms limit agricultural production. The impact of this fungus on farm yield and economic performance prompted the development of research that would bring measurable benefits to farms enduring epidemics induced by *S. sclerotiorum*. In our case, however, the occurrence of a small-scale *S. sclerotiorum*-induced disease in lettuce was fought off and silenced with a crop rotation practice (Appendix A, Figure A2). However, the management of lettuce drop is mainly based on a few fungicides [67], and biocontrol methods are used, such as spraying plants with bacteria such as *Streptomycetes* or *Arthrobacter* [68,69]. Unfortunately, those methods are insufficient for combatting large-scale farm epidemics triggered by *S. sclerotiorum* [70]. Manipulating the content and/or composition of phenolic compounds using contemporary approaches may provide the foundation for modern management and defense strategy for successful lettuce cultivation [71].

Supplementary Materials: The following supporting information can be downloaded at: https://www.mdpi.com/article/10.3390/pathogens12121416/s1. Table S1: Soil temperature and pH, and air temperature and humidity measured at 3 p.m. daily during the first crop from 7 July 2016 to 18 August 2016; Table S2: Soil temperature and pH, and air temperature and humidity measured at 3 p.m. daily during second crop from 3 September 2016 to 23 October 2016; Table S3. ANOVA of the lettuce survival during vegetation seasons; Table S4: Pearson's correlations for the survival of lettuce cultivars with air and soil parameters in the infected field; Table S5: Pearson's correlation coefficients for the survival of lettuce cultivars with air and soil parameters in the control field.

Author Contributions: Conceptualization, V.K.M. and P.M.; methodology, V.K.M.; formal analysis, V.K.M.; investigation, P.M.; resources, V.K.M. and P.M.; writing—original draft preparation, V.K.M. and P.M.; writing—review and editing, V.K.M. and A.K.K.; visualization, V.K.M. and P.M.; supervision, A.K.K.; funding acquisition, A.K.K., V.K.M. and P.M. All authors have read and agreed to the published version of the manuscript.

Funding: This research was funded in part by the National Centre for Research and Development, Poland, grant no. ERA-CAPS II/1/2015 and the University of Lodz, Poland, grant no. B1611000000211.01. The APC was funded by the University of Bialystok, Poland.

Institutional Review Board Statement: Not applicable.

Informed Consent Statement: Not applicable.

Data Availability Statement: Data are contained within the article.

Acknowledgments: All the authors, and especially A.K.K. and V.K.M. thank wholeheartedly Wiesław and Izabela Królikiewicz (Horticulture Market Farm, Majków Średni, Poland) for their support during the realization of the project and enabling us to grow lettuce cultivars and perform the experiments in the field.

Conflicts of Interest: The authors declare no conflict of interest. The funders had no role in the design of the study; in the collection, analyses, or interpretation of data; in the writing of the manuscript; or in the decision to publish the results.

Appendix A

Non-Evaluated Crops (Spring Crop 2016, Spring and Summer Crops 2017, and Spring Crop 2018).

In spring on 4 April 2016, the experimental crop was established with two lettuce cultivars, Crispy Iceberg *Diamentinas* and Leafy red *Lollo Rossa*. However, the crop was longer than the summer and autumn ones and lasted until 20 June (81 days) due to relatively low temperatures on days and nights. The evaluation of the disease progression and the collection of samples was carried out on the fourth and eleventh weeks of cultivation. This crop's different establishment and longer cultivation period than the other 2016 crops made it impossible to correlate all three crops in 2016. Therefore, we have decided not to consider the spring crop in 2016 in our study (Figure A1).

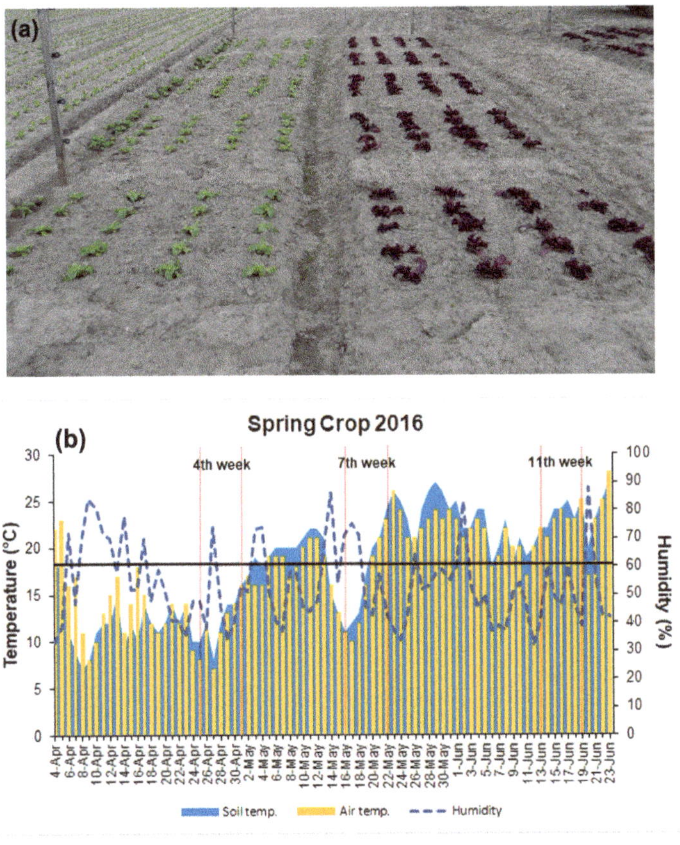

Figure A1. Establishment of the spring crops in 2016. (**a**) Seven-week-old seedlings in the infected field, from the left Crispy *Diamentinas* and Leafy *Lollo Rossa*; (**b**) soil and air temperature, and relative air humidity in the spring crop.

In the spring and summer of 2017, Brussel sprouts plants instead of lettuce have been planted as a crop rotation practice in experimental plots. In early autumn 2017, the experimental plots were established, and the plant material comprised three lettuce cultivars: iceberg lettuce *Templin*, green lettuce *Lollo Bionda*, and red lettuce *Lollo Rossa*. However, it was impossible to collect enough samples from the fungus-infected and the control fields due to unfavorable weather conditions at that time (heavy rain).

Another lettuce crop was established on 30 April 2018, similar to the crops in 2016 and 2017. This crop lasted until 20 June 2018. However, there was no sign of any disease symptoms on any lettuce plants (Figure A2). Therefore, no samples were harvested, and no data were collected.

Figure A2. The lettuce crop was established on 30 April 2018, without any signs of *S. sclerotiorum* infection. Photos of lettuce cultivars in the infected field were taken on 15 May: (**a**) *Templin*; (**b**) *Lollo Bionda*; (**c**) *Lollo Rossa*; photos of lettuce cultivars in the infected field taken on 5 June: (**d**) *Templin*; (**e**) *Lollo Bionda*; (**f**) *Lollo Rossa*; photos of lettuce cultivars in the infected field taken on 18 June: (**g**) *Templin*; (**h**) *Lollo Bionda*; (**i**) *Lollo Rossa*.

References

1. Křístková, E.; Doležalová, I.; Lebeda, A.; Vinter, V.; Novotná, A. Description of morphological characters of lettuce (*Lactuca sativa* L.) genetic resources. *Hortic. Sci.* **2008**, *35*, 113–129. [CrossRef]
2. Han, R.; Truco, M.J.; Lavelle, D.O.; Michelmore, R.W. A Composite Analysis of Flowering Time Regulation in Lettuce. *Front. Plant Sci.* **2021**, *12*, 632708. [CrossRef] [PubMed]
3. Mou, B. *Lettuce. Vegetables I—Asteraceae, Brassicaceae, Chenopodiaceae, and Cucurbitaceae*; Prohens, J., Nuez, F., Eds.; Springer: New York, NY, USA, 2008; pp. 75–116.
4. Rolnik, A.; Olas, B. The Plants of the *Asteraceae* Family as Agents in the Protection of Human Health. *Int. J. Mol. Sci.* **2021**, *22*, 3009. [CrossRef] [PubMed]
5. Akbar, S. *Lactuca sativa* L. (*Asteraceae/Compositae*). In *Handbook of 200 Medicinal Plants*; Akbar, S., Ed.; Springer: Cham, Switzerland, 2020; pp. 1067–1075. [CrossRef]
6. Ryder, E.J. Lettuce. In *Leafy Salad Vegetables*; Ryder, E.J., Ed.; Springer: Dordrecht, The Netherlands, 1979; pp. 13–94.
7. Damerum, A.; Chapman, M.A.; Taylor, G. Innovative breeding technologies in lettuce for improved post-harvest quality. *Postharvest Biol. Technol.* **2020**, *168*, 111266. [CrossRef] [PubMed]

8. Mulabagal, V.; Ngouajio, M.; Nair, A.; Zhang, Y.; Gottumukkala, A.L.; Nair, M.G. In vitro evaluation of red and green lettuce (*Lactuca sativa*) for functional food properties. *Food Chem.* **2010**, *118*, 300–306. [CrossRef]
9. Shi, M.; Gu, J.; Wu, H.; Rauf, A.; Emran, T.B.; Khan, Z.; Mitra, S.; Aljohani, A.S.M.; Alhumaydhi, F.A.; Al-Awthan, Y.S.; et al. Phytochemicals, Nutrition, Metabolism, Bioavailability, and Health Benefits in Lettuce—A Comprehensive Review. *Antioxidants* **2022**, *11*, 1158. [CrossRef] [PubMed]
10. Jeong, S.W.; Kik, G.-S.; Lee, W.S.; Kim, Y.-H.; Kang, N.J.; Jin, J.S.; Lee, G.M.; Kim, S.T.; El-Aty, A.M.; Shim, J.H.; et al. The effects of different night-time temperatures and cultivation durations on the polyphenolic contents of lettuce: Application of principal component analysis. *J. Adv. Res.* **2015**, *6*, 493–499. [CrossRef] [PubMed]
11. Kim, M.J.; Moon, Y.; Tou, J.C.; Mou, B.; Waterland, N.L. Nutritional value, bioactive compounds and health benefits of lettuce (*Lactuca sativa* L.). *J. Food Compos. Anal.* **2016**, *49*, 19–34. [CrossRef]
12. Brazaitytė, A.; Vaštakaitė-Kairienė, V.; Sutulienė, R.; Rasiukevičiūtė, N.; Viršilė, A.; Miliauskienė, J.; Laužikė, K.; Valiuškaitė, A.; Dėnė, L.; Chrapačienė, S.; et al. Phenolic Compounds Content Evaluation of Lettuce Grown under Short-Term Preharvest Daytime or Nighttime Supplemental LEDs. *Plants* **2022**, *11*, 1123. [CrossRef]
13. de Souza, A.S.N.; de Oliveira Schmidt, H.; Pagno, C.; Rodrigues, E.; Silva da Silva, M.A.; Hickmann Flôres, S.; de Oliveira Rios, A. Influence of cultivar and season on carotenoids and phenolic compounds from red lettuce influence of cultivar and season on lettuce. *Food Res. Int.* **2022**, *155*, 111110. [CrossRef]
14. Pérez-López, U.; Sgherri, C.; Miranda-Apodaca, J.; Micaelli, F.; Lacuesta, M.; Mena-Petite, A.; Frank Quartacci, M.; Muñoz-Rued, A. Concentration of phenolic compounds is increased in lettuce grown under high light intensity and elevated CO_2. *Plant Physiol. Biochem.* **2018**, *123*, 233–241. [CrossRef] [PubMed]
15. Liu, X.; Ardo, S.; Bunning, M.; Parry, J.; Zhou, K.; Stushnoff, C.; Stoniker, F.; Yu, L.; Kendall, P. Total phenolic content and DPPH radical scavenging activity of lettuce (*Lactuca sativa* L.) grown in Colorado. *LWT-Food Sci. Technol.* **2007**, *40*, 552–557. [CrossRef]
16. Llorach, R.; Martınez-Sanchez, A.; Tomas-Barberan, F.A.; Gil, M.I.; Ferreres, F. Characterisation of polyphenols and antioxidant properties of five lettuce varieties and escarole. *Food Chem.* **2008**, *108*, 1028–1038. [CrossRef]
17. Sinkovič, L.; Sinkovič, D.K.; Ugrinović, K. Yield and nutritional quality of soil-cultivated crisphead lettuce (*Lactuca sativa* L. var. *capitata*) and corn salad (*Valerianella* spp.) harvested at different growing periods. *Food Sci. Nutr.* **2023**, *11*, 1755–1769. [CrossRef] [PubMed]
18. Turini, T.A.; Joseph, S.V.; Baldwin, R.A.; Koike, S.T.; Natwick, E.T.; Ploeg, A.T.; Smith, R.F.; Dara, S.K.; Fennimore, S.A.; LeStrange, M.; et al. UC IPM Pest Management Guidelines: Lettuce. UC ANR Publication 3450. Davis, CA, USA. Available online: https://ipm.ucanr.edu/agriculture/lettuce/ (accessed on 30 April 2023).
19. Fall, M.L.; Van der Heyden, H.; Carisse, O. *Bremia lactucae* infection efficiency in lettuce is modulated by temperature and leaf wetness duration under Quebec field conditions. *Plant Dis.* **2015**, *99*, 1010–1019. [CrossRef] [PubMed]
20. Shim, C.K.; Kim, M.J.; Kim, Y.K.; Jee, H.J. Evaluation of lettuce germplasm resistance to gray mold disease for organic cultivations. *Plant Pathol. J.* **2014**, *30*, 90–95. [CrossRef]
21. Iwaniuk, P.; Lozowicka, B. Biochemical compounds and stress markers in lettuce upon exposure to pathogenic *Botrytis cinerea* and fungicides inhibiting oxidative phosphorylation. *Planta* **2022**, *255*, 61. [CrossRef]
22. Pink, H.; Talbot, A.; Graceson, A.; Graham, J.; Higgins, G.; Taylor, A.; Jackson, A.C.; Truco, M.; Michelmore, R.; Yao, C.; et al. Identification of genetic loci in lettuce mediating quantitative resistance to fungal pathogens. *Theor. Appl. Genet.* **2022**, *135*, 2481–2500. [CrossRef]
23. Wu, B.M.; Subbarao, K.V. Analyses of Lettuce Drop Incidence and Population Structure of *Sclerotinia sclerotiorum* and *S. minor*. *Phytopathology* **2006**, *96*, 1322–1329. [CrossRef]
24. Liang, X.; Rollins, J.A. Mechanisms of Broad Host Range Necrotrophic Pathogenesis in *Sclerotinia sclerotiorum*. *Phytopathology* **2018**, *108*, 1128–1140. [CrossRef]
25. Bolton, M.D.; Thomma, B.P.H.J.; Nelson, B.D. *Sclerotinia sclerotiorum* (Lib.) de Bary: Biology and molecular traits of a cosmopolitan pathogen. *Mol. Plant Pathol.* **2006**, *7*, 1–16. [CrossRef]
26. Saharan, G.S.; Mehta, N. History and Host Range. In *Sclerotinia Diseases of Crop Plants: Biology, Ecology and Disease Management*; Saharan, G.S., Mehta, N., Eds.; Springer: Dordrecht, The Netherlands, 2008; pp. 19–39. [CrossRef]
27. Derbyshire, M.C.; Newman, T.E.; Khentry, Y.; Taiwo, A.O. The evolutionary and molecular features of the broad-host-range plant pathogen *Sclerotinia sclerotiorum*. *Mol. Plant Pathol.* **2022**, *23*, 1075–1090. [CrossRef] [PubMed]
28. Durman, S.B.; Menendez, A.B.; Godeas, A.M. Variation in oxalic acid production and mycelial compatibility within field populations of *Sclerotinia sclerotiorum*. *Soil Biol. Biochem.* **2005**, *37*, 2180–2184. [CrossRef]
29. Williams, B.; Kabbage, M.; Kim, H.-J.; Britt, R.; Dickman, M.B. Tipping the balance: *Sclerotinia sclerotiorum* secreted oxalic acid suppresses host defenses by manipulating the host redox environment. *PLoS Pathog.* **2011**, *7*, e1002107. [CrossRef] [PubMed]
30. Erental, A.; Dickman, M.B.; Yarden, O. Sclerotial development in *Sclerotinia sclerotiorum*: Awakening molecular analysis of a "Dormant" structure. *Fungal Biol. Rev.* **2008**, *22*, 6–16. [CrossRef]
31. Mamo, B.E.; Eriksen, R.L.; Adhikari, N.D.; Hayes, R.J.; Mou, B.; Simko, I. Epidemiological Characterization of Lettuce Drop (*Sclerotinia* spp.) and Biophysical Features of the Host Identify Soft Stem as a Susceptibility Factor. *PhytoFrontiers* **2021**, *1*, 182–204. [CrossRef]

32. Mbengue, M.; Navaud, O.; Peyraud, R.; Barascud, M.; Badet, T.; Vincent, R.; Barbacci, A.; Raffaele, S. Emerging trends in molecular interactions between plants and the broad host range fungal pathogens *Botrytis cinerea* and *Sclerotinia sclerotiorum*. *Front. Plant Sci.* **2016**, *7*, 422. [CrossRef]
33. Hayes, R.J.; Wu, B.M.; Pryor, B.M.; Chitrampalam, P.; Subbarao, K.V. Assessment of Resistance in Lettuce (*Lactuca sativa* L.) to Mycelial and Ascospore Infection by *Sclerotinia minor* Jagger and *S. sclerotiorum* (Lib.) de Bary. *HortScience* **2010**, *45*, 333–341. [CrossRef]
34. Mamo, B.E.; Hayes, R.J.; Truco, M.J.; Puri, K.D.; Michelmore, R.W.; Subbarao, K.; Simko, I. The genetics of resistance to lettuce drop (*Sclerotinia* spp.) in lettuce in a recombinant inbred line population from Reine des Glaces × Eruption. *Theor. Appl. Genet.* **2019**, *132*, 2439–2460. [CrossRef]
35. Macioszek, V.K.; Wielanek, M.; Morkunas, I.; Ciereszko, I.; Kononowicz, A.K. Leaf position-dependent effect of *Alternaria brassicicola* development on host cell death, photosynthesis and secondary metabolites in *Brassica juncea*. *Physiol. Plant.* **2020**, *168*, 601–616. [CrossRef]
36. Lee, J.; Durst, R.W.; Wrolstad, R.E. Determination of Total Monomeric Anthocyanin Pigment Content of Fruit Juices, Beverages, Natural Colorants, and Wines by the pH Differential Method: Collaborative Study. *J. AOAC Int.* **2005**, *88*, 1269–1278. [CrossRef] [PubMed]
37. Hameed, M.K.; Umar, W.; Razzaq, A.; Aziz, T.; Maqsood, M.A.; Wei, S.; Niu, Q.; Huang, D.; Chang, L. Differential Metabolic Responses of Lettuce Grown in Soil, Substrate and Hydroponic Cultivation Systems under NH_4^+/NO_3^- Application. *Metabolites* **2022**, *12*, 444. [CrossRef] [PubMed]
38. Nagano, S.; Mori, N.; Tomari, Y.; Mitsugi, N.; Deguchi, A.; Kashima, M.; Tezuka, A.; Nagano, A.J.; Usami, H.; Tanabata, T.; et al. Effect of differences in light source environment on transcriptome of leaf lettuce (*Lactuca sativa* L.) to optimize cultivation conditions. *PLoS ONE* **2022**, *17*, e0265994. [CrossRef] [PubMed]
39. Lei, C.; Engeseth, N.J. Comparison of growth characteristics, functional qualities, and texture of hydroponically grown and soil-grown lettuce. *LWT—Food Sci. Technol.* **2021**, *150*, 111931. [CrossRef]
40. Becker, C.; Urlić, B.; Špika, M.J.; Kläring, H.P.; Krumbein, A.; Baldermann, S.; Ban, S.G.; Perica, S.; Schwarz, D. Nitrogen limited red and green leaf lettuce accumulate flavonoid glycosides, caffeic acid derivatives, and sucrose while losing chlorophylls, β-carotene and xanthophylls. *PLoS ONE* **2015**, *10*, e0142867. [CrossRef] [PubMed]
41. Martínez-Ispizua, E.; Calatayud, Á.; Marsal, J.I.; Cannata, C.; Basile, F.; Abdelkhalik, A.; Soler, S.; Valcárcel, J.V.; Martínez-Cuenca, M.-R. The Nutritional Quality Potential of Microgreens, Baby Leaves, and Adult Lettuce: An Underexploited Nutraceutical Source. *Foods* **2022**, *11*, 423. [CrossRef] [PubMed]
42. Medina-Lozano, I.; Bertolín, J.R.; Díaz, A. Nutritional value of commercial and traditional lettuce (*Lactuca sativa* L.) and wild relatives: Vitamin C and anthocyanin content. *Food Chem.* **2021**, *359*, 129864. [CrossRef]
43. Myers, S.S.; Smith, M.R.; Guth, S.; Golden, C.D.; Vaitla, B.; Mueller, N.D.; Dangour, A.D.; Huybers, P. Climate Change and Global Food Systems: Potential Impacts on Food Security and Undernutrition. *Annu. Rev. Public Health* **2017**, *38*, 259–277. [CrossRef]
44. Duchenne-Moutien, R.A.; Neetoo, H. Climate Change and Emerging Food Safety Issues: A Review. *J. Food Prot.* **2021**, *84*, 1884–1897. [CrossRef]
45. Raza, M.M.; Bebber, D.P. Climate change and plant pathogens. *Curr. Opin. Microbiol.* **2022**, *70*, 102233. [CrossRef]
46. Singh, B.K.; Delgado-Baquerizo, M.; Egidi, E.; Guirado, E.; Leach, J.E.; Liu, H.; Trivedi, P. Climate change impacts on plant pathogens, food security and paths forward. *Nat. Rev. Microbiol.* **2023**, *21*, 640–656. [CrossRef] [PubMed]
47. Chaloner, T.M.; Gurr, S.J.; Bebber, D.P. Plant pathogen infection risk tracks global crop yields under climate change. *Nat. Clim. Chang.* **2021**, *11*, 710–715. [CrossRef]
48. Miranda-Apodaca, J.; Artetxe, U.; Aguado, I.; Martin-Souto, L.; Ramirez-Garcia, A.; Lacuesta, M.; Becerril, J.M.; Estonba, A.; Ortiz-Barredo, A.; Hernández, A.; et al. Stress Response to Climate Change and Postharvest Handling in Two Differently Pigmented Lettuce Genotypes: Impact on *Alternaria alternata* Invasion and Mycotoxin Production. *Plants* **2023**, *12*, 1304. [CrossRef] [PubMed]
49. Shahoveisi, F.; Manesh, M.R.; Del Río Mendoza, L.E. Modeling risk of *Sclerotinia sclerotiorum*-induced disease development on canola and dry bean using machine learning algorithms. *Sci. Rep.* **2022**, *12*, 864. [CrossRef] [PubMed]
50. Shahoveisi, F.; Del Río Mendoza, L.E. Effect of Wetness Duration and Incubation Temperature on Development of Ascosporic Infections by *Sclerotinia sclerotiorum*. *Plant Dis.* **2020**, *104*, 1817–1823. [CrossRef] [PubMed]
51. Clarkson, J.P.; Fawcett, L.; Anthony, S.D.; Young, C. A model for *Sclerotinia sclerotiorum* infection and disease development in lettuce, based on the effects of temperature, relative humidity and ascospore density. *PLoS ONE* **2014**, *9*, e94049. [CrossRef] [PubMed]
52. Wójtowicz, A.; Jajor, E.; Wójtowicz, M.; Pasternak, M. Influence of temperature on mycelial growth and number of sclerotia of *Sclerotinia sclerotiorum* the cause of *Sclerotinia* stem rot. *Prog. Plant Prot.* **2016**, *56*, 241–244. [CrossRef]
53. Grube, R.; Ryder, E.J. Identification of lettuce (*Lactuca sativa* L.) germplasm with genetic resistance to drop caused by *Sclerotinia minor*. *J. Am. Soc. Hortic. Sci.* **2004**, *129*, 70–76. [CrossRef]
54. Tak, Y.; Kumar, M. Phenolics: A Key Defence Secondary Metabolite to Counter Biotic Stress. In *Plant Phenolics in Sustainable Agriculture*; Lone, R., Shuab, R., Kamili, A., Eds.; Springer: Singapore, 2020; pp. 309–329. [CrossRef]
55. Kulbat, K. The role of phenolic compounds in plant resistance. *Biotechnol. Food Sci.* **2016**, *80*, 97–108.

56. Romani, A.; Pinelli, P.; Galardi, C.; Sani, G.; Cimato, A.; Heimler, D. Polyphenols in greenhouse and open-air-grown lettuce. *Food Chem.* **2002**, *79*, 337–342. [CrossRef]
57. Gan, Y.Z.; Azrina, A. Antioxidant properties of selected varieties of lettuce (*Lactuca sativa* L.) commercially available in Malaysia. *Int. Food Res. J.* **2016**, *23*, 2357–2362.
58. Kang, H.-M.; Saltveit, M.-E. Antioxidant capacity of lettuce leaf tissue increases after wounding. *J. Agric. Food Chem.* **2002**, *50*, 7536–7541. [CrossRef] [PubMed]
59. Chen, Y.; Huang, L.; Liang, X.; Dai, P.; Zhang, Y.; Li, B.; Lin, X.; Sun, C. Enhancement of polyphenolic metabolism as an adaptive response of lettuce (*Lactuca sativa*) roots to aluminum stress. *Environ. Pollut.* **2020**, *261*, 114230. [CrossRef] [PubMed]
60. Jain, A.; Singh, A.; Singh, S.; Singh, H.B. Phenols enhancement effect of microbial consortium in pea plants restrains *Sclerotinia sclerotiorum*. *Biol. Control* **2015**, *89*, 23–32. [CrossRef]
61. Khoo, H.E.; Azlan, A.; Tang, S.T.; Lim, S.M. Anthocyanidins and anthocyanins: Colored pigments as food, pharmaceutical ingredients, and the potential health benefits. *Food Nutr. Res.* **2017**, *61*, 1361779. [CrossRef] [PubMed]
62. Gazula, A.; Kleinhenz, M.D.; Scheerens, J.C.; Ling, P.P. Anthocyanin Levels in Nine Lettuce (*Lactuca sativa*) Cultivars: Influence of Planting Date and Relations among Analytic, Instrumented, and Visual Assessments of Color. *HortScience* **2007**, *42*, 232–238. [CrossRef]
63. Chon, S.-U.; Boo, H.-O.; Heo, B.-G.; Gorinstein, S. Anthocyanin content and the activities of polyphenol oxidase, peroxidase and phenylalanine ammonia-lyase in lettuce cultivars. *Int. J. Food Sci. Nutr.* **2012**, *63*, 45–48. [CrossRef]
64. Gazula, A.; Kleinhenz, M.D.; Streeter, J.G.; Miller, A.R. Temperature and cultivar effects on anthocyanin and chlorophyll b concentrations in three related Lollo Rosso lettuce cultivars. *HortScience* **2005**, *40*, 1731–1733. [CrossRef]
65. Singh, M.; Avtar, R.; Lakra, N.; Pal, A.; Singh, V.K.; Punia, R.; Kumar, N.; Bishnoi, M.; Kumari, N.; Khedwal, R.S.; et al. Early oxidative burst and anthocyanin-mediated antioxidant defense mechanism impart resistance against *Sclerotinia sclerotiorum* in Indian mustard. *Physiol. Mol. Plant Pathol.* **2022**, *120*, 101847. [CrossRef]
66. Shalaby, S.; Larkov, O.; Lamdan, N.L.; Goldshmidt-Tran, O.; Horwitz, B.A. Plant phenolic acids induce programmed cell death of a fungal pathogen: MAPK signaling and survival of *Cochliobolus heterostrophus*. *Environ. Microbiol.* **2016**, *18*, 4188–4199. [CrossRef]
67. Świerczyńska, I.; Perek, A.; Pieczul, K.; Jajor, E. Sensitivity of *Sclerotinia sclerotiorum* to active substances of fungicides. *Prog. Plant Prot.* **2016**, *56*, 348–353. [CrossRef]
68. Chen, X.; Pizzatti, C.; Bonaldi, M.; Saracchi, M.; Erlacher, A.; Kunova, A.; Berg, G.; Cortesi, P. Biological Control of Lettuce Drop and Host Plant Colonization by Rhizospheric and Endophytic *Streptomycetes*. *Front. Microbiol.* **2016**, *7*, 714. [CrossRef] [PubMed]
69. Aggeli, F.; Ziogas, I.; Gkizi, D.; Fragkogeorgi, G.A.; Tjamos, S.E. Novel biocontrol agents against *Rhizoctonia solani* and *Sclerotinia sclerotiorum* in lettuce. *BioControl* **2020**, *65*, 763–773. [CrossRef]
70. Smolińska, U.; Kowalska, B. Biological control of the soil-borne fungal pathogen *Sclerotinia sclerotiorum*—A review. *J. Plant Pathol.* **2018**, *100*, 1–12. [CrossRef]
71. Wallis, C.M.; Galarneau, E.R.-A. Phenolic Compound Induction in Plant-Microbe and Plant-Insect Interactions: A Meta-Analysis. *Front. Plant Sci.* **2020**, *11*, 580753. [CrossRef] [PubMed]

pathogens MDPI

Review

The Ubiquitous Wilt-Inducing Pathogen *Fusarium oxysporum*—A Review of Genes Studied with Mutant Analysis

Edan Jackson [1,2,†], Josh Li [1,†], Thilini Weerasinghe [1,2,†] and Xin Li [1,2,*]

1 Michael Smith Laboratories, University of British Columbia, Vancouver, BC V6T 1Z4, Canada
2 Department of Botany, University of British Columbia, Vancouver, BC V6T 1Z4, Canada
* Correspondence: xinli@msl.ubc.ca
† These authors contributed equally to this work.

Abstract: *Fusarium oxysporum* is one of the most economically important plant fungal pathogens, causing devastating Fusarium wilt diseases on a diverse range of hosts, including many key crop plants. Consequently, *F. oxysporum* has been the subject of extensive research to help develop and improve crop protection strategies. The sequencing of the *F. oxysporum* genome 14 years ago has greatly accelerated the discovery and characterization of key genes contributing to *F. oxysporum* biology and virulence. In this review, we summarize important findings on the molecular mechanisms of *F. oxysporum* growth, reproduction, and virulence. In particular, we focus on genes studied through mutant analysis, covering genes involved in diverse processes such as metabolism, stress tolerance, sporulation, and pathogenicity, as well as the signaling pathways that regulate them. In doing so, we hope to present a comprehensive review of the molecular understanding of *F. oxysporum* that will aid the future study of this and related species.

Keywords: plant fungal pathogen; Fusarium wilt; *Fusarium oxysporum*; fungal pathogenesis; fungal growth; conidiation; mutant analysis; knockout

1. Introduction

The *Fusarium* genus contains a diverse range of filamentous ascomycete fungi widely distributed in the soil and associated with plants. While *Fusarium* strains are predominantly non-pathogenic and often form mutualistic relationships with host plants, the genus also includes some of the most important pathogens of both plants and animals. The genus is highly diverse, with approximately 300 phylogenetically distinct species, many of which have been grouped into monophyletic species complexes [1]. Amongst them, *Fusarium oxysporum* Schltdl. emend. Snyder & Hansen is considered the most economically important, containing the most plant pathogenic strains in the genus [2].

Plant pathogenic strains of *F. oxysporum* cause Fusarium wilt diseases in many crop species, including several economically important crops such as tomato, banana, sweet potato, onion, and legumes [3]. In addition, they can infect ornamental plants such as tulips, carnations, and orchids. Fusarium wilt displays characteristic symptoms, including browning of the vascular tissues and drooping of the older leaves, followed by necrosis, defoliation, and death of the entire plant.

The taxonomy of *F. oxysporum* has proved difficult to resolve. A very broad definition of the species has been adopted, encompassing significant genetic and morphological diversity [4]. Reflecting this, *F. oxysporum* is generally referred to as a species complex, and future taxonomic revisions within this complex are needed to resolve the complexity as more genetic evidence becomes available. Although the *F. oxysporum* species complex (FOSC) has a broad host range, individual strains tend to specialize on a small number of hosts, even to the level of a specific cultivar. Consequently, pathogenic strains of *F. oxysporum* are typically categorized based on host specificity by the informal taxonomic ranking *formae specialis* (f. sp.), with over 120 *formae speciales* described [1]. Genealogical studies have revealed that many *formae speciales* of *F. oxysporum* are not monophyletic, indicating that pathogenicity to certain host species has evolved independently multiple times [5–7]. This conclusion has important implications for the study of *F. oxysporum* biology and the development of disease control strategies, as different clonal lineages may depend upon different pathogenicity factors to infect the same host. Despite this, the *formae specialis* classification is widely used by plant pathologists as a useful and efficient system for describing and studying the pathogenicity of *F. oxysporum* isolates, although caution must be exercised when generalizing results to non-monophyletic groupings.

Even within the same *formae speciales*, different *F. oxysporum* strains can display differing virulence on different cultivars of the same host species. In these cases, the *formae specialis* is further subdivided into pathogenic races, with each race specializing on a different host cultivar [6]. *F. oxysporum* isolates are also categorized into vegetative compatibility groups (VCGs), defined as groupings of isolates that can fuse to form stable heterokaryons [7]. This distinct classification appears to better reflect phylogenetic relatedness than the *formae speciales* system [8].

F. oxysporum virulence is mediated through its diverse repertoire of pathogenicity factors that facilitate invasion and colonization, including cell wall-degrading enzymes (CWDEs), mycotoxins, and secreted effectors [9–11]. Many of these pathogenicity factors form host-specific gene-for-gene interactions, in which pathogen effectors interact with host immune receptors locked in an evolutionary arms race. These interactions often play important roles in determining the high degree of host specificity exhibited by *F. oxysporum* strains [12,13].

Like many *Fusarium* species, *F. oxysporum* is anamorphic, with no observed sexual stage [14]. It produces three types of asexual spores, macroconidia, microconidia, and chlamydospores, in a process known as conidiation [15]. Macroconidia are long, sickle-shaped, multinucleate with multiple septa, and are produced from conidiophores in specialized structures known as sporodochia. Microconidia are smaller, oval-shaped uninucleate spores produced by conidiophores on the mycelium. Chlamydospores are small, round, thick-walled spores produced by vegetative hyphae or from macroconidia [2]. While conidia and even mycelia may be able to survive in the soil for short periods, chlamydospores are the most resistant form; they are believed to be the main vessel of *F. oxysporum* persistence in the soil [16].

Spore germination is triggered upon favorable temperature and humidity, and they can grow into hyphae to penetrate the host root epidermis [9] (Figure 1). Once initial infection has been established, hyphae can grow intercellularly through the root cortex to the vascular tissues and enter the xylem. In the xylem, hyphae produce macro- and microconidia, which are then transported to the above-ground tissues to cause systemic infection [17]. Hyphae can grow to form a thick mycelium, blocking the xylem and leading to browning of the vasculature and wilting and chlorosis of the leaves and stems. The fungus can rapidly proliferate on dying host tissue, producing spores that can survive in soil to infect new plants. While primarily considered a soilborne pathogen, certain *F. oxysporum* strains produce airborne conidia from the surfaces of infected tissues, aiding systemic colonization and dispersal to new soils [18–20]. Spores have even been found to be dispersed via insect vectors such as shore flies, allowing dispersal over long distances [21].

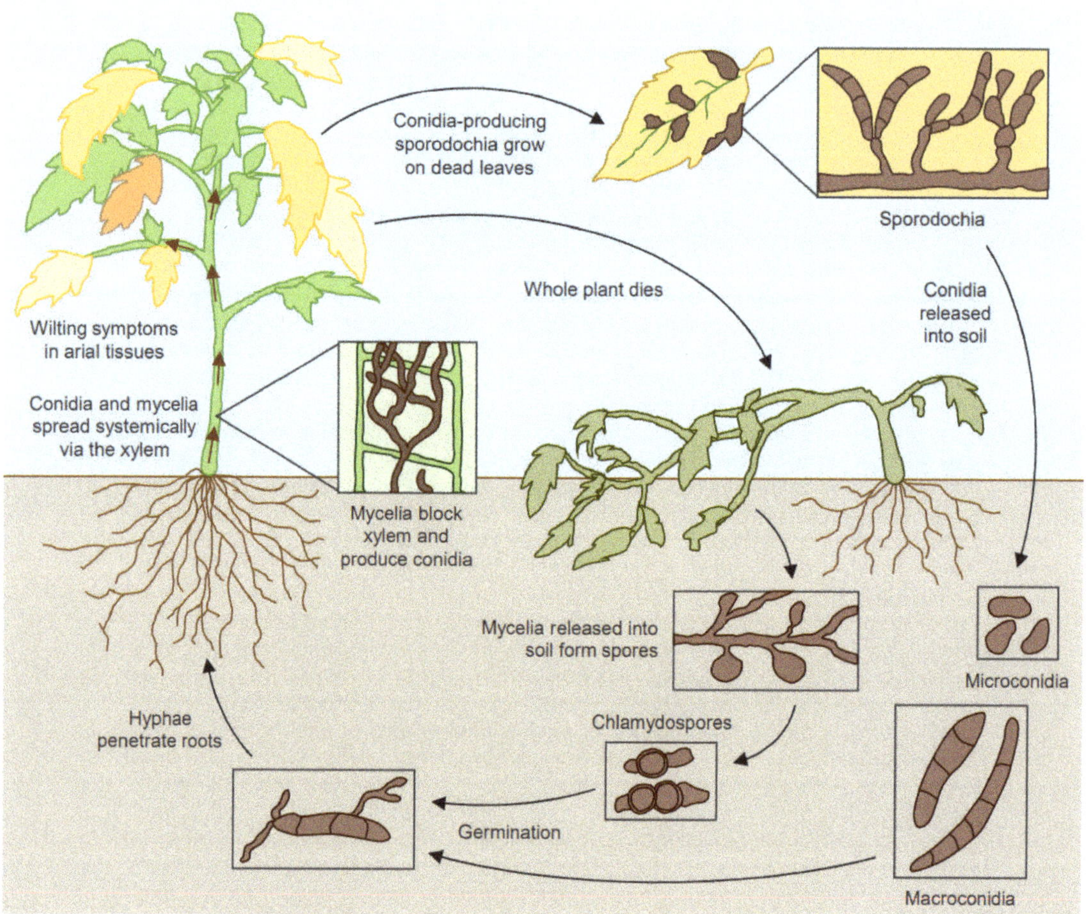

Figure 1. Lifecycle and disease cycle of *F. oxysporum*.

Due to its ability to persist in the soil for long periods, *F. oxysporum* is difficult to control. Whilst fungicides such as carbendazim, chloropicrin, and 1,3-dichloropropene have been effective, there are many concerns with their use, including harm to beneficial microbes, disruption of aquatic ecosystems, and the development of fungicide resistance [22]. Management practices such as crop rotation and destruction of infected plants have been used with some success in reducing soil inoculum, but the long persistence of chlamydospores makes such strategies difficult to sustain [23]. Future management will also find it challenging to deal with the development of fungicide resistant cultivars.

In this review, we provide an update on recent developments in the molecular biology of *F. oxysporum*, with a focus on mutant analysis. Mutant analysis is a powerful tool for the study of fungal biology and establishing a causal relationship between genes and their biological functions. Several techniques have been utilized for the generation of *F. oxysporum* mutants, including homologous recombination (HR)-based gene deletion, RNA interference (RNAi), and T-DNA insertion. We begin with an overview of the *F. oxysporum* genome and the insights provided by genomic, transcriptomic, and secretomic analyses. We then discuss key molecular findings from the *F. oxysporum* genes that have been studied through mutant analysis to provide a comprehensive synthesis of the molecular genetic work that has taken place in recent decades.

2. The Features of the *F. oxysporum* Genome

2.1. Genome Sequences

The reference genome of *F. oxysporum* f. sp. *lycopersici* (*Fol*) 4287 (race 2, VCG 0030), isolated from tomato (Murcia, Spain), was first generated with Sanger sequencing with 6.8× coverage [24]. Using 114 scaffolds with an N_{50} scaffold length of 1.98 Mb, the *Fol* strain was determined to have 15 chromosomes comprising approximately 59.9 Mb (GenBank accession number: AAXH01000000, https://www.ncbi.nlm.nih.gov/nuccore/AAXH00000000.1/, accessed on 1 August 2023). The genome was predicted to have 17,735 genes with an average gene length of 1292 bp.

At the same time, the closely related *Fusarium graminearum* genome (AACM00000000, https://www.ncbi.nlm.nih.gov/nuccore/AACM00000000.2/, accessed on 1 August 2023) was sequenced [25]. It was compared with the genomes of *Fusarium verticillioides* (AAIM02000000, https://www.ncbi.nlm.nih.gov/nuccore/AAIM00000000.2/, accessed on 1 August 2023) and *Fol* [24]. This revealed lineage-specific (LS) regions on chromosomes 3, 6, 14, and 15 of the *F. oxysporum* genome that were missing in *F. graminearum* and *F. verticillioides*. These LS regions contain genes involved in host cell wall degradation, ethylene and necrosis induction, and were shown to be upregulated during early tomato infection. As the transfer of these chromosomes could induce pathogenicity in non-pathogenic *F. oxysporum* strains, these LS regions were proposed to facilitate the transfer of pathogenicity factors among *Fusarium* species and subsequently broaden their host range [24].

More recently, the *Fol* genome was sequenced with higher coverage (66×) using Illumina HiSeq and PacBio [26]. This version of the whole-genome sequence is the current reference genome for *Fol* 4287 (QESU00000000, https://www.ncbi.nlm.nih.gov/nuccore/QESU00000000.1, accessed on 1 August 2023). This assembly is 53.9 Mb, with 499 contigs and an N_{50} scaffold length of 1.3 Mb. The largest contig is 5.7 Mb with a 47.7% GC content.

Several recent sequencing projects for other *F. oxysporum* strains with differing hosts such as flax [27], cucumber [28], cabbage [29], cowpea [30], and melon [31] have been reported. Further, Schmidt et al. described the sequencing of several strains of *F. oxysporum* f. sp. *melonis* (*Fom*), which revealed an avirulence protein that interacts with the melon resistance gene *Fom-2* [31]. The availability of *F. oxysporum* genomes for various strains has facilitated comparative genomic approaches to reveal genes involved in pathogenicity. Most recently, 35 genomes of different *Fusarium* species have been integrated into the *F. oxysporum* Pangenome Database (FoPGDB), allowing for efficient and comprehensive genomic analysis [32].

2.2. Transcriptomic and Secretomic Analyses

Transcriptomic and secretomic analyses of *F. oxysporum* during host infection have revealed the regulation of several pathogenicity-related genes. For instance, RNA-seq performed during host infection identified upregulated genes encoding cell wall-degrading enzymes, such as endo-polygalacturonases *PG1* and *PG5* [33]. Additionally, genes involved in synthesizing mycotoxins, such as trichothecene and fumonisin, were upregulated. Using a similar approach, Chang et al. discovered a β-lactamase-encoding gene that allowed *F. oxysporum* to infect soybean even in the presence of bacterial competitor *Burkholderia ambifaria* [34].

In a study comparing the secretomes of two strains of the banana pathogen *F. oxysporum* f. sp. *cubense* (*Foc* or *Focub*), 120 and 129 secreted proteins were identified during root infection in strains *Foc R1* and *Foc TR4*, respectively [35]. Specifically, in *Foc TR4*, a cysteine biosynthesis enzyme was found to be highly induced during root infection and was necessary for pathogenicity. In *Fol*, a secretomic analysis of acetylated proteins using LC-MS/MS revealed 32 genes that were not found in lineage-specific regions, 26 of which were upregulated during root infection [36]. These genes induced during infection provide a resource for potential reverse genetic studies to elucidate the mechanisms required for *F. oxysporum* pathogenicity.

Despite the broad range of -omic studies, an *F. oxysporum* deletion mutant for a potential gene should be generated and characterized to understand whether the gene is necessary for pathogenicity. Hence, the rest of this review will summarize the genes that have been studied in *F. oxysporum* using mutant analysis.

3. Molecular Dissection of *F. oxysporum* Biology

In this section, we describe the known functions *F. oxysporum* genes studied through mutant analysis. A comprehensive list of genes can be found in Supplementary Table S1, while an overview of the key phenotypes identified and their locations in the *F. oxysporum* genome are provided in Figures 2 and 3, respectively. We begin with the signaling pathways and gene expression regulators that are shared between many aspects of *F. oxysporum* biology (Figure 4). Next, we describe the core genes involved in vegetative growth and then consider genes specific to reproductive development and virulence. Protein, gene, and mutant nomenclatures vary amongst *Fusarium* researchers. For clarity, the most consistently used system has been adopted here: proteins are denoted with the first letter capitalized (Abc1), genes are italicized with all letters capitalized (*ABC1*), and mutant alleles are denoted in lower case (*abc1*).

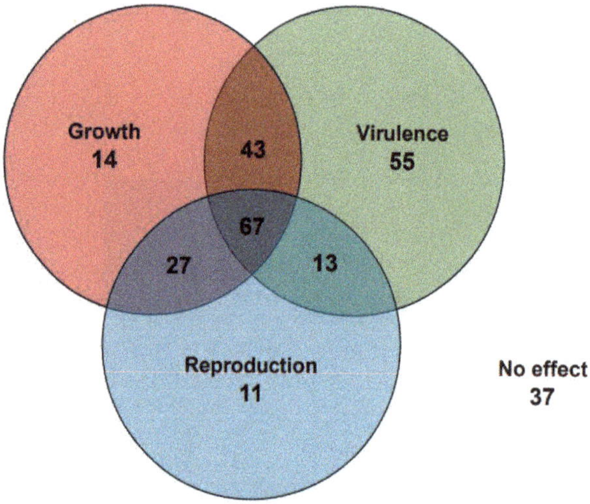

Figure 2. Venn diagram of *F. oxysporum* genes studied through mutant analysis. Growth includes genes affecting vegetative hyphal growth and stress tolerance. Reproduction includes genes affecting conidia and chlamydospore production. Virulence includes genes affecting pathogenicity, invasive growth, surface hydrophobicity, and fusaric acid production. A full list of genes is available in Supplementary Table S1.

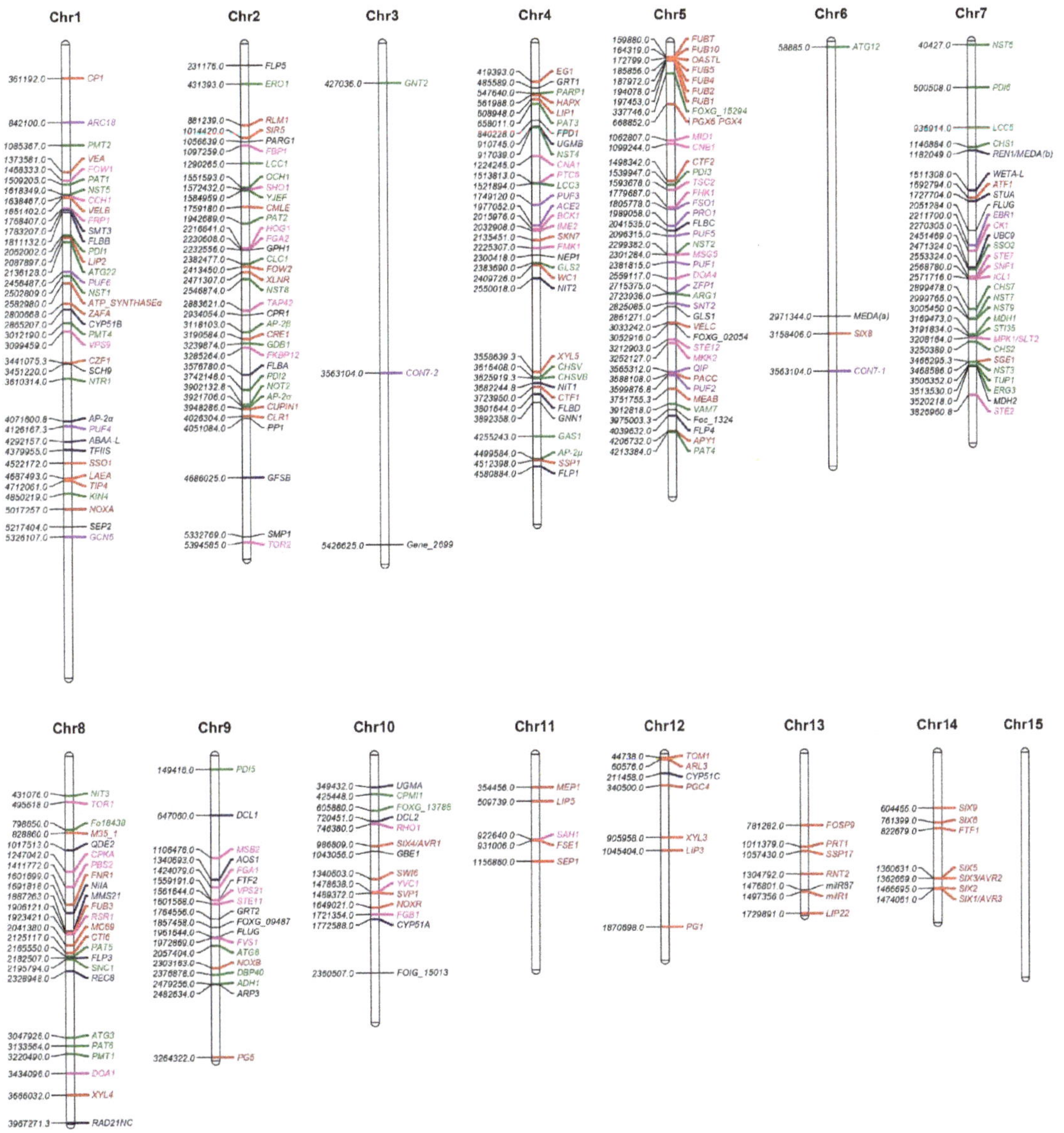

Figure 3. Chromosome map of *F. oxysporum* genes studied through mutant analysis. Components of central signaling pathways are labeled in pink, shared transcription factors are labeled in purple, genes involved in growth and stress tolerance are labeled in green, genes involved in reproduction are labeled in blue, and genes involved in virulence are labeled in red. The chromosomal map was drawn using MapChart. Full details are available in Supplementary Table S1.

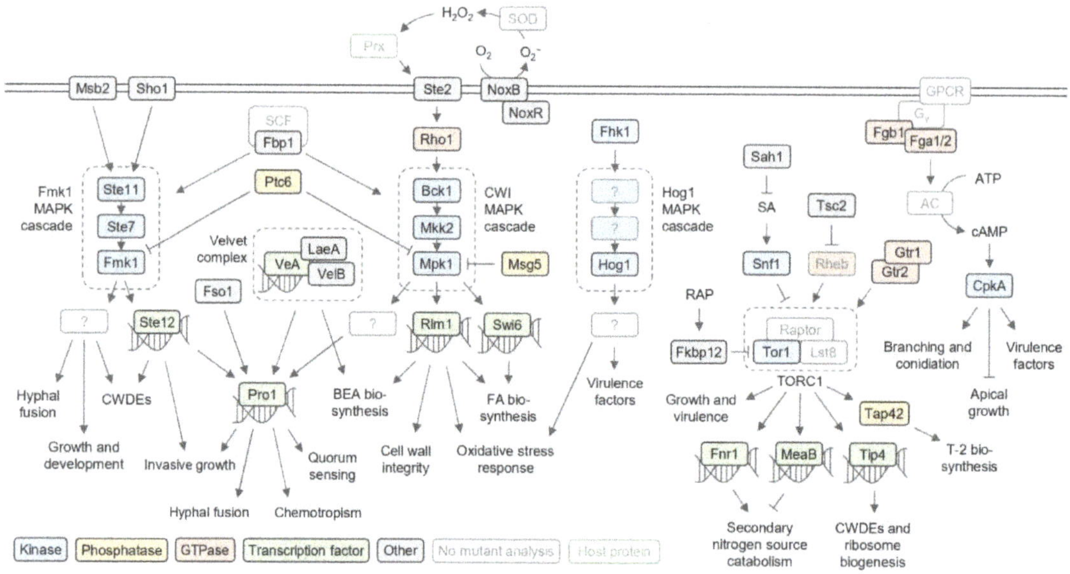

Figure 4. Major *F. oxysporum* signaling pathways identified through mutant analysis. CWDEs, cell wall-degrading enzymes; MAPK, mitogen-activated protein kinase; CWI, cell wall integrity; BEA, beauvericin; FA, fusaric acid; SA, salicylic acid; RAP, rapamycin; GPCR, G protein-coupled receptor; SOD, superoxide dismutase.

3.1. Central Signaling Pathways and Transcription Factors

3.1.1. MAP Kinase Signaling

Mitogen-activated protein kinase (MAPK) signaling cascades play a central role in multiple aspects of fungal biology (Figure 4). The *F. oxysporum* MAPK Fmk1 is required for growth, virulence, and development, with essential roles regulating invasive growth, expression of CWDEs, vegetative hyphal fusion, and surface hydrophobicity (a key determinant of virulence) [37–40]. The Fmk1 cascade involves Fmk1, the MAPKK Ste7, and the MAPKKK Ste11, and is regulated by the membrane proteins Msb2 and Sho1, which act upstream of Fmk1 to promote invasive growth and cell wall integrity [41–43].

Another MAPK involved in *F. oxysporum* virulence is Hog1. The Hog1 pathway is activated by the histidine kinase Fhk1 and is a key mediator of virulence and oxidative stress responses [39,40,44]. The cell wall integrity (CWI) MAPK cascade—consisting of the MAPK Mpk1/Slt2, the MAPKK Mkk2, and the MAPKKK Bck1 acting downstream of the Rho-type GTPase Rho1—is required for cell wall stress resistance, hyphal fusion, and chemotrophic growth [40,43,45]. Furthermore, the MAP kinase Pbs2 is required for host perception, colonization, and virulence [39]. The serine/threonine protein kinase Ime2 controls vegetative growth, hyphal branching, conidiation, pathogenicity, and stress responses [46]. Ime2 is thought to act upstream of MAPK cascades, though its placement requires further study.

Dephosphorylation of MAPK cascade components is key to regulating their signaling activity. In *F. oxysporum*, the type 2C protein phosphatase (PP2C) Ptc6 is involved in regulating MAPKs such as Mpk1 and Fmk1 and downstream growth and virulence activities [47]. The dual-specificity phosphatase Msg5 also dephosphorylates Mpk1, as well as the MAPK Fus3 [48]. Through this dephosphorylation, Msg5 functions as a regulator of pheromone responses and cell wall integrity.

3.1.2. TOR Signaling

A key regulator of growth and virulence in *F. oxysporum* is the protein complex TORC1, centered around the serine/threonine protein kinase target of rapamycin (TOR). TOR is highly conserved in eukaryotes, playing central roles in nutrient and hormone signaling networks [49,50]. Due to its importance, TOR knockout mutants are often lethal. siRNA-mediated knockdown of *TOR1* resulted in inhibited mycelial growth on potato leaves, while transgenic *TOR1* RNAi potato plants showed increased resistance to *F. oxysporum* infection, suggesting an important role for Tor1 in virulence [51]. However, chemical inhibition of TORC1 by rapamycin treatment—a process mediated by FK-506-binding protein (Fkbp12)—resulted in increased mycelial growth, suggesting a complex function of this complex [52,53]. Transcriptomic analysis has revealed that Tor1 regulates key growth and virulence-related pathways, including ribosome biogenesis and CWDEs [53].

The Gtr1–Gtr2 GTPase complex acts upstream of Tor1 to regulate its functions in growth and secondary metabolism in response to different nutrient conditions. In *F. oxysporum*, the presence of certain amino acids, most notably cysteine, causes the Gtr1–Gtr2 complex to recruit TORC1, which induces the production of T-2 toxin (T-2)—a mycotoxin harmful to humans—via the downstream phosphatase Tap42 [54].

The phytohormone salicylic acid (SA), a key modulator of the plant immune system, inhibits TOR1 by activating the adenosine monophosphate-activated protein kinase (AMPK) Snf1. Despite inhibiting TOR1, Snf1 is also required for the expression of CWDEs and is thus a key virulence factor in its own right [51,55]. To combat the inhibitory effects of SA, *F. oxysporum* produces salicylate hydroxylases (SAHs) such as Sah1 to degrade SA [51].

3.1.3. GTPase Signaling

G proteins are heterotrimeric GTP-binding proteins containing α, β, and γ subunits. They transduce external signals perceived by G protein-coupled receptors (GPCRs) to a range of intracellular targets, often through activating a cyclic AMP-protein kinase A (cAMP-PKA) cascade [56]. The *F. oxysporum* G protein α subunits Fga1 and Fga2 and the β subunit Fgb1 are required for development and pathogenesis [57–60]. G protein subunit mutants displayed defects including reduced intracellular cAMP levels, reduced pathogenicity, and altered physiological features such as heat resistance, colony morphology, conidia formation, and conidia germination. Downstream of this, cAMP-dependent protein kinase A (CpkA) affects growth, morphology, root attachment, penetration, and pathogenesis [61].

In addition to heterotrimeric G proteins, the Ras-related small GTPase Rsr1 has recently been identified as an important regulator of FA biosynthesis, conidiation, and secondary metabolism [62], and the Rab family small GTPase Vps21 is required for development and virulence [63]. RAS GTPases are activated by guanine nucleotide exchange factors (GEFs). The *F. oxysporum* GEF Vsp9 acts as a GEF for Vps21 and plays an important function in endocytosis and autophagy [63]. In contrast, GTPases are inactivated by GTPase-activating proteins (GAPs). The *F. oxysporum* GAP Tsc2 acts as a negative regulator of TORC1, with loss of Tsc2 resulting in reduced growth, stress tolerance, and virulence, indicating that constitutive activation of TORC1 negatively impacts these processes [64].

3.1.4. Ubiquitination

Ubiquitination is an important post-translational modification, regulating cellular processes across eukaryotes [65]. The addition of ubiquitin to target proteins is catalyzed by a series of enzymes: ubiquitin is activated by the ubiquitin-activating enzyme (E1), transferred to the ubiquitin-conjugating enzyme (E2), and finally transferred to the target protein through ubiquitin ligase (E3). Ubiquitination of substates can either regulate their activity or target them for degradation by the proteasome, forming the ubiquitin–proteasome system (UPS).

Several studies have demonstrated the importance of ubiquitination in *F. oxysporum*. The F-box protein Fbp1—a component of the SCF E3 complex—is required for pathogenesis, invasive growth, and cell wall integrity, and is believed to regulate multiple virulence-related MAPK signaling pathways [66]. Another F-box protein, Frp1, is also required for virulence, with a *frp1* mutant showing reduced expression of CWDEs and the glyoxylate cycle gene *ICL1* [67,68]. However, the direct targets of Frp1 remain unknown. Downstream of ubiquitination, Cdc48 is an essential ATPase that interacts with ubiquitinated proteins via ubiquitin-binding cofactors, including Doa1 of *Saccharomyces cerevisiae* [69]. Deletion of the *F. oxysporum* f. sp. *niveum* (*Fon*) homologue disrupts vegetative growth, conidiation, and stress tolerance [70]. Ubiquitination is a reversible modification, with ubiquitin tags removed by deubiquitinases. Deletion of Doa4, a putative deubiquitinase, also results in compromised vegetative growth, conidiation, and stress tolerance, reflecting the diverse roles ubiquitination plays in fungal biology [70].

3.1.5. Other Signaling Components

Temporal and spatial changes of cytoplasmic calcium (Ca^{2+}) ions play a key role in regulating the cellular and developmental responses in fungi [71]. Three Ca^{2+} channel genes, *CCH1*, *MID1*, and *YVC1*, have been identified with important roles in *F. oxysporum* biology. Cch1 and Mid1 are both required for normal vegetative growth, while Mid1 and Yvc1 contribute to sporulation [72].

In fungi, the calcium/calmodulin-dependent serine/threonine protein phosphatase complex calcineurin is involved in maintaining a diverse range of cellular processes such as growth, morphogenesis, cellular processes, stress response, and pathogenicity [73]. In *F. oxysporum*, deletion of the catalytic (Cna1) and regulatory (Cnb1) subunits demonstrated that calcineurin plays an important function in phosphatase activity and vegetative growth, virulence, and conidiation [74].

Another important regulator of *F. oxysporum* biology is Casein kinase 1 (Ck1), which negatively regulates the essential plasma membrane H^+-ATPase Pma1 to promote alkalization of the extracellular environment and regulate hyphal growth and conidiation [75]. Ck1 also controls the hyphal chemotropism toward plant roots and pathogenicity on host plants.

FVS1 encodes a protein with a sterile alpha motif (SAM) domain that is involved in protein–protein interactions related to signal transduction and gene regulation. Fvs1 is involved in the production of micro- and macroconidia, the development of conidiogenesis cells, conidiophores, and phialides, as well as in vegetative growth and virulence [76].

Interestingly, the mitochondrial carrier protein Fow1 is required for virulence and colonization of host tissues, but not for mycelial growth or development, in contrast to most mitochondria-localized proteins [77]. Similarly, the putative membrane protein Fpd1 contributes to virulence [78]. How Fow1 and Fpd1 fit into the known regulators of *F. oxysporum* virulence remains to be determined.

3.1.6. Shared Transcription Factors and Gene Regulation Components

While some transcription factors (discussed later) regulate specific biological processes, many appear to act as 'global regulators', controlling multiple aspects of *F. oxysporum* biology. Con7-1 regulates a diverse range of key processes including cell wall biogenesis and remodeling, cell division, and invasive growth [79]. The C2H2 zinc finger transcription factor Czf1 plays an important function in the production of fusaric acid (FA; a virulence factor of *F. oxysporum*), secondary metabolism, conidiation, and early host infection [45]. The BAH/PHD domain-containing transcription factor Snt2 is also important for conidia production, vegetative growth, and hyphal septation of *F. oxysporum* [80], and the Zn_2Cys_6 domain-containing transcription factor Ebr1 contributes to growth and virulence via the regulation of virulence factors and genes involved in diverse metabolic pathways [81].

The GATA-type transcription factor Pro1 has recently been characterized at the interface of multiple signaling pathways, integrating signals from the Fmk1 and CWI MAPK cascades and the fungal-specific velvet transcription factor complex to regulate quorum sensing, hyphal fusion, and chemotropism [82]. Pro1 also acts downstream of Fso1, a regulatory protein of unknown biochemical function that is required for hyphal fusion [38]. Furthermore, the C2H2 zinc finger transcription factor Zfp1 regulates growth, conidiation, stress tolerance, and pathogenicity on *Polygonatum kingianum*, and Ace2 regulates growth, conidiation and virulence on banana, at least in part through the regulation of cell well integrity [83,84].

The Ccr4–Not complex—a multi-functional complex that regulates both transcription and translation—is similarly involved in FA biosynthesis, as well as oxidative stress tolerance, cell wall integrity, conidiation, and vegetative growth [85]. Furthermore, the histone acetyltransferase (HAT) Gcn5 plays a key role in *F. oxysporum* biology through the regulation of gene expression. Gcn5 is a member of the GNATs family of type A HATs and regulates the apical deposition of the cell wall material, as well as tolerance to heat, salt, and cell wall inhibitors [86].

In fungi, the velvet complex is a key regulator of development and the biosynthesis of secondary metabolites, acting through the modulation of chromatin accessibility and gene expression [87,88]. The velvet family proteins VeA and VelB interact with the non-velvet protein LaeA in the absence of light to form the heterotrimeric velvet complex [87,89]. Mutation analysis indicates that VeA and LaeA have partially overlapping functions in the development of hyphae and the conidiation and light response of *F. oxysporum* [90–92].

Post-transcriptional regulation also plays a key role in the control of gene expression. The Pumilio protein family (PUF) of RNA-binding proteins is important for the regulation of mRNA stability and translation in eukaryotes [93]. PUF proteins have diverse roles in *F. oxysporum* biology, with Puf1-4 regulating vegetative growth, Puf1-6 involved in macroconidia development, and Puf1 required for full virulence [94]. In particular, Puf1 interacts with the actin-related protein 2/3 (ARP2/3) complex via the complex component Arc18. Arc18 itself plays an important role in *F. oxysporum* virulence and ATP generation in mitochondria.

Small RNAs also play important roles in the regulation of *F. oxysporum* virulence at the translational level. Fungi produce microRNA-like RNAs (milRNAs) that are similar to plant and animal microRNAs in structure, playing important functions in different biological processes [95]. Deletions of the Argonaute protein Qde2, the Dicer-like proteins Dcl1 and Dcl2, and the exonuclease Qip—all components of milRNA processing pathways—variously impacted growth, conidiation, and virulence [96,97]. In particular, Qde2 upregulates the expression of the milRNA gene *milR87*, which contributes to virulence by suppressing the avirulence gene *FOIG_15013* [97]. In addition, the *milR106* is important in promoting conidiation, oxidative stress tolerance, and virulence [98].

3.2. Genes Involved in Vegetative Hyphal Growth and Stress Tolerance

This sub-section deals with the genes involved in the vegetative hyphal growth and stress tolerance of *F. oxysporum*, although many of these genes also impact virulence and reproduction.

3.2.1. Protein Post-Translational Modifications (PTMs)

Post-translational modifications (PTMs) are vital for the activity and regulation of proteins in all aspects of cellular biology. Protein O-mannosylation is a PTM conserved in eukaryotes that is catalyzed by protein O-mannosyltransferases (PMTs). In *F. oxysporum* f. sp. *cucumerinum*, mutants of *PMT* genes such as *PMT1*, *PMT2*, and *PMT4* show retarded growth, reduced conidiation, cell wall defects, attenuated virulence, and altered ER stress response [99]. Pmt1 targets nuclear proteins and components of the protein folding machinery. Pmt2 also regulates protein folding as well as cell wall synthesis. Pmt4 acts on proteins in secretory pathways, notably the GPI anchoring pathway involved in polarized growth.

Protein palmitoylation, another PTM, is catalyzed by a group of palmitoyl transferases (PATs). In the *Fon* genome, six PAT genes play key roles in conidiation, conidial morphology, stress response, and vegetative growth [100]. Among them, *PAT1*, *PAT2*, and *PAT4* regulate virulence. In an in vivo assay, Pat2 palmitoylated subunits of the AP-2 complex, a heterotetrameric endocytic cargo-binding adaptor. This palmitoylation contributes to the interaction and stability of the core subunits and is required for vegetative growth, cell wall integrity, asexual reproduction, and virulence.

Glycosylation is another PTM with important roles. Nucleotide sugar transporters (NSTs) link the synthesis of nucleotide sugars and glycosylation in the ER or Golgi and function as antiporters of nucleotide monophosphates [101,102]. The *Fon* genome contains nine *NST* genes that show distinct functions in vegetative growth, cell wall stress response, asexual production, and virulence [103]. In particular, Nst2 and Nst3 are essential for virulence, with Nst2 mainly affecting host colonization. Nst2 acts as a UDP-galactose transporter and interacts with the protein disulfide isomerase Pdi1 and the oxidoreductase Ero1, important regulators of disulfide bond formation.

Poly(ADP-ribosyl)ation (PARylation) is another important PTM in eukaryotes and is catalyzed by poly(ADP-ribose) polymerases (PARPs) and hydrolyzed by poly(ADP-ribose) glycohydrolases (PARGs). *F. oxysporum* Parp1 is required for pathogenicity—while its targets remain unknown, Parp1 is phosphorylated in vitro by the kinase Kin4 to enhance PARP activity [104].

3.2.2. Vesicle Trafficking

Soluble N-ethylmaleimide-sensitive factor attachment protein receptors (SNAREs) are conserved in fungi, animals, and plants and have a vital function in vesicle trafficking. The *F. oxysporum* Vam7 protein contains SNARE and Phox homology (PX) domains [105]. Vesicle trafficking mediated by Vam7 is critical in vegetative growth, asexual reproduction, and host infection of *F. oxysporum*. Vam7 also regulates the sensitivity of the fungus to salt and osmotic stress and cell wall stresses. In addition, the SNARE proteins Sso1 and Sso2 contribute to growth, conidiation, and virulence by forming complexes with the SNARE proteins Sec9 and Snc1 [106]. Sso1 appears to regulate exocytosis at the growing hyphal apex, while Sso2 is mainly expressed in older hyphae.

3.2.3. Autophagy

Autophagy appears to play key roles in multiple aspects of *F. oxysporum* biology, with mutations in components of the autophagy pathway (*ATG* genes) resulting in severe defects. The ubiquitin-like protein Atg8—an autophagy component required for the formation of the autophagosome—mediates nuclear degradation after hyphal fusion and plays a general function in the control of nuclear functions [107]. Atg3 regulates the conidiation, hyphal growth, and virulence of *F. oxysporum* [108]. Atg22 is important in the formation of autophagosomes and regulates the hyphal development, conidiation, and pathogenicity of *F. oxysporum* [109]. Atg12 affects the expression of genes involved in pathogenicity, vegetative growth, and morphological features under various stresses [110].

3.2.4. Metabolism and Nutrient Acquisition

As expected, mutations in genes related to primary metabolic pathways can result in compromised growth, virulence, and development. Loss of the isocitrate lyase *ICL1*—an enzyme in the glyoxylate cycle—resulted in reduced growth on various carbon sources [68]. Expression of *ICL1* and other genes involved in glyoxylate metabolism is regulated by the CCCH-type zinc finger-containing protein Dbp40, the deletion of which also results in decreased growth and virulence [111]. Tup1, a component of the Tup1–Cyc8 transcriptional corepressor complex, is also involved in mycelial growth, conidia development, and virulence through the regulation of numerous metabolic pathways, including the tricarboxylic acid (TCA) cycle [112]. Similarly, deletion of genes encoding the TCA cycle enzyme malate dehydrogenase (*MDH1/2*) compromised mycelial growth, conidiation, and virulence [112].

Other genes related to diverse metabolic pathways have also been studied in *F. oxysporum*. Disruption of the putative argininosuccinate lyase Arg1, a key enzyme in arginine biosynthesis, resulted in reduced vegetative growth and virulence [113]. The alcohol dehydrogenase gene *adh1* is highly expressed during hypoxia, and a small deletion in the

ADH1 gene resulted in reduced growth in hypoxic conditions and delayed virulence on tomato [114]. Deletion of the predicted glycogen debranching enzyme-encoding gene *GDB1* resulted in compromised virulence and vegetative hyphal growth—although deletion of other genes involve in glycogen biosynthesis and catabolism caused no notable phenotype [115]. Thiamine (vitamin B1) is an essential vitamin that functions as a major cofactor of enzymes that are critical for carbohydrate metabolism, including the pentose–phosphate cycle, the citric acid cycle, and glycolysis [116]. The *F. oxysporum* stress-induced gene *STI35* functions in thiamine biosynthesis [117]. In yeast, orthologs of *STI35* are highly expressed and depend on transcriptional repression by thiamine [118]. Mutation analysis of *F. oxysporum STI35* has revealed that Sti35 plays an important role in thiamine biosynthesis and in oxidative stress tolerance [119]. Finally, deletion of YjeF, a homolog of an *Escherichia coli* cellular metabolism damage-repair enzyme, significantly reduced growth, sporulation, and virulence [120].

Nitrate assimilation is also central to fungal growth and development. *F. oxysporum* strains such as *Fol* can utilize nitrate as the only source of nitrogen. Mutation analysis has shown that the predicted nitrate reductase gene *NIT1* and high-affinity nitrate/nitrite transporter gene *NTR1* are critical for nitrate assimilation of *Fol* under aerobic and anaerobic conditions, but this is not essential for virulence [121].

3.2.5. Cell Wall Biogenesis, Integrity, and Remodeling

Fungal cell walls are essential for their survival. Enzymes involved in the synthesis and maintenance of the cell wall are crucial for *F. oxysporum* growth and development. Mutant analysis has also indicated that cell wall integrity plays a key role in *F. oxysporum* virulence: deletions of genes involved in cell wall architecture have resulted in reduced virulence, including those encoding chitin synthases (*CHS2*, *CHS7*, *CHSV*, and *CHSVB*), an ergosterol biosynthesis gene (*ERG3*), a β-1,3-glucanosyltransferase (*gas1*), an *N*-acetylglucosaminyl transferase (*GNT2*), protein O-mannosyltransferases (*PMT1*, *PMT2*, and *PMT4*), a phosphomannose isomerase (*CPMI1*), a β-1,3-glucan synthase subunit (*GLS2*), and a putative α-1,6-mannosyltransferase (*OCH1*) [99,122–130]. The putative UDP-galactopyranose mutase (UGM)-encoding genes *UGMA* and *UGMB* are involved in the production of galactofuranose-containing sugar chains, affecting vegetative growth, pathogenesis and conidiation of *F. oxysporum* [131]. These studies have revealed a number of mechanisms through which cell wall integrity may contribute to virulence, including tolerance to host-associated stresses and defense compounds, recognition of external signals, and invasive growth and reproduction.

3.2.6. Stress Tolerance and Defense

During pathogen colonization, plants produce reactive oxygen species (ROS) as both immune signals and direct anti-microbials. To protect themselves from ROS damage, fungal pathogens produce a variety of antioxidant enzymes to detoxify the host environment, including catalases and peroxidases. In *F. oxysporum*, deletion of predicted catalase- (FOXG_15294), peroxidase- (FOXG_13788) and catalase-peroxidase- (FOXG_17180) encoding genes resulted in reduced virulence on tomato [132]. Expression of antioxidant genes in *Fol* is controlled by the serine/arginine protein kinase Srpk1, which is deacetylated upon ROS exposure, allowing translocation of Srpk1 to the nucleus.

Laccases—copper-containing phenol oxidases that catalyze the oxidation of phenolic compounds and reduce molecular oxygen to water—are also important for the protection of fungal pathogens against toxic plant compounds [133]. In *F. oxysporum*, loss of the laccase genes *LCC1* and *LCC3*, though not *LCC5*, results in more sensitivity to toxic compounds and oxidative stress [134]. Laccase activity depends upon the predicted chloride channel Clc1, with deletion of *CLC1* resulting in reduced laccase activity and delayed virulence on tomato [135].

Fungal plant pathogens also face toxic compounds produced by microbial antagonists, including β-lactam antibiotics. During soybean infection, *F. oxysporum* employs the β-lactamase-encoding gene *Fo18438* to protect against β-lactam antibiotics secreted by microbial competitors [34]. *Fo18438* is upregulated in the presence of *Burkholderia ambifaria*—a plant growth-promoting bacterium that inhibits *F. oxysporum* colonization [136].

3.3. Genes Involved in Reproduction

Under favorable conditions, *F. oxysporum* can produce microconidia, macroconidia, or chlamydospores [137]. These spores facilitate efficient fungal propagation and reproduction. Upon host infection, single or double-celled microconidia formed inside host tissues from phialides can germinate and grow in adjacent cells, while multicellular macroconidia predominantly form on the surface of infected tissue from conidiophores and can become airborne to spread to other hosts [138]. Finally, single or double-celled chlamydospores can be formed at the tips of macroconidia under nutrient-limiting conditions. They can overwinter in soil for long periods [138]. This subsection highlights genes that are primarily necessary for *F. oxysporum* reproduction.

3.3.1. Cell Division

Knocking out genes involved in cell division is anticipated to negatively affect conidiation. Indeed, deleting components of the mitotic cohesin complex such as *RAD21* and *REC8* diminished conidial germination under cell cycle stress conditions [139]. However, conidiation was unaffected in these mutants, suggesting that other specific regulators are present to regulate conidiation.

3.3.2. Transcriptional Regulators of Conidiation

Several genes encoding transcriptional regulators of reproduction in other fungi also regulate conidiation in *F. oxysporum*. From an insertional mutagenesis screen, Ren1 was identified as the *rensa* mutant of *F. oxysporum*, which exhibited reduced micro- and macroconidia formation [140]. Ren1 shows close homology to transcriptional regulators of conidiation in *Magnaporthe grisea* (Acr1) and *Aspergillus nidulans* (MedA). Ren1 and Aba1, a homolog of *A. nidulans* transcription factor AbaA, were suggested to be regulated by the transcription elongation factor TFIIS in *F. oxysporum* [141]. TFIIS deletion strains exhibited reduced *REN1* and *ABA1* expression and overall reduced conidiation and virulence.

In *A. nidulans*, AbaA is a component of the central regulatory pathway (CRP) required for conidiophore development and subsequent spore germination [142]. The CRP also consists of the transcription factors BrlA and WetA. In *F. oxysporum*, the homologs Aba1 and WetA-L are required to produce conidia, phialides, and chlamydospores [92]. BrlA has no known homolog in *F. oxysporum*, but MedA(a), an ortholog of Ren1, has a similar function to BrlA, being required for conidiophore and phialide development and regulating *ABA1* expression. Similarly, deleting the developmental transcriptional regulator StuA in *F. oxysporum* reduced the expression of *MEDA(a)* and *ABA1*, leading to reduced micro- and macroconidia. In *A. nidulans*, the upstream developmental activators (UDAs) flbB, flbC, and flbD control conidiation by regulating the expression of *BrlA* [143]. However, the expression of CRP components in *F. oxysporum*, such as *MEDA(a)*, *ABA1*, and *WETA-L*, were not affected by UDA deletion, suggesting a species-specific function of the UDAs [92]. Nevertheless, deleting *F. oxysporum* FlbB, FlbC, and FlbD resulted in dysregulated conidiation.

3.3.3. Post-Transcriptional Regulators

Along with transcriptional regulation, *F. oxysporum* mRNA stability and protein modification can control reproduction. In *Neurospora crassa*, the argonaut protein Qde2 and dicer Dcl1 were found to regulate virulence and microconidiation. Deleting these RNAi components simultaneously in *F. oxysporum* led to decreased levels of conidiation-regulating genes such as *STUA* and *NIIA*, which suggests a role of mRNA degradation in promoting conidiation [97]. Additionally, the deletion of several RNA-binding Pumilio proteins (*PUF2-6*) required for RNA processing resulted in reduced vegetative growth and decreased macroconidiation [94]. Several genes of the SUMOylation pathway involved in protein modification are required for vegetative growth and tolerance to chemical stressors; deleting *UBC9, MMS21, SMT3*, and *AOS1* produced smaller macroconidia and decreased microconidiation [144]. Similarly, deleting palmitoylation pathway components *PAT3, PAT5*, and *PAT6* reduced mycelial growth and dysregulated macro- and microconidiation [100]. This was suggested to occur due to the dysregulation of palmitoylating AP-2 complex components, which are known regulators of fungal virulence, cell wall integrity, and conidiation.

3.3.4. Nutrient Metabolism

The availability of nutrients in the environment is critical in determining whether *Fusarium* species can reproduce [145]. Hence, genes involved in nutrient metabolism are necessary for regulating conidiation. Transcriptomic analyses in the previously mentioned *ren1* and *stuA* revealed that the nitrite reductase gene *NIIA* was downregulated in *F. oxysporum* [146]. Deleting *NIIA* resulted in reduced macroconidia production like *ren1* and *stuA* mutants. Interestingly, the nitrite-reducing activity of *NIIA* was not required for conidiation but was speculated to be necessary for producing nitric oxide, a byproduct that promotes conidiation. Expectedly, the nitrate reductase gene *NIT1* deletion mutant also exhibited decreased macroconidia production, exemplifying the importance of nitrogen metabolism toward conidiation [121].

Aside from nitrogen, the nucleotide sugar transporters (NSTs) involved in nucleotide sugar synthesis and glycosylation are required for vegetative growth, virulence, and conidiation [103]. Similarly, the sterol 14α-demethylases (*CYP51A* and *CYP51B*) necessary for ergosterol production seemed to negatively regulate conidiation in *F. oxysporum* [147]. Likewise, deleting the human lysine deacetylase *SIRT5* ortholog (*SIR5*) resulted in enhanced virulence and conidiation in the host [148]. In *F. oxysporum, SIR5* was suggested to reduce conidial germination through the restriction of ATP synthesis by inhibiting the pyruvate dehydrogenase complex and repressing the expression of citric acid components.

3.3.5. Cell Wall Production and Stability

The proper composition of fungal cell walls consisting of glucans, glycoproteins, and chitin is integral to fungal growth, virulence, and reproduction [149]. In microconidia, the content of mannose, a cell wall component involved in stability, is decreased compared to mycelium [150]. This likely suggests that genes regulating the production and stability of cell walls are important for *F. oxysporum* reproduction. Indeed, deleting the UDP-galactopyranose mutase (*UGMB*) and galf-transferase (*GFSB*), required for cell wall stability via facilitating galactofuranose biosynthesis, led to a reduction in virulence and conidiation [131]. The *ugmA gfsB* double mutant displayed stronger conidiation deficiencies and decreased vegetative growth. Further, RNAi of the fasciclin-like protein (FLP) family involved in cell wall stability in plants and virulence of *Magnaporthe oryzae* had greatly diminished conidiation and host colonization in *F. oxysporum* [151].

3.3.6. Virulence Genes Affecting Conidiation

In some instances, knocking out genes studied in the primary context of virulence also affected conidiation. For example, deleting the yeast MAPK pheromone response pathway components *STE12* and *FMK1* in *F. oxysporum* led to increased conidiation [152,153]. As *ste12 fmk1* also exhibited decreased virulence, the roles of Ste12 and Fmk1 likely vary depending on nutrient availability in the infected plant host. Further, when the velvet protein complex transcription factors involved in mycotoxin production, *VEA, VELB, VELC*, and *LAEA*, were deleted, microconidiation was also dysregulated [90]. Although all velvet deletion mutants had attenuated virulence, the number and morphology of microconidia differed among mutants: *velB* and *velC* had increased microconidia, while this was decreased in *laeA*. *veA* also had increased microconidia but had a strikingly elongated morphology. Seemingly, the velvet proteins regulate the conidiation of *F. oxysporum* in a complex manner. Finally, deleting the small, secreted protein (*SSP1*) enhanced virulence and conidiation [154]. At the same time, Ssp1 is secreted and is likely recognized by the host as a pathogen-associated molecular pattern (PAMP), negatively affecting *F. oxysporum* virulence.

3.4. Genes Involved in Virulence

F. oxysporum employs diverse signaling modules to facilitate invasion and colonization of host tissues. This sub-section deals with components involved in the pathogenesis mechanisms, detailing the regulated processes we understand, as well as the virulence-specific transcription factors controlling them.

3.4.1. Chemotrophic Growth

Chemotropism plays a key role in the initiation of fungal pathogenesis, allowing hyphae to elongate in the direction of potential host tissue. In *F. oxysporum*, chemotrophic growth depends on the NADPH oxidase B (NoxB) complex, with the complex subunits NoxB and NoxR both required for chemotrophic growth and full virulence [155]. The NoxB complex, in coordination with secreted superoxide dismutase (SOD), mediates the synthesis of ROS that activate peroxidases (Prx) secreted by the plant roots. Prx activation leads to the activation of the cell wall integrity (CWI) MAPK cascade via the G protein-coupled receptor Ste2, which coordinates growth along the Prx activity gradient [43].

3.4.2. Fusaric Acid (FA) and Nitric Oxide (NO) Production

An important *F. oxysporum* pathogenicity factor is the production of fusaric acid (FA or FSA), a phytotoxin produced by numerous *Fusarium* species. While the function of FA is not fully understood, evidence suggests that it plays a role in cell membrane damage and in chelating metal ions [156,157]. FA biosynthesis in *F. oxysporum* is controlled by the FA biosynthetic (*FUB*) cluster of twelve genes [158], which appears to be well conserved among *F. oxysporum* strains [159]. Targeted deletions of six genes (*FUB1-5* and *FUB10*) resulted in reduced FA biosynthesis, invasive growth, and disease severity, indicating their importance in *F. oxysporum* pathogenicity [157,159]. Another key component of FA production is the major facilitator superfamily (MFS) transporter protein FubT. *FUBT* is required for FA secretion in *F. oxysporum* f. sp. *vasinfectum* (*Fov*), but its disruption also resulted in reductions in FA biosynthesis and fungal resistance to exogenous FA [160], alluding to complex regulation of FA homeostasis.

NO has been widely implicated in plant–pathogen interactions, with important functions as a signaling molecule in growth, development, and stress responses of both plants and fungi [161]. Recently, NO was shown to play a key role in the pathogenicity of *F. oxysporum* f. sp. *cubense* (*Foc* or *Focub*) on banana [162]. Meta-transcriptomic analysis identified the upregulation of NO biosynthesis and detoxification genes upon infection, and deletion of two NO biosynthesis genes encoding putative NAD(+)-dependent formate dehydrogenase and nitrite reductase resulted in compromised virulence.

3.4.3. Cell Wall Degrading Enzymes (CWDEs)

As with many fungal plant pathogens, *F. oxysporum* secretes a large repertoire of cell wall-degrading enzymes (CWDEs), including polygalacturonases (PGs), xylanases, glycosidases, and proteases. Initial studies have shown these enzymes to be secreted during infection, but their contributions to virulence have been difficult to determine. The PG genes *PG1*, *PGX4*, *PG5*, and *PGX6*, the xylanase genes *XYL3*, *XYL4*, and *XYL5*, the protease gene *PRT1*, and the lipase genes *LIP1*, *LIP2*, *LIP3*, *LIP5*, and *LIP22* have all been deleted in *Fol* without any impact on virulence [163–169]. The lack of virulence phenotypes may be in part due to the functional redundancy of many CWDE genes—single deletions of *PG1* and *PGX6* had no impact of virulence, but a *pg1pgx6* double mutant showed reduced pathogenicity on tomato [163]. Furthermore, knockouts of genes regulating multiple CWDEs generally have a clearer impact on virulence, such as the carbon catabolite repressor *SNF1* and the lipase and cutinase transcriptional regulators *CTF1* and *CTF2* [55,168]. Genetic redundancy poses a major challenge for studies utilizing mutant analysis, particularly in forward genetic screens that largely characterize single mutants. Future studies will require alternative strategies to overcome redundancy, including generating higher-order mutants, employing RNAi or CRISPR for efficient knockdown/knockout of multiple genes, and utilizing overexpression of target genes.

In some cases, however, a single deletion was sufficient to reduce virulence, such as for the PG-encoding gene *PGC4* in *Focub* and the glycosidase-encoding gene *EG1* in *Fol* [170,171], suggesting their key roles in host cell wall degradation. 3-carboxy-cis,cis-muconate lactonizing enzyme (Cmle) is also required for full virulence and is believed to play an important role in the breakdown of lignin and phenolic compounds secreted by the host during infection to strengthen its cell wall [172]. Interestingly, deletion of the glycosidase gene *FOIG_15013* increased virulence, potentially because the enzyme releases fragments from the host cell wall that are recognized by the host as damage-associated molecular patterns (DAMPs), inducing an immune response [97]. To combat this, *FOIG_15013* is suppressed during infection by the microRNA-like RNA *milR87*.

3.4.4. Xylem Effectors: Secreted in Xylem (SIX) Proteins

Numerous secreted effector proteins have been identified from the xylem sap of tomato plants infected with *Fol* and are thus named secreted in xylem (SIX) proteins [173–175]. Homologous proteins have since been identified in other strains, though the number of *SIX* genes varies between and even within *formae speciales* [176]. Mutant analysis has shown that many SIX proteins function as key virulence factors—deletions of *SIX1*, *SIX3*, *SIX4*, *SIX5*, *SIX6* and *SIX8* resulted in reduced pathogenicity of various *formae speciales*, although *SIX2* and *SIX9* knockouts produced no virulence defects [162,177–188]. While the specific functions of SIX proteins are largely unknown, some studies point to roles in inhibiting host defense pathways, such as jasmonic acid signaling and the hypersensitive response (HR) [182,184]. In *F. oxysporum* f. sp. *conglutinans* (*Focon*), *SIX8* is physically linked with the effector gene *PSE1* [188]. The *SIX8–PSE1* linkage pair is required for virulence on *Arabidopsis* and is thought to act by suppressing the phytoalexin camalexin.

Despite a clear contribution to virulence, certain *SIX* genes are known to participate in gene-for-gene interactions with resistance (*R*) genes in the respective host plants, causing recognition by specific host cultivars. In *Fol*, *SIX4* triggers resistance mediated by the tomato *R* gene *IMMUNITY* (*I*), *SIX3* and *SIX5* trigger *I-2*-mediated resistance, and *SIX1* triggers *I-3* resistance [12,173,175,183]. These *SIX* genes are thus also referred to as avirulence (*AVR*) genes (*SIX4* is *AVR1*, *SIX3* is *AVR2*, and *SIX1* is *AVR3*). Similarly, *SIX6* deletion in *Fon* increased virulence on watermelon, although the host resistance gene involved is yet to be identified [185]. These avirulent properties have also played an important role in race discrimination. *Fol* race 2 is thought to have evolved from race 1 through a deletion of the *SIX4/AVR1* gene, allowing it to evade *I*-mediated resistance. Race 3 then emerged through a mutation in *SIX3/AVR2*, circumventing *I-2*-mediated resistance.

3.4.5. Effectors in Pathogenicity and Suppression of Host Defenses

In addition to CWDEs and SIX proteins, mutant analysis has identified several other secreted proteins with key roles in *F. oxysporum* virulence, many of which operate by inhibiting host immunity. In *Fol*, the metalloprotease Mep1 and serine protease Sep1 contribute to virulence by cleaving host chitinases, which function in host resistance by attacking the fungal cell wall [189]. Similarly, the *Focub* metalloprotease M35_1 suppresses chitinase activity, as well as inhibiting the hypersensitive response (HR) [190]. The secreted enzyme O-acetylhomoserine (thiol)-lyase (Oastl) is thought to interfere with host biosynthesis of cysteine, a precursor to numerous defense compounds [35], while the Cupin domain-containing protein Cupin1, the α-pheromone-like protein Pp1, and the small secreted protein Ssp17 also inhibit host immune responses [191–193]. *Fol* also secretes the tomatinase enzyme Tom1, which degrades the tomato antimicrobial defense compound α-tomatine [194]. The *Focub* effector Fse1 regulates virulence via a direct interaction with a host MYB transcription factor involved in the induction of cell death [195]. Furthermore, the *Fol* secreted protein SVP1 suppresses host defenses by relocating the tomato defense protein SlPR1 from the apoplast to the nucleus, preventing its defense signaling activity [196]. Interestingly, SVP1 is protected from ubiquitin-mediated degradation in both the fungal and plant cells by acetylation catalyzed by the sine acetyltransferase ARD1.

Like other pathogenic fungi, *F. oxysporum* employs cross-kingdom RNA interference during pathogenesis. The microRNA-like RNA *milR1* acts as a secreted effector in *Fol*, contributing to virulence by suppressing a tomato resistance gene [197]. Extracellular ATP also plays a role in suppressing host immune responses—mutants lacking the ATP synthase gene *ATP SYNTHASEα* were unable to suppress expression of host sugar transporters, which are believed to contribute to host immunity by starving invading pathogens [198].

A number of other putative secreted effectors contribute to virulence through unknown mechanisms, including the aminopeptidase Apy1 [187], the ribonuclease Rnt2 [199], the secreted proteins Fosp9 and Foc 1324 [200,201], and the small secreted proteins Mc69 and Cep28 [202,203]. Cp1, a secreted cerato-platanin (CP) protein, is required for penetration of *Focub* [204]. The secretion of these effectors requires the creation of secretory vesicles in a process orchestrated by the Arf family proteins, including Arf, Arl, and Sar proteins. In *F. oxysporum*, *ARL3* deletion resulted in reduced virulence, while viable cells could not be obtained from *ARF1* deletion, indicating the essential roles of these proteins [205].

3.4.6. Virulence-Specific Transcription Factors and Small RNA Regulation

F. oxysporum employs many transcription factors to regulate the expression of virulence factors. While several transcription factors regulating growth and development are also required for full virulence, this section will focus on those specific to pathogenesis.

CWDEs are regulated by a range of transcription factors. Ctf1 and Ctf2 exert both positive and negative regulation of lipase and cutinase genes [168], while XlnR is a transcriptional activator of xylanase genes, but is not required for virulence [206]. Tip4 acts downstream of Tor1 to promote expression of CWDEs, as well as genes related to ribosome biogenesis [53]. Clr1 activates expression of cellulolytic enzymes, but its deletion also led to increased expression of other virulence factors, including other CWDEs, resulting in increased virulence [207]. PacC and Cre1 act as negative regulators of virulence, with PacC suppressing the expression of polygalacturonase and pH-responsive virulence genes, and Cre1 suppressing CWDE and nutrient acquisition genes [208,209].

FA production is controlled in part by the C2H2 zinc finger transcription factor Czf1, which activates transcription of *FUB* genes [210]. Downstream of the CWI MAPK cascade, Rlm1 and Swi6 also regulate FA biosynthesis, and Rlm1 further functions in cell wall integrity, beauvericin biosynthesis, and oxidative stress responses [211,212]. The core genome transcription factor Sge1 regulates *SIX* gene expression and is required for pathogenicity and conidiation, but not for colonization or penetration [213]. *SGE1* itself is regulated by Ftf1 and 2, which positively regulate virulence during early infection [214,215]. The homeodomain transcription factor Ste12 acts downstream of the Fmk1 cascade to regulate invasive growth and expression of CWDEs [152], while unknown transcription factors must affect the other functions of the Fmk1 cascade. In addition, the Zn(II)2Cys6 transcription factor Fow2 and the PHD finger-containing transcription factor Cti6 are also important regulators of *F. oxysporum* virulence, although their targets remain unknown [216,217].

Many transcription factors have characteristic roles in responding to environmental conditions during infection. The transcription factor Wc1 mediates photoreception in *F. oxysporum*, regulating surface hydrophobicity, carotenoid biosynthesis, and hyphal growth in response to UV light [218]. While *wc1* is not required for pathogenesis on tomato, it is a key virulence determinant during infection of immunocompromised mice. During infection, phytopathogenic fungi experience nutrient limitation, including nitrogen and metal irons. The GATA transcription factor Fnr1/AreA and the bZIP transcription factor MeaB act downstream of Tor1 to coordinate the regulation of nitrogen catabolism during infection [52,219]. Fnr1 promotes the catabolism of secondary nitrogen sources under the nitrogen-poor conditions present during early infection, while MeaB suppresses this response when preferred nitrogen sources are present. The bZIP transcription factor HapX contributes to virulence by regulating iron homeostasis [220], while ZafA is induced in response to zinc-poor conditions to induce expression of zinc transporters [221]. The transcription factors Atf1 and Skn7 are involved in the oxidative stress response, likely by inducing expression of antioxidant genes such as catalases and peroxidases to protect against ROS damage in the host tissues [222,223].

4. Conclusions and Future Directions

Fusarium wilt is a widespread and devastating disease that causes huge damage to crop plants globally. The last two decades have greatly expanded our molecular understanding of *Fusarium oxysporum* biology and its interactions with host species. Approximately 200 genes have now been studied through mutant analysis, identifying key components of diverse biological processes and elucidating the signaling framework. However, with over 17,000 predicted genes in the *F. oxysporum* genome [24], much work remains to further our understanding of this important fungal pathogen. Most signaling pathways remain incomplete, and many key components in fundamental cellular processes have yet to be studied. In particular, the regulation of *F. oxysporum* development and reproduction remains poorly understood. Furthermore, while the virulence mechanisms of *F. oxysporum* have been better studied, many of the genetic components are still missing.

Enhancing our molecular knowledge of these processes will be vital for the development of novel disease control strategies. In particular, mutant analysis provides a useful approach to identifying effective targets of RNAi silencing-based technologies such as host- and spray-induced gene silencing (HIGS/SIGS) to inhibit pathogen growth or virulence. Furthermore, a deeper understanding of *F. oxysporum* pathogenesis will assist the development of targeted crop engineering strategies to enhance disease resistance. However, such strategies will have to consider the considerable genetic diversity within and between *F. oxysporum formae speciales* and the diverse virulence strategies employed by different strains.

Recent advances in genetic engineering technologies such as CRISPR are now being applied to accelerate gene discovery and characterization. However, challenges remain. The intricate crosstalk among different signaling pathways and genetic redundancy make it difficult to dissect gene function through single mutant analysis. Mutant lethality is also a major difficulty in studying genes with vital roles in *F. oxysporum* biology, many of which may be of particular interest for disease control. Future work will need to overcome these challenges by employing alternative approaches, including multiple gene knockout, RNAi-based gene knockdown, and overexpression analysis. Furthermore, detailed studies must combine genetic analysis with cell biology and biochemical tools to develop a more complete understanding of *F. oxysporum*.

Supplementary Materials: The following supporting information can be downloaded at: https://www.mdpi.com/article/10.3390/pathogens13100823/s1, Supplementary Table S1: List of *Fusarium oxysporum* genes studied by mutant analysis. References [224–230] are cited in the Supplementary Materials.

Author Contributions: X.L. conceptualized the review. J.L., T.W. and E.J. collated Supplementary Table S1. E.J. was the primary author of the Abstract, Introduction, Genes Involved in Virulence, and Conclusions and Future Directions sections. J.L. was the primary author of the Features of the *F. oxysporum* Genome and Genes Involved in Reproduction sections. T.W. was the primary author of the Genes Involved in Vegetative Hyphal Growth and Stress Tolerance section. E.J. and T.W. contributed to the Central Signaling Pathways and Transcriptional Factors section. All authors reviewed and revised the manuscript. All authors have read and agreed to the published version of the manuscript.

Funding: This research was supported by funds from the Natural Sciences and Engineering Research Council of Canada (NSERC)-Discovery and NSERC-CREATE-PRoTECT funds.

Data Availability Statement: No new data were created or analyzed in this study.

Acknowledgments: We apologize for the works that are not cited in the manuscript. We would like to thank the NSERC-Discovery and NSERC-CREATE-PRoTECT programs for financial support of the laboratory of X.L.

Conflicts of Interest: The authors declare no conflicts of interest. The funders had no role in the design or the writing of the manuscript.

References

1. Aoki, T.; O'Donnell, K.; Geiser, D.M. Systematics of key phytopathogenic *Fusarium* species: Current status and future challenges. *J. Gen. Plant Pathol.* **2014**, *80*, 189–201. [CrossRef]
2. Michielse, C.B.; Rep, M. Pathogen profile update: *Fusarium oxysporum*. *Mol. Plant Pathol.* **2009**, *10*, 311–324. [CrossRef]
3. Edel-Hermann, V.; Lecomte, C. Current Status of *Fusarium oxysporum* Formae Speciales and Races. *Phytopathology* **2019**, *109*, 512–530. [CrossRef] [PubMed]
4. Snyder, W.C.; Hansen, H.N. The Species Concept in Fusarium. *Am. J. Bot.* **1940**, *27*, 64–67. [CrossRef]
5. O'Donnell, K.; Kistler, H.C.; Cigelnik, E.; Ploetz, R.C. Multiple evolutionary origins of the fungus causing Panama disease of banana: Concordant evidence from nuclear and mitochondrial gene genealogies. *Proc. Natl. Acad. Sci. USA* **1998**, *95*, 2044–2049. [CrossRef]
6. Baayen, R.P.; O'Donnell, K.; Bonants, P.J.; Cigelnik, E.; Kroon, L.P.; Roebroeck, E.J.; Waalwijk, C. Gene Genealogies and AFLP Analyses in the *Fusarium oxysporum* Complex Identify Monophyletic and Nonmonophyletic Formae Speciales Causing Wilt and Rot Disease. *Phytopathology* **2000**, *90*, 891–900. [CrossRef]
7. Skovgaard, K.; Nirenberg, H.I.; O'Donnell, K.; Rosendahl, S. Evolution of *Fusarium oxysporum* f. sp. *vasinfectum* Races Inferred from Multigene Genealogies. *Phytopathology* **2001**, *91*, 1231–1237. [CrossRef]
8. Kistler, H.C.; Alabouvette, C.; Baayen, R.P.; Bentley, S.; Brayford, D.; Coddington, A.; Correll, J.; Daboussi, M.J.; Elias, K.; Fernandez, D.; et al. Systematic Numbering of Vegetative Compatibility Groups in the Plant Pathogenic Fungus *Fusarium oxysporum*. *Phytopathology* **1998**, *88*, 30–32. [CrossRef] [PubMed]
9. Beckman, C.H. *The Nature of Wilt Diseases of Plants*; APS Press: St. Paul, MN, USA, 1987.
10. Covey, P.A.; Kuwitzky, B.; Hanson, M.; Webb, K.M. Multilocus analysis using putative fungal effectors to describe a population of *Fusarium oxysporum* from sugar beet. *Phytopathology* **2014**, *104*, 886–896. [CrossRef]
11. Munkvold, G.P. Fusarium Species and Their Associated Mycotoxins. *Methods Mol. Biol.* **2017**, *1542*, 51–106. [CrossRef]
12. Houterman, P.M.; Cornelissen, B.J.; Rep, M. Suppression of plant resistance gene-based immunity by a fungal effector. *PLoS Pathog.* **2008**, *4*, e1000061. [CrossRef] [PubMed]
13. Ma, L.J.; Geiser, D.M.; Proctor, R.H.; Rooney, A.P.; O'Donnell, K.; Trail, F.; Gardiner, D.M.; Manners, J.M.; Kazan, K. Fusarium pathogenomics. *Annu. Rev. Microbiol.* **2013**, *67*, 399–416. [CrossRef] [PubMed]
14. Booth, C. *The Genus Fusarium*; Commonwealth Mycological Institute: Kew, UK, 1971.
15. Toussoun, T.A.; Marasas, W.F.O. *Fusarium Species: An Illustrated Manual for Identification*; Pennsylvania State University Press: Pennsylvania, PA, USA, 1983.
16. Gordon, T.R. *Fusarium oxysporum* and the Fusarium Wilt Syndrome. *Annu. Rev. Phytopathol.* **2017**, *55*, 23–39. [CrossRef] [PubMed]
17. Bishop, C.D.; Cooper, R.M. An ultrastructural study of root invasion in three vascular wilt diseases. *Physiol. Plant Pathol.* **1983**, *22*, 15–27. [CrossRef]
18. Katan, T.; Shlevin, E.; Katan, J. Sporulation of*Fusarium oxysporum* f. sp. *lycopersici* on Stem Surfaces of Tomato Plants and Aerial Dissemination of Inoculum. *Phytopathology* **1997**, *87*, 712–719. [CrossRef]
19. Rekah, Y.; Shtienberg, D.; Katan, J. Disease Development Following Infection of Tomato and Basil Foliage by Airborne Conidia of the Soilborne Pathogens *Fusarium oxysporum* f. sp. *radicis-lycopersici* and *F. oxysporum* f. sp. *basilici*. *Phytopathology* **2000**, *90*, 1322–1329. [CrossRef]
20. Rowe, R.C.; Farley, J.D.; Coplin, D.L. Airborne spore dispersal and recolonization of steamed soil by *Fusarium oxysporum* in tomato greenhouses. *Phytopathology* **1977**, *67*, 1513–1517. [CrossRef]
21. Scarlett, K.; Tesoriero, L.; Daniel, R.; Guest, D. Sciarid and shore flies as aerial vectors of *Fusarium oxysporum* f. sp. *cucumerinum* in greenhouse cucumbers. *J. Appl. Entomol.* **2013**, *138*, 368–377. [CrossRef]
22. Sampaio, A.M.; Araújo, S.d.S.; Rubiales, D.; Vaz Patto, M.C. Fusarium Wilt Management in Legume Crops. *Agronomy* **2020**, *10*, 1073. [CrossRef]
23. Jiménez-Díaz, R.M.; Castillo, P.; del Mar Jiménez-Gasco, M.; Landa, B.B.; Navas-Cortés, J.A. Fusarium wilt of chickpeas: Biology, ecology and management. *Crop Prot.* **2015**, *73*, 16–27. [CrossRef]
24. Ma, L.J.; van der Does, H.C.; Borkovich, K.A.; Coleman, J.J.; Daboussi, M.J.; Di Pietro, A.; Dufresne, M.; Freitag, M.; Grabherr, M.; Henrissat, B.; et al. Comparative genomics reveals mobile pathogenicity chromosomes in Fusarium. *Nature* **2010**, *464*, 367–373. [CrossRef]

25. Cuomo, C.A.; Güldener, U.; Xu, J.R.; Trail, F.; Turgeon, B.G.; Di Pietro, A.; Walton, J.D.; Ma, L.J.; Baker, S.E.; Rep, M.; et al. The Fusarium graminearum genome reveals a link between localized polymorphism and pathogen specialization. *Science* **2007**, *317*, 1400–1402. [CrossRef]
26. Ayhan, D.H.; López-Díaz, C.; Di Pietro, A.; Ma, L.J. Improved Assembly of Reference Genome*Fusarium oxysporum* f. sp. *lycopersici* Strain Fol4287. *Microbiol. Resour. Announc.* **2018**, *7*. [CrossRef]
27. Dvorianinova, E.M.; Pushkova, E.N.; Novakovskiy, R.O.; Povkhova, L.V.; Bolsheva, N.L.; Kudryavtseva, L.P.; Rozhmina, T.A.; Melnikova, N.V.; Dmitriev, A.A. Nanopore and illumina genome sequencing of *Fusarium oxysporum* f. sp. *lini* strains of different virulence. *Front. Genet.* **2021**, *12*, 662928. [CrossRef]
28. Yang, J.; Mao, A.; Zhang, J.; Zhang, X.; Xia, C.; Zhao, H.; Wang, Y.; Wen, C.; Liu, H.; Wang, Q. Whole-genome sequencing of *Fusarium oxysporum* f. sp. *cucumerinum* strain race-4 infecting cucumber in China. *Plant Dis.* **2023**, *107*, 1210–1213. [CrossRef]
29. Yu, F.; Zhang, W.; Wang, S.; Wang, H.; Yu, L.; Zeng, X.; Fei, Z.; Li, J. Genome sequence of *Fusarium oxysporum* f. sp. conglutinans, the etiological agent of cabbage Fusarium wilt. *Mol. Plant Microbe Interact.* **2021**, *34*, 210–213. [CrossRef]
30. Hao, Y.; Li, Y.; Ping, X.; Yang, Q.; Mao, Z.; Zhao, J.; Lu, X.; Xie, B.; Yang, Y.; Ling, J. The genome of *Fusarium oxysporum* f. sp. phaseoli provides insight into the evolution of genomes and effectors of *Fusarium oxysporum* species. *Int. J. Mol. Sci.* **2023**, *24*, 963. [CrossRef]
31. Schmidt, S.M.; Lukasiewicz, J.; Farrer, R.; van Dam, P.; Bertoldo, C.; Rep, M. Comparative genomics of *Fusarium oxysporum* f. sp. *melonis* reveals the secreted protein recognized by the Fom-2 resistance gene in melon. *New Phytol.* **2016**, *209*, 307–318. [CrossRef]
32. Meng, T.; Jiao, H.; Zhang, Y.; Zhou, Y.; Chen, S.; Wang, X.; Yang, B.; Sun, J.; Geng, X.; Ayhan, D.H.; et al. FoPGDB: A pangenome database of *Fusarium oxysporum*, a cross-kingdom fungal pathogen. *Database* **2024**, *2024*. [CrossRef]
33. Luo, J.; Zhang, A.; Tan, K.; Yang, S.; Ma, X.; Bai, X.; Hou, Y.; Bai, J. Study on the interaction mechanism between Crocus sativus and *Fusarium oxysporum* based on dual RNA-seq. *Plant Cell Rep.* **2023**, *42*, 91–106. [CrossRef]
34. Chang, H.X.; Noel, Z.A.; Chilvers, M.I. A β-lactamase gene of *Fusarium oxysporum* alters the rhizosphere microbiota of soybean. *Plant J.* **2021**, *106*, 1588–1604. [CrossRef]
35. Wang, D.; Peng, C.; Zheng, X.; Chang, L.; Xu, B.; Tong, Z. Secretome analysis of the banana Fusarium wilt fungi Foc R1 and Foc TR4 reveals a new effector OASTL required for full pathogenicity of Foc TR4 in banana. *Biomolecules* **2020**, *10*, 1430. [CrossRef]
36. Li, J.; Gao, M.; Gabriel, D.W.; Liang, W.; Song, L. Secretome-wide analysis of lysine acetylation in fusarium oxysporum f. sp. lycopersici provides novel insights into infection-related proteins. *Front. Microbiol.* **2020**, *11*, 559440. [CrossRef]
37. Di Pietro, A.; García-MacEira, F.I.; Méglecz, E.; Roncero, M.I. A MAP kinase of the vascular wilt fungus *Fusarium oxysporum* is essential for root penetration and pathogenesis. *Mol. Microbiol.* **2001**, *39*, 1140–1152. [CrossRef]
38. Prados Rosales, R.C.; Di Pietro, A. Vegetative hyphal fusion is not essential for plant infection by *Fusarium oxysporum*. *Eukaryot. Cell* **2008**, *7*, 162–171. [CrossRef]
39. Pareek, M.; Rajam, M.V. RNAi-mediated silencing of MAP kinase signalling genes (Fmk1, Hog1, and Pbs2) in *Fusarium oxysporum* reduces pathogenesis on tomato plants. *Fungal Biol.* **2017**, *121*, 775–784. [CrossRef]
40. Segorbe, D.; Di Pietro, A.; Pérez-Nadales, E.; Turrà, D. Three *Fusarium oxysporum* mitogen-activated protein kinases (MAPKs) have distinct and complementary roles in stress adaptation and cross-kingdom pathogenicity. *Mol. Plant Pathol.* **2016**, *18*, 912–924. [CrossRef]
41. Pérez-Nadales, E.; Di Pietro, A. The membrane mucin Msb2 regulates invasive growth and plant infection in *Fusarium oxysporum*. *Plant Cell* **2011**, *23*, 1171–1185. [CrossRef]
42. Perez-Nadales, E.; Di Pietro, A. The transmembrane protein Sho1 cooperates with the mucin Msb2 to regulate invasive growth and plant infection in *Fusarium oxysporum*. *Mol. Plant Pathol.* **2015**, *16*, 593–603. [CrossRef]
43. Turrà, D.; El Ghalid, M.; Rossi, F.; Di Pietro, A. Fungal pathogen uses sex pheromone receptor for chemotropic sensing of host plant signals. *Nature* **2015**, *527*, 521–524. [CrossRef]
44. Rispail, N.; Di Pietro, A. The two-component histidine kinase Fhk1 controls stress adaptation and virulence of *Fusarium oxysporum*. *Mol. Plant Pathol.* **2010**, *11*, 395–407. [CrossRef]
45. Ding, Z.; Li, M.; Sun, F.; Xi, P.; Sun, L.; Zhang, L.; Jiang, Z. Mitogen-activated protein kinases are associated with the regulation of physiological traits and virulence in *Fusarium oxysporum* f. sp. *cubense*. *PLoS ONE* **2015**, *10*, e0122634. [CrossRef]
46. Xiao, J.; Zhang, Y.; Yang, K.; Tang, Y.; Wei, L.; Liu, E.; Liang, Z. Protein kinase Ime2 is associated with mycelial growth, conidiation, osmoregulation, and pathogenicity in *Fusarium oxysporum*. *Arch. Microbiol.* **2022**, *204*, 455. [CrossRef]
47. Nunez-Rodriguez, J.C.; Ruiz-Roldán, C.; Lemos, P.; Membrives, S.; Hera, C. The phosphatase Ptc6 is involved in virulence and MAPK signalling in *Fusarium oxysporum*. *Mol. Plant Pathol.* **2020**, *21*, 206–217. [CrossRef]
48. Fernandes, T.R.; Sánchez Salvador, E.; Tapia, Á.; Di Pietro, A. Dual-specificity protein phosphatase Msg5 controls cell wall integrity and virulence in *Fusarium oxysporum*. *Fungal Genet. Biol.* **2021**, *146*, 103486. [CrossRef]
49. De Virgilio, C.; Loewith, R. Cell growth control: Little eukaryotes make big contributions. *Oncogene* **2006**, *25*, 6392–6415. [CrossRef]
50. Yuan, H.X.; Xiong, Y.; Guan, K.L. Nutrient sensing, metabolism, and cell growth control. *Mol. Cell* **2013**, *49*, 379–387. [CrossRef]
51. Li, L.; Zhu, T.; Song, Y.; Feng, L.; Kear, P.J.; Riseh, R.S.; Sitohy, M.; Datla, R.; Ren, M. Salicylic acid fights against Fusarium wilt by inhibiting target of rapamycin signaling pathway in *Fusarium oxysporum*. *J. Adv. Res.* **2022**, *39*, 1–13. [CrossRef]
52. López-Berges, M.S.; Rispail, N.; Prados-Rosales, R.C.; Di Pietro, A. A nitrogen response pathway regulates virulence functions in *Fusarium oxysporum* via the protein kinase TOR and the bZIP protein MeaB. *Plant Cell* **2010**, *22*, 2459–2475. [CrossRef]

53. Li, L.; Zhu, T.; Song, Y.; Luo, X.; Datla, R.; Ren, M. Target of rapamycin controls hyphal growth and pathogenicity through FoTIP4 in *Fusarium oxysporum*. *Mol. Plant Pathol.* **2021**, *22*, 1239–1255. [CrossRef]
54. Qiu, M.; Deng, Y.; Deng, Q.; Sun, L.; Fang, Z.; Wang, Y.; Huang, X.; Zhao, J. Cysteine Inhibits the Growth of *Fusarium oxysporum* and Promotes T-2 Toxin Synthesis through the Gtr/Tap42 Pathway. *Microbiol. Spectr.* **2022**, *10*, e0368222. [CrossRef]
55. Ospina-Giraldo, M.D.; Mullins, E.; Kang, S. Loss of function of the *Fusarium oxysporum* SNF1 gene reduces virulence on cabbage and Arabidopsis. *Curr. Genet.* **2003**, *44*, 49–57. [CrossRef]
56. Neer, E.J. Heterotrimeric G proteins: Organizers of transmembrane signals. *Cell* **1995**, *80*, 249–257. [CrossRef]
57. Jain, S.; Akiyama, K.; Mae, K.; Ohguchi, T.; Takata, R. Targeted disruption of a G protein alpha subunit gene results in reduced pathogenicity in *Fusarium oxysporum*. *Curr. Genet.* **2002**, *41*, 407–413. [CrossRef]
58. Jain, S.; Akiyama, K.; Kan, T.; Ohguchi, T.; Takata, R. The G protein beta subunit FGB1 regulates development and pathogenicity in *Fusarium oxysporum*. *Curr. Genet.* **2003**, *43*, 79–86. [CrossRef]
59. Jain, S.; Akiyama, K.; Takata, R.; Ohguchi, T. Signaling via the G protein alpha subunit FGA2 is necessary for pathogenesis in *Fusarium oxysporum*. *FEMS Microbiol. Lett.* **2005**, *243*, 165–172. [CrossRef]
60. Delgado-Jarana, J.; Martínez-Rocha, A.L.; Roldán-Rodriguez, R.; Roncero, M.I.; Di Pietro, A. *Fusarium oxysporum* G-protein beta subunit Fgb1 regulates hyphal growth, development, and virulence through multiple signalling pathways. *Fungal Genet. Biol.* **2005**, *42*, 61–72. [CrossRef]
61. Kim, H.S.; Park, S.Y.; Lee, S.; Adams, E.L.; Czymmek, K.; Kang, S. Loss of cAMP-dependent protein kinase A affects multiple traits important for root pathogenesis by *Fusarium oxysporum*. *Mol. Plant Microbe Interact.* **2011**, *24*, 719–732. [CrossRef]
62. Yang, C.; Sun, J.; Wu, Z.; Jiang, M.; Li, D.; Wang, X.; Zhou, C.; Liu, X.; Ren, Z.; Wang, J.; et al. FoRSR1 is important for conidiation, fusaric acid production, and pathogenicity in *Fusarium oxysporum*. *Phytopathology* **2023**, *113*, 1244–1253. [CrossRef]
63. Li, B.; Mao, H.Y.; Zhang, Z.Y.; Chen, X.J.; Ouyang, S.Q. FolVps9, a Guanine Nucleotide Exchange Factor for FolVps21, Is Essential for Fungal Development and Pathogenicity in *Fusarium oxysporum* f. sp. *lycopersici*. *Front. Microbiol.* **2019**, *10*, 2658. [CrossRef]
64. Navarro-Velasco, G.Y.; Di Pietro, A.; López-Berges, M.S. Constitutive activation of TORC1 signalling attenuates virulence in the cross-kingdom fungal pathogen *Fusarium oxysporum*. *Mol. Plant Pathol.* **2023**, *24*, 289–301. [CrossRef]
65. Hershko, A.; Ciechanover, A. The ubiquitin system. *Annu. Rev. Biochem.* **1998**, *67*, 425–479. [CrossRef]
66. Miguel-Rojas, C.; Hera, C. The F-box protein Fbp1 functions in the invasive growth and cell wall integrity mitogen-activated protein kinase (MAPK) pathways in *Fusarium oxysporum*. *Mol. Plant Pathol.* **2016**, *17*, 55–64. [CrossRef]
67. Duyvesteijn, R.G.; van Wijk, R.; Boer, Y.; Rep, M.; Cornelissen, B.J.; Haring, M.A. Frp1 is a *Fusarium oxysporum* F-box protein required for pathogenicity on tomato. *Mol. Microbiol.* **2005**, *57*, 1051–1063. [CrossRef]
68. Jonkers, W.; Rodrigues, C.D.; Rep, M. Impaired colonization and infection of tomato roots by the Δ*frp1* mutant of *Fusarium oxysporum* correlates with reduced CWDE gene expression. *Mol. Plant Microbe Interact.* **2009**, *22*, 507–518. [CrossRef]
69. Mullally, J.E.; Chernova, T.; Wilkinson, K.D. Doa1 is a Cdc48 adapter that possesses a novel ubiquitin binding domain. *Mol. Cell Biol.* **2006**, *26*, 822–830. [CrossRef]
70. Noman, M.; Azizullah; Ahmed, T.; Gao, Y.; Wang, H.; Xiong, X.; Wang, J.; Lou, J.; Li, D.; Song, F. Degradation of α-Subunits, Doa1 and Doa4, are Critical for Growth, Development, Programmed Cell Death Events, Stress Responses, and Pathogenicity in the Watermelon Fusarium Wilt Fungus *Fusarium oxysporum* f. sp. *niveum*. *J. Agric. Food Chem.* **2023**, *71*, 11667–11679. [CrossRef]
71. Dodd, A.N.; Kudla, J.; Sanders, D. The language of calcium signaling. *Annu. Rev. Plant Biol.* **2010**, *61*, 593–620. [CrossRef]
72. Kim, H.S.; Kim, J.E.; Frailey, D.; Nohe, A.; Duncan, R.; Czymmek, K.J.; Kang, S. Roles of three *Fusarium oxysporum* calcium ion (Ca(2+)) channels in generating Ca(2+) signatures and controlling growth. *Fungal Genet. Biol.* **2015**, *82*, 145–157. [CrossRef]
73. Park, H.S.; Lee, S.C.; Cardenas, M.E.; Heitman, J. Calcium-Calmodulin-Calcineurin Signaling: A Globally Conserved Virulence Cascade in Eukaryotic Microbial Pathogens. *Cell Host Microbe* **2019**, *26*, 453–462. [CrossRef]
74. Hou, Y.H.; Hsu, L.H.; Wang, H.F.; Lai, Y.H.; Chen, Y.L. Calcineurin regulates conidiation, chlamydospore formation and virulence in*Fusarium oxysporum* f. sp. *lycopersici*. *Front. Microbiol.* **2020**, *11*, 539702. [CrossRef]
75. Mariscal, M.; Miguel-Rojas, C.; Hera, C.; Fernandes, T.R.; Di Pietro, A. *Fusarium oxysporum* casein kinase 1, a negative regulator of the plasma membrane H+-ATPase Pma1, is required for development and pathogenicity. *J. Fungi* **2022**, *8*, 1300. [CrossRef] [PubMed]
76. Iida, Y.; Fujiwara, K.; Yoshioka, Y.; Tsuge, T. Mutation of FVS1, encoding a protein with a sterile alpha motif domain, affects asexual reproduction in the fungal plant pathogen *Fusarium oxysporum*. *FEMS Microbiol. Lett.* **2014**, *351*, 104–112. [CrossRef] [PubMed]
77. Inoue, I.; Namiki, F.; Tsuge, T. Plant colonization by the vascular wilt fungus *Fusarium oxysporum* requires FOW1, a gene encoding a mitochondrial protein. *Plant Cell* **2002**, *14*, 1869–1883. [CrossRef]
78. Kawabe, M.; Mizutani, K.; Yoshida, T.; Teraoka, T.; Yoneyama, K.; Yamaguchi, I.; Arie, T. Cloning of the pathogenicity-related gene FPD1 in*Fusarium oxysporum* f. sp. *lycopersici*. *J. Gen. Plant Pathol.* **2004**, *70*, 16–20. [CrossRef]
79. Ruiz-Roldán, C.; Pareja-Jaime, Y.; González-Reyes, J.A.; Roncero, M.I. The Transcription Factor Con7-1 Is a Master Regulator of Morphogenesis and Virulence in *Fusarium oxysporum*. *Mol. Plant Microbe Interact.* **2015**, *28*, 55–68. [CrossRef]
80. Denisov, Y.; Freeman, S.; Yarden, O. Inactivation of Snt2, a BAH/PHD-containing transcription factor, impairs pathogenicity and increases autophagosome abundance in *Fusarium oxysporum*. *Mol. Plant Pathol.* **2011**, *12*, 449–461. [CrossRef] [PubMed]
81. Jonkers, W.; Xayamongkhon, H.; Haas, M.; Olivain, C.; van der Does, H.C.; Broz, K.; Rep, M.; Alabouvette, C.; Steinberg, C.; Kistler, H.C. EBR1 genomic expansion and its role in virulence of Fusarium species. *Env. Microbiol.* **2014**, *16*, 1982–2003. [CrossRef]

82. Palos-Fernández, R.; Turrà, D.; Pietro, A.D. The Gal4-Type Transcription Factor Pro1 Integrates Inputs from Two Different MAPK Cascades to Regulate Development in the Fungal Pathogen *Fusarium oxysporum*. *J. Fungi* **2022**, *8*, 1242. [CrossRef]
83. Ding, Z.; Lin, H.; Liu, L.; Lu, T.; Xu, Y.; Peng, J.; Ren, Y.; Xu, T.; Zhang, X. Transcription factor FoAce2 regulates virulence, vegetative growth, conidiation, and cell wall homeostasis in *Fusarium oxysporum* f. sp. *cubense*. *Fungal Biol.* **2024**, *128*, 1960–1967. [CrossRef]
84. Su, J.; Wang, J.; Tang, J.; Yu, W.; Liu, J.; Dong, X.; Dong, J.; Chai, X.; Ji, P.; Zhang, L. Zinc finger transcription factor ZFP1 is associated with growth, conidiation, osmoregulation, and virulence in the Polygonatum kingianum pathogen *Fusarium oxysporum*. *Sci. Rep.* **2024**, *14*, 16061. [CrossRef]
85. Dai, Y.; Cao, Z.; Huang, L.; Liu, S.; Shen, Z.; Wang, Y.; Wang, H.; Zhang, H.; Li, D.; Song, F. CCR4-Not complex subunit Not2 plays critical roles in vegetative growth, conidiation and virulence in watermelon Fusarium wilt pathogen *Fusarium oxysporum* f. sp. *niveum*. *Front. Microbiol.* **2016**, *7*, 1449. [CrossRef] [PubMed]
86. Liu, J.; An, B.; Luo, H.; He, C.; Wang, Q. The histone acetyltransferase FocGCN5 regulates growth, conidiation, and pathogenicity of the banana wilt disease causal agent *Fusarium oxysporum* f. sp. *cubense* tropical race 4. *Res. Microbiol.* **2022**, *173*, 103902. [CrossRef] [PubMed]
87. Bayram, O.; Krappmann, S.; Ni, M.; Bok, J.W.; Helmstaedt, K.; Valerius, O.; Braus-Stromeyer, S.; Kwon, N.J.; Keller, N.P.; Yu, J.H.; et al. VelB/VeA/LaeA complex coordinates light signal with fungal development and secondary metabolism. *Science* **2008**, *320*, 1504–1506. [CrossRef] [PubMed]
88. Reyes-Dominguez, Y.; Bok, J.W.; Berger, H.; Shwab, E.K.; Basheer, A.; Gallmetzer, A.; Scazzocchio, C.; Keller, N.; Strauss, J. Heterochromatic marks are associated with the repression of secondary metabolism clusters in *Aspergillus nidulans*. *Mol. Microbiol.* **2010**, *76*, 1376–1386. [CrossRef] [PubMed]
89. Bok, J.W.; Noordermeer, D.; Kale, S.P.; Keller, N.P. Secondary metabolic gene cluster silencing in *Aspergillus nidulans*. *Mol. Microbiol.* **2006**, *61*, 1636–1645. [CrossRef] [PubMed]
90. López-Berges, M.S.; Hera, C.; Sulyok, M.; Schäfer, K.; Capilla, J.; Guarro, J.; Di Pietro, A. The velvet complex governs mycotoxin production and virulence of *Fusarium oxysporum* on plant and mammalian hosts. *Mol. Microbiol.* **2013**, *87*, 49–65. [CrossRef]
91. Li, P.; Pu, X.; Feng, B.; Yang, Q.; Shen, H.; Zhang, J.; Lin, B. FocVel1 influences asexual production, filamentous growth, biofilm formation, and virulence in *Fusarium oxysporum* f. sp. *cucumerinum*. *Front. Plant Sci.* **2015**, *6*, 312. [CrossRef]
92. Lu, S.; Deng, H.; Lin, Y.; Huang, M.; You, H.; Zhang, Y.; Zhuang, W.; Lu, G.; Yun, Y. A Network of Sporogenesis-Responsive Genes Regulates the Growth, Asexual Sporogenesis, Pathogenesis and Fusaric Acid Production of *Fusarium oxysporum* f. sp. *cubense*. *J. Fungi* **2023**, *10*, 1. [CrossRef]
93. Miller, M.A.; Olivas, W.M. Roles of Puf proteins in mRNA degradation and translation. *Wiley Interdiscip. Rev. RNA* **2011**, *2*, 471–492. [CrossRef]
94. Gao, Y.; Xiong, X.; Wang, H.; Bi, Y.; Wang, J.; Yan, Y.; Li, D.; Song, F. *Fusarium oxysporum* f. sp. *niveum* Pumilio 1 Regulates Virulence on Watermelon through Interacting with the ARP2/3 Complex and Binding to an A-Rich Motif in the 3′ UTR of Diverse Transcripts. *mBio* **2023**, *14*, e0015723. [CrossRef]
95. Lee, H.C.; Li, L.; Gu, W.; Xue, Z.; Crosthwaite, S.K.; Pertsemlidis, A.; Lewis, Z.A.; Freitag, M.; Selker, E.U.; Mello, C.C.; et al. Diverse pathways generate microRNA-like RNAs and Dicer-independent small interfering RNAs in fungi. *Mol. Cell* **2010**, *38*, 803–814. [CrossRef] [PubMed]
96. Cheng, L.; Ling, J.; Liang, L.; Luo, Z.; Zhang, J.; Xie, B. *Qip* gene in *Fusarium oxysporum* is required for normal hyphae morphology and virulence. *Mycology* **2015**, *6*, 130–137. [CrossRef] [PubMed]
97. Li, M.; Xie, L.; Wang, M.; Lin, Y.; Zhong, J.; Zhang, Y.; Zeng, J.; Kong, G.; Xi, P.; Li, H.; et al. FoQDE2-dependent milRNA promotes *Fusarium oxysporum* f. sp. *cubense* virulence by silencing a glycosyl hydrolase coding gene expression. *PLoS Pathog.* **2022**, *18*, e1010157. [CrossRef] [PubMed]
98. Xie, L.; Bi, Y.; He, C.; Situ, J.; Wang, M.; Kong, G.; Xi, P.; Jiang, Z.; Li, M. Unveiling microRNA-like small RNAs implicated in the initial infection of *Fusarium oxysporum* f. sp. *cubense* through small RNA sequencing. *Mycology* **2024**, 1–16. [CrossRef]
99. Xu, Y.; Zhou, H.; Zhao, G.; Yang, J.; Luo, Y.; Sun, S.; Wang, Z.; Li, S.; Jin, C. Genetical and O-glycoproteomic analyses reveal the roles of three protein O-mannosyltransferases in phytopathogen *Fusarium oxysporum* f. sp. *cucumerinum*. *Fungal Genet. Biol.* **2020**, *134*, 103285. [CrossRef]
100. Xiong, X.; Gao, Y.; Wang, J.; Wang, H.; Lou, J.; Bi, Y.; Yan, Y.; Li, D.; Song, F. Palmitoyl Transferase FonPAT2-Catalyzed Palmitoylation of the FonAP-2 Complex Is Essential for Growth, Development, Stress Response, and Virulence in *Fusarium oxysporum* f. sp. *Microbiol. Spectr.* **2023**, *11*, e0386122. [CrossRef]
101. Caffaro, C.E.; Hirschberg, C.B. Nucleotide sugar transporters of the Golgi apparatus: From basic science to diseases. *Acc. Chem. Res.* **2006**, *39*, 805–812. [CrossRef] [PubMed]
102. Hadley, B.; Maggioni, A.; Ashikov, A.; Day, C.J.; Haselhorst, T.; Tiralongo, J. Structure and function of nucleotide sugar transporters: Current progress. *Comput. Struct. Biotechnol. J.* **2014**, *10*, 23–32. [CrossRef]
103. Gao, Y.; Xiong, X.; Wang, H.; Wang, J.; Bi, Y.; Yan, Y.; Cao, Z.; Li, D.; Song, F. Ero1-Pdi1 module-catalysed dimerization of a nucleotide sugar transporter, FonNst2, regulates virulence of *Fusarium oxysporum* on watermelon. *Env. Microbiol.* **2022**, *24*, 1200–1220. [CrossRef]

104. Wang, J.; Gao, Y.; Xiong, X.; Yan, Y.; Lou, J.; Noman, M.; Li, D.; Song, F. The Ser/Thr protein kinase FonKin4-poly(ADP-ribose) polymerase FonPARP1 phosphorylation cascade is required for the pathogenicity of watermelon fusarium wilt fungus. *Front. Microbiol.* **2024**, *15*, 1397688. [CrossRef]
105. Li, B.; Gao, Y.; Mao, H.Y.; Borkovich, K.A.; Ouyang, S.Q. The SNARE protein FolVam7 mediates intracellular trafficking to regulate conidiogenesis and pathogenicity in*Fusarium oxysporum* f. sp. *lycopersici*. *Env. Microbiol.* **2019**, *21*, 2696–2706. [CrossRef] [PubMed]
106. Fang, Z.; Zhao, Q.; Yang, S.; Cai, Y.; Fang, W.; Abubakar, Y.S.; Lin, Y.; Yun, Y.; Zheng, W. Two distinct SNARE complexes mediate vesicle fusion with the plasma membrane to ensure effective development and pathogenesis of *Fusarium oxysporum* f. sp. *cubense*. *Mol. Plant Pathol.* **2024**, *25*, e13443. [CrossRef] [PubMed]
107. Corral-Ramos, C.; Roca, M.G.; Di Pietro, A.; Roncero, M.I.; Ruiz-Roldán, C. Autophagy contributes to regulation of nuclear dynamics during vegetative growth and hyphal fusion in *Fusarium oxysporum*. *Autophagy* **2015**, *11*, 131–144. [CrossRef]
108. Khalid, A.R.; Lv, X.; Naeem, M.; Mehmood, K.; Shaheen, H.; Dong, P.; Qiu, D.; Ren, M. Autophagy related gene (ATG3) is a key regulator for cell growth, development, and virulence of *Fusarium oxysporum*. *Genes* **2019**, *10*, 658. [CrossRef]
109. Khalid, A.R.; Zhang, S.; Luo, X.; Mehmood, K.; Rahim, J.; Shaheen, H.; Dong, P.; Qiu, D.; Ren, M. Role of autophagy-related gene *atg22* in developmental process and virulence of *Fusarium oxysporum*. *Genes* **2019**, *10*, 365. [CrossRef]
110. Khalid, A.R.; Zhang, S.; Luo, X.; Shaheen, H.; Majeed, A.; Maqbool, M.; Zahid, N.; Rahim, J.; Ren, M.; Qiu, D. Functional analysis of autophagy-related gene *ATG12* in potato dry rot fungus *Fusarium oxysporum*. *Int. J. Mol. Sci.* **2021**, *22*, 4932. [CrossRef] [PubMed]
111. Zhao, B.; He, D.; Gao, S.; Zhang, Y.; Wang, L. Hypothetical protein FoDbp40 influences the growth and virulence of *Fusarium oxysporum* by regulating the expression of isocitrate lyase. *Front. Microbiol.* **2022**, *13*, 1050637. [CrossRef]
112. Huang, Z.; Lou, J.; Gao, Y.; Noman, M.; Li, D.; Song, F. FonTup1 functions in growth, conidiogenesis and pathogenicity of *Fusarium oxysporum* f. sp. *niveum* through modulating the expression of the tricarboxylic acid cycle genes. *Microbiol. Res.* **2023**, *272*, 127389. [CrossRef]
113. Namiki, F.; Matsunaga, M.; Okuda, M.; Inoue, I.; Nishi, K.; Fujita, Y.; Tsuge, T. Mutation of an arginine biosynthesis gene causes reduced pathogenicity in *Fusarium oxysporum* f. sp. *melonis*. *Mol. Plant Microbe Interact.* **2001**, *14*, 580–584. [CrossRef]
114. Corrales Escobosa, A.R.; Rangel Porras, R.A.; Meza Carmen, V.; Gonzalez Hernandez, G.A.; Torres Guzman, J.C.; Wrobel, K.; Roncero, M.I.; Gutierrez Corona, J.F. *Fusarium oxysporum* Adh1 has dual fermentative and oxidative functions and is involved in fungal virulence in tomato plants. *Fungal Genet. Biol.* **2011**, *48*, 886–895. [CrossRef]
115. Corral-Ramos, C.; Roncero, M.I. Glycogen catabolism, but not its biosynthesis, affects virulence of *Fusarium oxysporum* on the plant host. *Fungal Genet. Biol.* **2015**, *77*, 40–49. [CrossRef] [PubMed]
116. Hohmann, S.; Meacock, P.A. Thiamin metabolism and thiamin diphosphate-dependent enzymes in the yeast Saccharomyces cerevisiae: Genetic regulation. *Biochim. Biophys. Acta* **1998**, *1385*, 201–219. [CrossRef] [PubMed]
117. Choi, G.H.; Marek, E.T.; Schardl, C.L.; Richey, M.G.; Chang, S.Y.; Smith, D.A. sti35, a stress-responsive gene in Fusarium spp. *J. Bacteriol.* **1990**, *172*, 4522–4528. [CrossRef] [PubMed]
118. Maundrell, K. nmt1 of fission yeast. A highly transcribed gene completely repressed by thiamine. *J. Biol. Chem.* **1990**, *265*, 10857–10864. [CrossRef] [PubMed]
119. Ruiz-Roldán, C.; Puerto-Galán, L.; Roa, J.; Castro, A.; Di Pietro, A.; Roncero, M.I.; Hera, C. The *Fusarium oxysporum* sti35 gene functions in thiamine biosynthesis and oxidative stress response. *Fungal Genet. Biol.* **2008**, *45*, 6–16. [CrossRef]
120. Wei, C.; Wen, C.; Zhang, Y.; Du, H.; Zhong, R.; Guan, Z.; Wang, M.; Qin, Y.; Wang, F.; Song, L.; et al. The FomYjeF Protein Influences the Sporulation and Virulence of *Fusarium oxysporum* f. sp. momordicae. *Int. J. Mol. Sci.* **2023**, *24*, 7260. [CrossRef]
121. Gomez-Gil, L.; Camara Almiron, J.; Rodriguez Carrillo, P.L.; Olivares Medina, C.N.; Bravo Ruiz, G.; Romo Rodriguez, P.; Corrales Escobosa, A.R.; Gutierrez Corona, F.; Roncero, M.I. Nitrate assimilation pathway (NAP): Role of structural (nit) and transporter (ntr1) genes in*Fusarium oxysporum* f. sp. *lycopersici* growth and pathogenicity. *Curr. Genet.* **2018**, *64*, 493–507. [CrossRef]
122. Caracuel, Z.; Martínez-Rocha, A.L.; Di Pietro, A.; Madrid, M.P.; Roncero, M.I. *Fusarium oxysporum* gas1 encodes a putative beta-1,3-glucanosyltransferase required for virulence on tomato plants. *Mol. Plant Microbe Interact.* **2005**, *18*, 1140–1147. [CrossRef]
123. López-Fernández, L.; Ruiz-Roldán, C.; Pareja-Jaime, Y.; Prieto, A.; Khraiwesh, H.; Roncero, M.I. The *Fusarium oxysporum* gnt2, encoding a putative N-acetylglucosamine transferase, is involved in cell wall architecture and virulence. *PLoS ONE* **2013**, *8*, e84690. [CrossRef]
124. Madrid, M.P.; Di Pietro, A.; Roncero, M.I. Class V chitin synthase determines pathogenesis in the vascular wilt fungus *Fusarium oxysporum* and mediates resistance to plant defence compounds. *Mol. Microbiol.* **2003**, *47*, 257–266. [CrossRef]
125. Martín-Udíroz, M.; Madrid, M.P.; Roncero, M.I. Role of chitin synthase genes in *Fusarium oxysporum*. *Microbiology* **2004**, *150*, 3175–3187. [CrossRef] [PubMed]
126. Martín-Urdíroz, M.; Roncero, M.I.; González-Reyes, J.A.; Ruiz-Roldán, C. ChsVb, a class VII chitin synthase involved in septation, is critical for pathogenicity in *Fusarium oxysporum*. *Eukaryot. Cell* **2008**, *7*, 112–121. [CrossRef] [PubMed]
127. Li, M.H.; Xie, X.L.; Lin, X.F.; Shi, J.X.; Ding, Z.J.; Ling, J.F.; Xi, P.G.; Zhou, J.N.; Leng, Y.; Zhong, S.; et al. Functional characterization of the gene FoOCH1 encoding a putative α-1,6-mannosyltransferase in *Fusarium oxysporum* f. sp. *cubense*. *Fungal Genet. Biol.* **2014**, *65*, 1–13. [CrossRef] [PubMed]
128. Han, S.; Sheng, B.; Zhu, D.; Chen, J.; Cai, H.; Zhang, S.; Guo, C. Role of FoERG3 in Ergosterol Biosynthesis by *Fusarium oxysporum* and the Associated Regulation by *Bacillus subtilis* HSY21. *Plant Dis.* **2023**, *107*, 1565–1575. [CrossRef]

129. Usman, S.; Ge, X.; Xu, Y.; Qin, Q.; Xie, J.; Wang, B.; Jin, C.; Fang, W. Loss of Phosphomannose Isomerase Impairs Growth, Perturbs Cell Wall Integrity, and Reduces Virulence of *Fusarium oxysporum* f. sp. *cubense* on Banana Plants. *J. Fungi* **2023**, *9*, 478. [CrossRef]
130. Wei-na, G.A.O.; Zhan-ming, H.O.U. Research on the function of FolGLS2 gene in *Fusarium oxysporum* f. sp. *lini*. *Chin. J. Oil Crop Sci.* **2024**, *46*. [CrossRef]
131. Zhou, H.; Xu, Y.; Ebel, F.; Jin, C. Galactofuranose (Galf)-containing sugar chain contributes to the hyphal growth, conidiation and virulence of *F. oxysporum* f. sp. *cucumerinum*. *PLoS ONE* **2021**, *16*, e0250064. [CrossRef]
132. Zhang, N.; Lv, F.; Qiu, F.; Han, D.; Xu, Y.; Liang, W. Pathogenic fungi neutralize plant-derived ROS via Srpk1 deacetylation. *EMBO J.* **2023**, *42*, e112634. [CrossRef]
133. Thurston, C.F. The structure and function of fungal laccases. *Microbiology* **1994**, *140*, 19–26. [CrossRef]
134. Cañero, D.C.; Roncero, M.I. Functional analyses of laccase genes from *Fusarium oxysporum*. *Phytopathology* **2008**, *98*, 509–518. [CrossRef]
135. Cañero, D.C.; Roncero, M.I.G. Influence of the chloride channel of *Fusarium oxysporum* on extracellular laccase activity and virulence on tomato plants. *Microbiology* **2008**, *154*, 1474–1481. [CrossRef] [PubMed]
136. Li, W.; Roberts, D.P.; Dery, P.D.; Meyer, S.L.F.; Lohrke, S.; Lumsden, R.D.; Hebbar, K.P. Broad spectrum anti-biotic activity and disease suppression by the potential biocontrol agent Burkholderia ambifaria BC-F. *Crop Prot.* **2002**, *21*, 129–135. [CrossRef]
137. Nelson, P.E.; Dignani, M.C.; Anaissie, E.J. Taxonomy, biology, and clinical aspects of Fusarium species. *Clin. Microbiol. Rev.* **1994**, *7*, 479–504. [CrossRef] [PubMed]
138. Agrios, G.N. *Plant Pathology*, 5th ed.; Elsevier Academic Press: Amsterdam, The Netherlands, 2005.
139. Pareek, M.; Almog, Y.; Bari, V.K.; Hazkani-Covo, E.; Onn, I.; Covo, S. Alternative functional *rad21* paralogs in *Fusarium oxysporum*. *Front. Microbiol.* **2019**, *10*, 1370. [CrossRef]
140. Ohara, T.; Inoue, I.; Namiki, F.; Kunoh, H.; Tsuge, T. REN1 is required for development of microconidia and macroconidia, but not of chlamydospores, in the plant pathogenic fungus *Fusarium oxysporum*. *Genetics* **2004**, *166*, 113–124. [CrossRef]
141. Qian, H.; Song, L.; Wang, L.; Yang, Q.; Wu, R.; Du, J.; Zheng, B.; Liang, W. FolIws1-driven nuclear translocation of deacetylated FolTFIIS ensures conidiation of *Fusarium oxysporum*. *Cell Rep.* **2024**, *43*, 114588. [CrossRef]
142. Tao, L.; Yu, J.H. AbaA and WetA govern distinct stages of Aspergillus fumigatus development. *Microbiology* **2011**, *157*, 313–326. [CrossRef]
143. Etxebeste, O.; Garzia, A.; Espeso, E.A.; Ugalde, U. *Aspergillus nidulans* asexual development: Making the most of cellular modules. *Trends Microbiol.* **2010**, *18*, 569–576. [CrossRef]
144. Azizullah; Noman, M.; Gao, Y.; Wang, H.; Xiong, X.; Wang, J.; Li, D.; Song, F. The SUMOylation Pathway Components Are Required for Vegetative Growth, Asexual Development, Cytotoxic Responses, and Programmed Cell Death Events in. *J. Fungi* **2023**, *9*, 94. [CrossRef]
145. Moura, R.D.; de Castro, L.A.M.; Culik, M.P.; Fernandes, A.A.R.; Fernandes, P.M.B.; Ventura, J.A. Culture medium for improved production of conidia for identification and systematic studies of Fusarium pathogens. *J. Microbiol. Methods* **2020**, *173*, 105915. [CrossRef]
146. Iida, Y.; Kurata, T.; Harimoto, Y.; Tsuge, T. Nitrite reductase gene upregulated during conidiation is involved in macroconidium formation in *Fusarium oxysporum*. *Phytopathology* **2008**, *98*, 1099–1106. [CrossRef] [PubMed]
147. Zheng, B.; Yan, L.; Liang, W.; Yang, Q. Paralogous Cyp51s mediate the differential sensitivity of *Fusarium oxysporum* to sterol demethylation inhibitors. *Pest. Manag. Sci.* **2019**, *75*, 396–404. [CrossRef]
148. Zhang, N.; Song, L.; Xu, Y.; Pei, X.; Luisi, B.F.; Liang, W. The decrotonylase FoSir5 facilitates mitochondrial metabolic state switching in conidial germination of *Fusarium oxysporum*. *eLife* **2021**, *10*. [CrossRef]
149. Garcia-Rubio, R.; de Oliveira, H.C.; Rivera, J.; Trevijano-Contador, N. The Fungal Cell Wall: Candida, Cryptococcus, and Aspergillus Species. *Front. Microbiol.* **2019**, *10*, 2993. [CrossRef]
150. Henry, C.; Fontaine, T.; Heddergott, C.; Robinet, P.; Aimanianda, V.; Beau, R.; Beauvais, A.; Mouyna, I.; Prevost, M.C.; Fekkar, A.; et al. Biosynthesis of cell wall mannan in the conidium and the mycelium of Aspergillus fumigatus. *Cell Microbiol.* **2016**, *18*, 1881–1891. [CrossRef] [PubMed]
151. Chauhan, S.; Rajam, M.V. RNAi-mediated down-regulation of fasciclin-like proteins (FoFLPs) in*Fusarium oxysporum* f. sp. *lycopersici* results in reduced pathogenicity and virulence. *Microbiol. Res.* **2022**, *260*, 127033. [CrossRef] [PubMed]
152. Rispail, N.; Di Pietro, A. *Fusarium oxysporum* Ste12 controls invasive growth and virulence downstream of the Fmk1 MAPK cascade. *Mol. Plant-Microbe Interact.* **2009**, *22*, 830–839. [CrossRef]
153. Asunción García-Sánchez, M.; Martín-Rodrigues, N.; Ramos, B.; de Vega-Bartol, J.J.; Perlin, M.H.; Díaz-Mínguez, J.M. fost12, the *Fusarium oxysporum* homolog of the transcription factor Ste12, is upregulated during plant infection and required for virulence. *Fungal Genet. Biol.* **2010**, *47*, 216–225. [CrossRef] [PubMed]
154. Wang, Y.; Zhang, X.; Wang, T.; Zhou, S.; Liang, X.; Xie, C.; Kang, Z.; Chen, D.; Zheng, L. The Small Secreted Protein FoSsp1 Elicits Plant Defenses and Negatively Regulates Pathogenesis in *Fusarium oxysporum* f. sp. *cubense* (Foc4). *Front. Plant Sci.* **2022**, *13*, 873451. [CrossRef]
155. Nordzieke, D.E.; Fernandes, T.R.; El Ghalid, M.; Turrà, D.; Di Pietro, A. NADPH oxidase regulates chemotropic growth of the fungal pathogen *Fusarium oxysporum* towards the host plant. *New Phytol.* **2019**, *224*, 1600–1612. [CrossRef]
156. Dong, X.; Ling, N.; Wang, M.; Shen, Q.; Guo, S. Fusaric acid is a crucial factor in the disturbance of leaf water imbalance in Fusarium-infected banana plants. *Plant Physiol. Biochem.* **2012**, *60*, 171–179. [CrossRef]

157. López-Díaz, C.; Rahjoo, V.; Sulyok, M.; Ghionna, V.; Martín-Vicente, A.; Capilla, J.; Di Pietro, A.; López-Berges, M.S. Fusaric acid contributes to virulence of *Fusarium oxysporum* on plant and mammalian hosts. *Mol. Plant Pathol.* **2018**, *19*, 440–453. [CrossRef] [PubMed]
158. Brown, D.W.; Lee, S.H.; Kim, L.H.; Ryu, J.G.; Lee, S.; Seo, Y.; Kim, Y.H.; Busman, M.; Yun, S.H.; Proctor, R.H.; et al. Identification of a 12-gene Fusaric Acid Biosynthetic Gene Cluster in Fusarium Species Through Comparative and Functional Genomics. *Mol. Plant Microbe Interact.* **2015**, *28*, 319–332. [CrossRef] [PubMed]
159. Liu, S.; Li, J.; Zhang, Y.; Liu, N.; Viljoen, A.; Mostert, D.; Zuo, C.; Hu, C.; Bi, F.; Gao, H.; et al. Fusaric acid instigates the invasion of banana by *Fusarium oxysporum* f. sp. *cubense* TR4. *New Phytol.* **2020**, *225*, 913–929. [CrossRef]
160. Crutcher, F.K.; Liu, J.; Puckhaber, L.S.; Stipanovic, R.D.; Bell, A.A.; Nichols, R.L. FUBT, a putative MFS transporter, promotes secretion of fusaric acid in the cotton pathogen *Fusarium oxysporum* f. sp. *vasinfectum*. *Microbiology* **2015**, *161*, 875–883. [CrossRef] [PubMed]
161. Arasimowicz-Jelonek, M.; Floryszak-Wieczorek, J. Nitric oxide: An effective weapon of the plant or the pathogen? *Mol. Plant Pathol.* **2014**, *15*, 406–416. [CrossRef] [PubMed]
162. Zhang, Y.; Liu, S.; Mostert, D.; Yu, H.; Zhuo, M.; Li, G.; Zuo, C.; Haridas, S.; Webster, K.; Li, M.; et al. Virulence of banana wilt-causing fungal pathogen *Fusarium oxysporum* tropical race 4 is mediated by nitric oxide biosynthesis and accessory genes. *Nat. Microbiol.* **2024**. [CrossRef]
163. Bravo Ruiz, G.; Di Pietro, A.; Roncero, M.I. Combined action of the major secreted exo- and endopolygalacturonases is required for full virulence of *Fusarium oxysporum*. *Mol. Plant Pathol.* **2016**, *17*, 339–353. [CrossRef]
164. García-Maceira, F.I.; Di Pietro, A.; Roncero, M.I. Cloning and disruption of pgx4 encoding an in planta expressed exopolygalacturonase from *Fusarium oxysporum*. *Mol. Plant Microbe Interact.* **2000**, *13*, 359–365. [CrossRef]
165. García-Maceira, F.I.; Di Pietro, A.; Huertas-González, M.D.; Ruiz-Roldán, M.C.; Roncero, M.I. Molecular characterization of an endopolygalacturonase from *Fusarium oxysporum* expressed during early stages of infection. *Appl. Env. Microbiol.* **2001**, *67*, 2191–2196. [CrossRef]
166. Gómez-Gómez, E.; Isabel, M.; Roncero, G.; Di Pietro, A.; Hera, C. Molecular characterization of a novel endo-beta-1,4-xylanase gene from the vascular wilt fungus *Fusarium oxysporum*. *Curr. Genet.* **2001**, *40*, 268–275. [CrossRef] [PubMed]
167. Di Pietro, A.; Huertas-González, M.D.; Gutierrez-Corona, J.F.; Martínez-Cadena, G.; Méglecz, E.; Roncero, M.I. Molecular characterization of a subtilase from the vascular wilt fungus *Fusarium oxysporum*. *Mol. Plant Microbe Interact.* **2001**, *14*, 653–662. [CrossRef] [PubMed]
168. Bravo-Ruiz, G.; Ruiz-Roldán, C.; Roncero, M.I. Lipolytic system of the tomato pathogen*Fusarium oxysporum* f. sp. *lycopersici*. *Mol. Plant Microbe Interact.* **2013**, *26*, 1054–1067. [CrossRef]
169. Gómez-Gómez, E.; Ruíz-Roldán, M.C.; Di Pietro, A.; Roncero, M.I.; Hera, C. Role in pathogenesis of two endo-beta-1,4-xylanase genes from the vascular wilt fungus *Fusarium oxysporum*. *Fungal Genet. Biol.* **2002**, *35*, 213–222. [CrossRef]
170. Dong, Z.; Luo, M.; Wang, Z. An Exo-polygalacturonase Pgc4 regulates aerial hyphal growth and virulence in *Fusarium oxysporum* f. sp. *cubense* race 4. *Int. J. Mol. Sci.* **2020**, *21*, 5886. [CrossRef]
171. Zhang, L.; Yan, J.; Fu, Z.; Shi, W.; Ninkuu, V.; Li, G.; Yang, X.; Zeng, H. FoEG1, a secreted glycoside hydrolase family 12 protein from *Fusarium oxysporum*, triggers cell death and modulates plant immunity. *Mol. Plant Pathol.* **2021**, *22*, 522–538. [CrossRef]
172. Michielse, C.B.; Reijnen, L.; Olivain, C.; Alabouvette, C.; Rep, M. Degradation of aromatic compounds through the β-ketoadipate pathway is required for pathogenicity of the tomato wilt pathogen*Fusarium oxysporum* f. sp. *lycopersici*. *Mol. Plant Pathol.* **2012**, *13*, 1089–1100. [CrossRef]
173. Houterman, P.M.; Speijer, D.; Dekker, H.L.; DE Koster, C.G.; Cornelissen, B.J.; Rep, M. The mixed xylem sap proteome of *Fusarium oxysporum*-infected tomato plants. *Mol. Plant Pathol.* **2007**, *8*, 215–221. [CrossRef] [PubMed]
174. Lievens, B.; Houterman, P.M.; Rep, M. Effector gene screening allows unambiguous identification of*Fusarium oxysporum* f. sp. *lycopersici* races and discrimination from other formae speciales. *FEMS Microbiol. Lett.* **2009**, *300*, 201–215. [CrossRef]
175. Rep, M.; van der Does, H.C.; Meijer, M.; van Wijk, R.; Houterman, P.M.; Dekker, H.L.; de Koster, C.G.; Cornelissen, B.J. A small, cysteine-rich protein secreted by *Fusarium oxysporum* during colonization of xylem vessels is required for I-3-mediated resistance in tomato. *Mol. Microbiol.* **2004**, *53*, 1373–1383. [CrossRef]
176. Jangir, P.; Mehra, N.; Sharma, K.; Singh, N.; Rani, M.; Kapoor, R. Secreted in Xylem Genes: Drivers of Host Adaptation in *Fusarium oxysporum*. *Front. Plant Sci.* **2021**, *12*, 628611. [CrossRef] [PubMed]
177. Li, E.; Wang, G.; Xiao, J.; Ling, J.; Yang, Y.; Xie, B. A SIX1 Homolog in *Fusarium oxysporum* f. sp. conglutinans Is Required for Full Virulence on Cabbage. *PLoS ONE* **2016**, *11*, e0152273. [CrossRef]
178. Rep, M.; Meijer, M.; Houterman, P.M.; van der Does, H.C.; Cornelissen, B.J. *Fusarium oxysporum* evades I-3-mediated resistance without altering the matching avirulence gene. *Mol. Plant Microbe Interact.* **2005**, *18*, 15–23. [CrossRef]
179. Widinugraheni, S.; Niño-Sánchez, J.; van der Does, H.C.; van Dam, P.; García-Bastidas, F.A.; Subandiyah, S.; Meijer, H.J.G.; Kistler, H.C.; Kema, G.H.J.; Rep, M. A SIX1 homolog in *Fusarium oxysporum* f. sp. *cubense* tropical race 4 contributes to virulence towards Cavendish banana. *PLoS ONE* **2018**, *13*, e0205896. [CrossRef] [PubMed]
180. Houterman, P.M.; Ma, L.; van Ooijen, G.; de Vroomen, M.J.; Cornelissen, B.J.; Takken, F.L.; Rep, M. The effector protein Avr2 of the xylem-colonizing fungus *Fusarium oxysporum* activates the tomato resistance protein I-2 intracellularly. *Plant J.* **2009**, *58*, 970–978. [CrossRef]

181. Kashiwa, T.; Inami, K.; Fujinaga, M.; Ogiso, H.; Yoshida, T.; Teraoka, T.; Arie, T. An avirulence gene homologue in the tomato wilt fungus*Fusarium oxysporum* f. sp. *lycopersici* race 1 functions as a virulence gene in the cabbage yellows fungus *F. oxysporum* f. sp. conglutinans. *J. Gen. Plant Pathol.* **2013**, *71*, 412–421. [CrossRef]
182. Thatcher, L.F.; Gardiner, D.M.; Kazan, K.; Manners, J.M. A highly conserved effector in *Fusarium oxysporum* is required for full virulence on Arabidopsis. *Mol. Plant Microbe Interact.* **2012**, *25*, 180–190. [CrossRef] [PubMed]
183. Ma, L.; Houterman, P.M.; Gawehns, F.; Cao, L.; Sillo, F.; Richter, H.; Clavijo-Ortiz, M.J.; Schmidt, S.M.; Boeren, S.; Vervoort, J.; et al. The AVR2-SIX5 gene pair is required to activate I-2-mediated immunity in tomato. *New Phytol.* **2015**, *208*, 507–518. [CrossRef]
184. Gawehns, F.; Houterman, P.M.; Ichou, F.A.; Michielse, C.B.; Hijdra, M.; Cornelissen, B.J.; Rep, M.; Takken, F.L. The *Fusarium oxysporum* effector Six6 contributes to virulence and suppresses I-2-mediated cell death. *Mol. Plant Microbe Interact.* **2014**, *27*, 336–348. [CrossRef]
185. Niu, X.; Zhao, X.; Ling, K.S.; Levi, A.; Sun, Y.; Fan, M. The FonSIX6 gene acts as an avirulence effector in the *Fusarium oxysporum* f. sp. *niveum*—Watermelon pathosystem. *Sci. Rep.* **2016**, *6*, 28146. [CrossRef]
186. van Dam, P.; Fokkens, L.; Ayukawa, Y.; van der Gragt, M.; Ter Horst, A.; Brankovics, B.; Houterman, P.M.; Arie, T.; Rep, M. A mobile pathogenicity chromosome in *Fusarium oxysporum* for infection of multiple cucurbit species. *Sci. Rep.* **2017**, *7*, 9042. [CrossRef]
187. An, B.; Hou, X.; Guo, Y.; Zhao, S.; Luo, H.; He, C.; Wang, Q. The effector SIX8 is required for virulence of *Fusarium oxysporum* f. sp. *cubense* tropical race 4 to Cavendish banana. *Fungal Biol.* **2019**, *123*, 423–430. [CrossRef]
188. Ayukawa, Y.; Asai, S.; Gan, P.; Tsushima, A.; Ichihashi, Y.; Shibata, A.; Komatsu, K.; Houterman, P.M.; Rep, M.; Shirasu, K.; et al. A pair of effectors encoded on a conditionally dispensable chromosome of *Fusarium oxysporum* suppress host-specific immunity. *Commun. Biol.* **2021**, *4*, 707. [CrossRef]
189. Jashni, M.K.; Dols, I.H.; Iida, Y.; Boeren, S.; Beenen, H.G.; Mehrabi, R.; Collemare, J.; de Wit, P.J. Synergistic Action of a Metalloprotease and a Serine Protease from*Fusarium oxysporum* f. sp. *lycopersici* Cleaves Chitin-Binding Tomato Chitinases, Reduces Their Antifungal Activity, and Enhances Fungal Virulence. *Mol. Plant Microbe Interact.* **2015**, *28*, 996–1008. [CrossRef]
190. Zhang, X.; Huang, H.; Wu, B.; Xie, J.; Viljoen, A.; Wang, W.; Mostert, D.; Xie, Y.; Fu, G.; Xiang, D.; et al. The M35 Metalloprotease Effector FocM35_1 Is Required for Full Virulence of. *Pathogens* **2021**, *10*, 670. [CrossRef]
191. Liu, L.; Huang, Y.; Song, H.; Luo, M.; Dong, Z. α-Pheromone Precursor Protein Foc4-PP1 Is Essential for the Full Virulence of *Fusarium oxysporum* f. sp. *cubense* Tropical Race 4. *J. Fungi* **2023**, *9*, 365. [CrossRef]
192. Yan, T.; Zhou, X.; Li, J.; Li, G.; Zhao, Y.; Wang, H.; Li, H.; Nie, Y.; Li, Y. FoCupin1, a Cupin_1 domain-containing protein, is necessary for the virulence of *Fusarium oxysporum* f. sp. *cubense* tropical race 4. *Front. Microbiol.* **2022**, *13*, 1001540. [CrossRef]
193. Wang, T.; Xu, Y.; Zhao, Y.; Liang, X.; Liu, S.; Zhang, Y.; Kang, Z.; Chen, D.; Zheng, L. Systemic screening of *Fusarium oxysporum* candidate effectors reveals FoSSP17 that suppresses plant immunity and contributes to virulence. *Phytopathol. Res.* **2023**, *5*. [CrossRef]
194. Pareja-Jaime, Y.; Roncero, M.I.; Ruiz-Roldán, M.C. Tomatinase from*Fusarium oxysporum* f. sp. *lycopersici* is required for full virulence on tomato plants. *Mol. Plant Microbe Interact.* **2008**, *21*, 728–736. [CrossRef]
195. Yang, Y.; An, B.; Guo, Y.; Luo, H.; He, C.; Wang, Q. A Novel Effector, FSE1, Regulates the Pathogenicity of *Fusarium oxysporum* f. sp. *cubense* Tropical Race 4 to Banana by Targeting the MYB Transcription Factor MaEFM-Like. *J. Fungi* **2023**, *9*, 472. [CrossRef]
196. Li, J.; Ma, X.; Wang, C.; Liu, S.; Yu, G.; Gao, M.; Qian, H.; Liu, M.; Luisi, B.F.; Gabriel, D.W.; et al. Acetylation of a fungal effector that translocates host PR1 facilitates virulence. *eLife* **2022**, *11*. [CrossRef] [PubMed]
197. Ji, H.M.; Mao, H.Y.; Li, S.J.; Feng, T.; Zhang, Z.Y.; Cheng, L.; Luo, S.J.; Borkovich, K.A.; Ouyang, S.Q. Fol-milR1, a pathogenicity factor of *Fusarium oxysporum,* confers tomato wilt disease resistance by impairing host immune responses. *New Phytol.* **2021**, *232*, 705–718. [CrossRef] [PubMed]
198. Gai, X.; Li, S.; Jiang, N.; Sun, Q.; Xuan, Y.H.; Xia, Z. Comparative transcriptome analysis reveals that ATP synthases regulate *Fusarium oxysporum* virulence by modulating sugar transporter gene expressions in tobacco. *Front. Plant Sci.* **2022**, *13*, 978951. [CrossRef] [PubMed]
199. Qian, H.; Wang, L.; Wang, B.; Liang, W. The secreted ribonuclease T2 protein FoRnt2 contributes to *Fusarium oxysporum* virulence. *Mol. Plant Pathol.* **2022**, *23*, 1346–1360. [CrossRef]
200. Chang, W.; Li, H.; Chen, H.; Qiao, F.; Zeng, H. Identification of mimp-associated effector genes in *Fusarium oxysporum* f. sp. *cubense* race 1 and race 4 and virulence confirmation of a candidate effector gene. *Microbiol. Res.* **2020**, *232*, 126375. [CrossRef]
201. Guo, L.; Wang, J.; Liang, C.; Yang, L.; Zhou, Y.; Liu, L.; Huang, J. Fosp9, a Novel Secreted Protein, Is Essential for the Full Virulence of *Fusarium oxysporum* f. sp. *Appl. Env. Microbiol.* **2022**, *88*, e0060421. [CrossRef]
202. Sasaki, K.; Ito, Y.; Hamada, Y.; Dowaki, A.; Jogaiah, S.; Ito, S.I. *FoMC69* Gene in *Fusarium oxysporum* f. sp. *radicis-lycopersici* Is Essential for Pathogenicity by Involving Normal Function of Chlamydospores. *Pathogens* **2022**, *11*, 1433. [CrossRef]
203. Sakane, K.; Akiyama, M.; Ando, A.; Shigyo, M.; Ito, S.I.; Sasaki, K. Identification of a novel effector gene and its functional tradeoff in *Fusarium oxysporum* f. sp. cepae that infects Welsh onion. *Physiol. Mol. Plant Pathol.* **2023**, *123*, 101939. [CrossRef]
204. Liu, S.; Wu, B.; Yang, J.; Bi, F.; Dong, T.; Yang, Q.; Hu, C.; Xiang, D.; Chen, H.; Huang, H.; et al. A cerato-platanin family protein FocCP1 is essential for the penetration and virulence of *Fusarium oxysporum* f. sp. *cubense* tropical race 4. *Int. J. Mol. Sci.* **2019**, *20*, 3785. [CrossRef]

205. Araiza-Cervantes, C.A.; Valle-Maldonado, M.I.; Patiño-Medina, J.A.; Alejandre-Castañeda, V.; Guzmán-Pérez, J.B.; Ramírez-Díaz, M.I.; Macias-Sánchez, K.L.; López-Berges, M.S.; Meza-Carmen, V. Transcript pattern analysis of Arf-family genes in the phytopathogen*Fusarium oxysporum* f. sp. *lycopersici* reveals the role of Arl3 in the virulence. *Antonie Van. Leeuwenhoek* **2021**, *114*, 1619–1632. [CrossRef]
206. Calero-Nieto, F.; Di Pietro, A.; Roncero, M.I.; Hera, C. Role of the transcriptional activator xlnR of *Fusarium oxysporum* in regulation of xylanase genes and virulence. *Mol. Plant Microbe Interact.* **2007**, *20*, 977–985. [CrossRef] [PubMed]
207. Gámez-Arjona, F.M.; Vitale, S.; Voxeur, A.; Dora, S.; Müller, S.; Sancho-Andrés, G.; Montesinos, J.C.; Di Pietro, A.; Sánchez-Rodríguez, C. Impairment of the cellulose degradation machinery enhances *Fusarium oxysporum* virulence but limits its reproductive fitness. *Sci. Adv.* **2022**, *8*, eabl9734. [CrossRef]
208. Caracuel, Z.; Roncero, M.I.; Espeso, E.A.; González-Verdejo, C.I.; García-Maceira, F.I.; Di Pietro, A. The pH signalling transcription factor PacC controls virulence in the plant pathogen *Fusarium oxysporum*. *Mol. Microbiol.* **2003**, *48*, 765–779. [CrossRef]
209. Jonkers, W.; Rep, M. Mutation of CRE1 in *Fusarium oxysporum* reverts the pathogenicity defects of the FRP1 deletion mutant. *Mol. Microbiol.* **2009**, *74*, 1100–1113. [CrossRef] [PubMed]
210. Yun, Y.; Zhou, X.; Yang, S.; Wen, Y.; You, H.; Zheng, Y.; Norvienyeku, J.; Shim, W.B.; Wang, Z. *Fusarium oxysporum* f. sp. *lycopersici* C2H2 transcription factor FolCzf1 is required for conidiation, fusaric acid production, and early host infection. *Curr. Genet.* **2019**, *65*, 773–783. [CrossRef]
211. Ding, Z.-j.; Qi, Y.-x.; Zeng, F.-y.; Peng, J.; Xie, Y.-x.; Zhang, X. A transcription factor FoSwi6 regulates physiology traits and virulence in *Fusarium oxysporum* f. sp. *cubense*. *Acta Phytopathol. Sin.* **2018**, *48*, 601–610. [CrossRef]
212. Ding, Z.; Xu, T.; Zhu, W.; Li, L.; Fu, Q. A MADS-box transcription factor FoRlm1 regulates aerial hyphal growth, oxidative stress, cell wall biosynthesis and virulence in *Fusarium oxysporum* f. sp. *cubense*. *Fungal Biol.* **2020**, *124*, 183–193. [CrossRef] [PubMed]
213. Michielse, C.B.; van Wijk, R.; Reijnen, L.; Manders, E.M.; Boas, S.; Olivain, C.; Alabouvette, C.; Rep, M. The nuclear protein Sge1 of *Fusarium oxysporum* is required for parasitic growth. *PLoS Pathog.* **2009**, *5*, e1000637. [CrossRef]
214. Niño-Sánchez, J.; Casado-Del Castillo, V.; Tello, V.; De Vega-Bartol, J.J.; Ramos, B.; Sukno, S.A.; Díaz Mínguez, J.M. The FTF gene family regulates virulence and expression of SIX effectors in *Fusarium oxysporum*. *Mol. Plant Pathol.* **2016**, *17*, 1124–1139. [CrossRef]
215. Casado-Del Castillo, V.; Benito, E.P.; Díaz-Mínguez, J.M. The Role of the *Fusarium oxysporum* FTF2 Transcription Factor in Host Colonization and Virulence in Common Bean Plants (*Phaseolus vulgaris* L.). *Pathogens* **2023**, *12*, 380. [CrossRef]
216. Imazaki, I.; Kurahashi, M.; Iida, Y.; Tsuge, T. Fow2, a Zn(II)2Cys6-type transcription regulator, controls plant infection of the vascular wilt fungus *Fusarium oxysporum*. *Mol. Microbiol.* **2007**, *63*, 737–753. [CrossRef] [PubMed]
217. Michielse, C.B.; van Wijk, R.; Reijnen, L.; Cornelissen, B.J.; Rep, M. Insight into the molecular requirements for pathogenicity of*Fusarium oxysporum* f. sp. *lycopersici* through large-scale insertional mutagenesis. *Genome Biol.* **2009**, *10*, R4. [CrossRef]
218. Ruiz-Roldán, M.C.; Garre, V.; Guarro, J.; Mariné, M.; Roncero, M.I. Role of the white collar 1 photoreceptor in carotenogenesis, UV resistance, hydrophobicity, and virulence of *Fusarium oxysporum*. *Eukaryot. Cell* **2008**, *7*, 1227–1230. [CrossRef] [PubMed]
219. Divon, H.H.; Ziv, C.; Davydov, O.; Yarden, O.; Fluhr, R. The global nitrogen regulator, FNR1, regulates fungal nutrition-genes and fitness during *Fusarium oxysporum* pathogenesis. *Mol. Plant Pathol.* **2006**, *7*, 485–497. [CrossRef]
220. López-Berges, M.S.; Capilla, J.; Turrà, D.; Schafferer, L.; Matthijs, S.; Jöchl, C.; Cornelis, P.; Guarro, J.; Haas, H.; Di Pietro, A. HapX-mediated iron homeostasis is essential for rhizosphere competence and virulence of the soilborne pathogen *Fusarium oxysporum*. *Plant Cell* **2012**, *24*, 3805–3822. [CrossRef]
221. López-Berges, M.S. ZafA-mediated regulation of zinc homeostasis is required for virulence in the plant pathogen *Fusarium oxysporum*. *Mol. Plant Pathol.* **2020**, *21*, 244–249. [CrossRef] [PubMed]
222. Qi, X.; Guo, L.; Yang, L.; Huang, J. Foatf1, a bZIP transcription factor of *Fusarium oxysporum* f. sp. *cubense*, is involved in pathogenesis by regulating the oxidative stress responses of *Cavendish banana* (*Musa* spp.). *Physiol. Mol. Plant Pathol.* **2013**, *84*, 76–85. [CrossRef]
223. Qi, X.; Liu, L.; Wang, J. Stress response regulator FoSkn7 participates in the pathogenicity of *Fusarium oxysporum* f. sp. *cubense* race 4 by conferring resistance to exogenous oxidative stress. *J. Gen. Plant Pathol.* **2019**, *85*, 382–394. [CrossRef]
224. Bailey, B.A.; Apel-Birkhold, P.C.; Luster, D.G. Expression of NEP1 by *Fusarium oxysporum* f. sp. erythroxyli After Gene Replacement and Overexpression Using Polyethylene Glycol-Mediated Transformation. *Phytopathology* **2002**, *92*, 833–841. [CrossRef]
225. Ohara, T.; Tsuge, T. FoSTUA, encoding a basic helix-loop-helix protein, differentially regulates development of three kinds of asexual spores, macroconidia, microconidia, and chlamydospores, in the fungal plant pathogen *Fusarium oxysporum*. *Eukaryot. Cell* **2004**, *3*, 1412–1422. [CrossRef]
226. Martínez-Rocha, A.L.; Roncero, M.I.; López-Ramirez, A.; Mariné, M.; Guarro, J.; Martínez-Cadena, G.; Di Pietro, A. Rho1 has distinct functions in morphogenesis, cell wall biosynthesis and virulence of *Fusarium oxysporum*. *Cell Microbiol.* **2008**, *10*, 1339–1351. [CrossRef] [PubMed]
227. Reyes-Medina, M.A.; Macías-Sánchez, K.L. GTPase Rho1 regulates the expression of xyl3 and laccase genes in *Fusarium oxysporum*. *Biotechnol. Lett.* **2015**, *37*, 679–683. [CrossRef] [PubMed]
228. Gurdaswani, V.; Ghag, S.B.; Ganapathi, T.R. FocSge1 in *Fusarium oxysporum* f. sp. *cubense* race 1 is essential for full virulence. *BMC Microbiol.* **2020**, *20*, 255. [CrossRef] [PubMed]

229. Hou, X.; An, B.; Wang, Q.; Guo, Y.; Luo, H.; He, C. SGE1 is involved in conidiation and pathogenicity of *Fusarium oxysporum* f. sp. *cubense*. *Can. J. Microbiol.* **2018**, *64*, 349–357. [CrossRef]
230. He, D.; Feng, Z.; Gao, S.; Wei, Y.; Han, S.; Wang, L. Contribution of NADPH-cytochrome P450 Reductase to Azole Resistance in *Fusarium oxysporum*. *Front. Microbiol.* **2021**, *12*, 709942. [CrossRef]

pathogens

MDPI

Review

Parasitic Plants—Potential Vectors of Phytopathogens

Stefan Savov, Bianka Marinova, Denitsa Teofanova, Martin Savov, Mariela Odjakova and Lyuben Zagorchev *

Department of Biochemistry, Faculty of Biology, Sofia University "St. Kliment Ohridski", 8 Dragan Tsankov blvd., 1164 Sofia, Bulgaria; stefan_savovidis@abv.bg (S.S.); bianca.marinova@gmail.com (B.M.); teofanova@biofac.uni-sofia.bg (D.T.); martin_savovidis@abv.bg (M.S.); modjakova@gmail.com (M.O.)
* Correspondence: lzagorchev@biofac.uni-sofia.bg; Tel.: +359-898211635

Abstract: Parasitic plants represent a peculiar group of semi- or fully heterotrophic plants, possessing the ability to extract water, minerals, and organic compounds from other plants. All parasitic plants, either root or stem, hemi- or holoparasitic, establish a vascular connection with their host plants through a highly specialized organ called haustoria. Apart from being the organ responsible for nutrient extraction, the haustorial connection is also a highway for various macromolecules, including DNA, proteins, and, apparently, phytopathogens. At least some parasitic plants are considered significant agricultural pests, contributing to enormous yield losses worldwide. Their negative effect is mainly direct, by the exhaustion of host plant fitness and decreasing growth and seed/fruit formation. However, they may pose an additional threat to agriculture by promoting the trans-species dispersion of various pathogens. The current review aims to summarize the available information and to raise awareness of this less-explored problem. We further explore the suitability of certain phytopathogens to serve as specific and efficient methods of control of parasitic plants, as well as methods for control of the phytopathogens.

Keywords: haustoria; parasitic plants; plant virus; phytoplasma

1. Parasitism in Plants

1.1. Variety of Parasitic Plants

Parasitic flowering plants are a highly specialized group of vascular plants, which switch to a partially or fully heterotrophic lifestyle. Depending on the degree of loss of photosynthetic ability, they are commonly divided into hemiparasites (photosynthetic or partially photosynthetic) and holoparasites (non-photosynthetic), but the terms facultative and obligate are also used [1], and parasitic plants are also classified as obligate (cannot complete lifecycle without a host) and facultative. The other common classification is based on the site of vascular connection with the host plant, giving either root or stem parasitic plants. Of the nearly 5000 known species [2,3], the holoparasites account for about 10%, while the distribution between root and stem parasites is more even, 60 to 40%, respectively [1]. There are a total of 12 parasitic plant clades, corresponding to 12 independent evolutional events. However, over 90% of all parasitic plants fall into two nearly equal in species number (with over 2000 species) clades—the root hemi- and holoparasites of the family Orobanchaceae and the stem hemiparasites of the order Santalales [3]. The third in species number is the genus *Cuscuta*, family Convolvulaceae, with a little over 200 species. The other clades contain a few to several dozen species, often rare and highly specialized. Some exotic examples include the Rafflesiaceae, characterized by enormously large flowers, but a highly reduced vegetative part [4]. However, this list is far from exhaustive, and new species are described every year, mainly because root hemiparasites may remain unnoticed, or because some species are very difficult to distinguish morphologically [5].

Parasitic plants also differ significantly in their host preference. While some are highly specialized in a single, or several, host plant species, others are generalists and infect tens, or hundreds, of host plants from different families. Some notable examples of single-host specialists are *Orobanche cumana* Wallr., specifically infecting sunflower (*Helianthus annuus* L.) due to specific requirements of germination stimulants of the strigolactone group [6]. However, a switch in strigolactone responsiveness may expand the host range of *O. cumana* to other plant species [7]. The juniper dwarf mistletoe (*Arceuthobium oxycedri* (DC.) M. Bieb) is highly specific to several juniper species but is occasionally found on other tree species [8]. Therefore, it is highly unlikely that a parasitic plant is restricted to a single host species, but still, many are restricted to several closely related hosts. Well-known generalists are the members of the *Cuscuta* genus, as some were reported to have 200 (*Cuscuta europaea* L., *Cuscuta campestris* Yunck.) to nearly 350 (*Cuscuta epithymum* L.) host species in a single study [9]. At the individual level, Orobanchaceae and mistletoes tend to expand to a single or few host plant individuals; a single *Cuscuta* individual may simultaneously spread and infect multiple hosts. From a practical point of view, parasitic plants that are generalists and infect multiple hosts simultaneously would be more efficient pathogen vectors than host specialists, and such species that infect a single host (e.g., mistletoes).

Some parasitic plants are dangerous pathogens on their own. Probably the most devastating are the members of the *Striga* (witchweeds) genus (Orobanchaceae), namely *Striga hermonthica* (Delile) Benth., *Striga asiatica* (L.) Kuntze, and *Striga gesnerioides* (Willd.) Vatke. They cause nearly entire yield loss in various cultures, such as rice, sorghum, cowpea, maize, finger, and pearl millet in over 40 countries, mainly in Sub-Saharan Africa, causing over USD 10 billion in economic losses [10,11]. Broomrapes (*Orobanche* and *Phelypanche* genera) of the same family are probably second in terms of agricultural impact. At least nine species were reported as dangerous weeds in Europe, Asia, and North Africa [12]. They were reported to cause yield losses, anywhere from 10 to 100% in various legumes, carrots, tomatoes, etc. Out of over 200 *Cuscuta* species, relatively few are also considered important pests, with *C. campestris* being the top villain, causing significant losses in agriculture worldwide [13]. Some other parasitic plant species are also dangerous pathogens, but to a much lesser degree than the above-mentioned [13]. For the current overview, we focused mainly on representatives of Orobanchaceae and *Cuscuta* spp. (root and stem parasites, respectively), due to their economic impact.

The primary cause of the negative impact of parasitic plants on their hosts is the direct extraction of nutrients (both minerals and organic compounds) and water, thus decreasing the biomass and seed production of their hosts [14]. However, their impact may be much broader, including the modulation of below-ground communities [15], inhibition of host photosynthesis [16–18], etc. The most striking effect is the "bewitching" effect of *Striga* spp. on their hosts, consisting of wilting and chlorosis of the host in the very early stages of *Striga* infection [19]. This effect is not entirely understood, suggested to be caused by excessive amounts of exuded abscisic acid [20], and the disruption of other hormones [19], but is certainly far more devastating than the exhaustion of nutrients. Finally, parasitic plants may also represent a significant vector of other pathogens, such as viruses, phytoplasma, bacteria, and fungi, which they may acquire and transfer from one host to another, facilitated by the vascular connection.

1.2. Haustorium Properties

Regardless of their taxonomic position, host specificity, or evolutionary history, all parasitic plants require a vascular connection with their hosts, called the haustorium. Haustoria represent multicellular invasive organs, able to attach and penetrate host tissues, sometimes overcoming tissue incompatibilities, as many parasitic plants are able to parasitize non-related, taxonomically distant host species [21]. Haustoria of root parasites are divided into two types, lateral haustoria and terminal haustoria, which are specifically formed by the apical root meristem [22]. While lateral haustoria do not interfere with root tip elongation and allow the formation of multiple haustoria, the terminal haustoria lead to the termination of root growth [23]. Some plants can form both lateral and terminal haustoria, like it was shown in *Phelipanche ramosa* (L.) Pomel (=*Orobanche ramosa* L.) (Orobanchaceae), while others are capable of only lateral haustoria formation like *Phtheirospermum japonicum*

(Thunb.) Kanitz (Orobanchaceae) [24]. To some extent, the presence of terminal haustoria is associated with obligate parasites, while lateral haustoria are present in hemiparasites and are regarded as evolutionarily older than terminal haustoria [25,26]. In the case of obligate parasites, like *Striga*, the terminal haustorium is formed first, immediately after germination, followed by the growth of adventitious roots and lateral haustoria [27]. In an older study [28], but still in use [27], terminal haustoria are also denoted as primary, while lateral haustoria are denoted as secondary. Haustoria of stem parasites are considered distinct from root haustoria and although similar in function, differ in molecular mechanisms of formation, like in *Cuscuta* spp. [26]. Other stem parasites may possess different haustoria in terms of structure and formation [1].

Roughly three distinct stages of haustorium formation can be defined—initiation (or prehaustorium), penetration into host tissues, and vascular connection establishment [21]. Before initiation, both root and stem parasites need to locate/recognize host tissue. In root parasites, this occurs through chemical signaling—the recognition of strigolactones, released in host root exudates to attract symbiotic mycorrhizal fungi [29–31]. Going further, obligate root parasites require strigolactone detection to germinate only in the presence of a suitable host [31], and are also definitive for the host specificity of certain root parasites [6,7]. Unlike them, members of the genus Cuscuta were not proven to require chemical stimuli for germination and employ light [32,33] and probably chemical [34] stimuli in host recognition. The initiation of the root haustorium (in Orobanchaceae) requires specific chemical compounds, released by the host, collectively called haustorium-inducing factors (HIFs), which were proved to be quite diverse [35,36]. Unlike them, haustorium initiation in *Cuscuta* further relies on specific light quality in combination with tactile stimuli [37]. The stage of penetration, regardless of the parasite taxonomy, involves a combination of hydrolytic and cell-wall-modifying enzymes [21,26,37]. Finally, the vascular connection is established, which is primarily xylem–xylem (xylem bridge), observed in all parasitic plants [21]. A linkage to the phloem of the host, however, seems to be less common and is observed in stem parasites *Cuscuta* spp., but also root parasites *Orobanche* spp., while not being presented in *Striga* spp. [38].

The main function of haustoria per se is the transfer of water, mineral nutrients, and photosynthates from the host to the parasite. It is often compared to grafting [39] and raises the problem of the complicated interaction between two genomes, connected via a non-interrupted connection [40]. In this context, the haustorium was reported on numerous occasions as a bidirectional transport highway for numerous different molecules, far more complicated than a simple source of nutrients for the parasite. In *Cuscuta* spp., this involves an open xylem–xylem connection, membrane-mediated phloem transport, and plasmodesmata connections between contacting parasite and host cell walls [41]. This allows extensive transport of macromolecules, including DNA [42], mRNAs [43,44], and proteins [45]. Similar results were reported and summarized for Orobanchaceae [46,47]. Concerning phytopathogens, there is an important question about the haustorial connection—whether it is a putative passage for such passengers, which it is [41,47], or if it could serve as a selective barrier, limiting their distribution. Although there is evidence that this connection may be selective at least for some chemical entities [48], overall, it is a putative passage for different phytopathogens.

2. Phytopathogens and Parasitic Plants

2.1. Major Phytopathogens

2.1.1. Viruses

The group of plant viruses is a constantly growing group of non-cellular infectious agents, containing DNA or RNA and a protein coat, that cause multiple diseases in plants, with visual expression as leaf discoloration, stunted growth, mottling, and necrosis. The estimated economic effect of plant viral diseases reaches USD 30 billion per year in crop yield losses [49]. The list of known plant viruses exceeds thousands [50] and will further expand in the future, but not all are equally devastating and agriculturally important.

Some efforts to classify plant viruses resulted in surveys, like the "Top 10 plant viruses in molecular plant pathology" [51] and "Top 10 economically important plant viruses" [52], based on their importance as agricultural pests and molecular study objects. While some are comparatively restricted in the host range, others are characterized by an extremely wide host range and wide range of vectors—for *Cucumber Mosaic Virus* (CMV), Bromoviridae, over 1200 host plant species from over 100 families, and 80 aphid species from 33 genera, were reported [51].

Therefore, it is not surprising that viruses were reported in parasitic plants. Historically, *Cuscuta* spp. was reported to be infected and transmit over 50 different viruses [53]. More recent reports are comparatively scarce, but confirm dodders as usual hosts for a variety of plant viruses. Of the six enlisted viruses (Table 1), one is a DNA virus of the Geminiviridae family (TYLCV) and all others are positive-strand RNA viruses from four different families. Reports on viruses in Orobanchaceae are even more scarce, although some of the viruses, found in *Cuscuta* [54,55], were also proved to be acquired by *Phelipanche aegyptiaca* (Pers.) Pomel (=*Orobanche aegyptiaca* Pers.) from host plants [56]. Overall, the recent literature lacks substantial studies of virus distribution among parasitic plants, also shown by the fact that some viruses were discovered accidentally during transcriptomics analyses [57].

Table 1. Recent examples of plant viruses, detected in parasitic plants.

Species	Parasitic Plant	Reference
TYLCV, CMV	*Cuscuta campestris*	[54]
TRV	*Cuscuta* spp.	[58]
GLRaV-7	*Cuscuta* spp.	[59]
LChV-1	*Cuscuta europaea*	[60]
PVY	*Cuscuta reflexa*	[55]
SaPlV1/2	*Striga hermonthica*	[57]
CMV, ToMV, PVY, TYLCV	*Phelipanche aegyptiaca*	[56]
PSTVd	*Orobanche ramosa*	[61,62]

Abbreviations: TYLCV—Tomato Yellow Leaf Curl Virus; CMV—Cucumber Mosaic Virus; TRV—*Tobacco Rattle Virus*; GLRaV-7—Grapevine Leafroll-Associated Virus-7; LChV-1—Little Cherry Virus-1; PVY—Potato Virus Y; SaPlV—Striga-Associated Poty-Like Virus; ToMV—Tomato Mosaic Virus; PSTVd—Potato Spindle Tuber Viroid.

2.1.2. Phytoplasma

Phytoplasmas (*Candidatus Phytoplasma*: Mollicutes) are obligate intracellular bacteria that lack cell walls [63] and cause a variety of diseases [64]. They are phloem-mobile, e.g., they move within plants by the phloem traffic [65]. Common disease symptoms include little or yellow leaves, phyllody, witches' brooms, etc., and could cause between 30% and 100% yield loss, depending on the crop plant [66]. Being phloem-mobile, it is not surprising that dodders are efficient reservoirs and vectors of several phytoplasma [67,68]. Phytoplasmas were also reported from several root parasites—*Ph. ramosa* as a host for tomato stolbur disease [69] and *Orobanche* spp. as a host for tomato big bud [70] and several *Orobanche*-specific phytoplasmas [71].

2.1.3. Bacteria

Phytopathogenic bacteria are among the most devastating disease-causing pests, contributing to enormous yield losses in crop plants [72]. Unlike viruses and phytoplasma, they are not fully dependent on hosts to survive, and more often are soilborne pathogens, infecting plants through wounds, and moving through the xylem vessels. Several of the most destructible pathovars belong to the *Pseudomonas* and *Xanthomonas* genera, as well as notorious pathogens like *Ralstonia solanacearum* (Smith 1896) Yabuuchi et al., 1995, single-handedly contributing to over USD 1 billion annual losses, and *Erwinia amylovora* (Burrill 1882) Winslow et al., 1920 [73]. Important pathogens among Actinomycetes include members of the *Clavibacter* [74] and *Curtobacterium* [75]. Most of the studies involving phytopathogenic bacteria and parasitic plants are related to methods of control of the parasitic plants [76] and will be further discussed in detail hereinafter.

Interestingly, extracts from the members of the genus *Cuscuta* were shown to possess strong antibacterial activity against major phytopathogenic bacteria. For example, water extracts of *Cuscuta pedicellata* Ledeb. were proven in vitro as efficient control agents for fruit lesions, caused by *Xanthomonas campestris* (Pammel 1895) Dowson 1939 [77]. Similarly, organic solvent extracts of *Cuscuta reflexa* Roxb. were effective in the inhibition of *X. campestris* and several human pathogenic bacteria [78]. Ethanol extracts of several *Orobanche* species were also shown as efficient antibacterial agents against *Agrobacterium* and *Erwinia* [79]. *Cuscuta* spp. are particularly rich in bioactive compounds [80], as well as members of Orobanchaceae [81,82], so their in vitro antibacterial activity is not surprising, and additionally, they are regarded as highly tolerant to phytopathogenic bacteria [79].

2.1.4. Fungi

The fourth major phytopathogenic group consists of fungi. According to some estimates, they account for nearly 80% of yield losses, caused by microbial pathogens [83]. Some of the most damaging species include *Magnaporthe oryzae* (T.T. Hebert) M.E. Barr, *Botrytis cinerea* Pers., and several *Fusarium* species, among others [84]. However, phytopathogenic fungi are a comparatively small portion of many fungi that live in symbiotic relations with plants, such as arbuscular mycorrhizal fungi, which are important and even critical for healthy plants [85], and endophytic fungi, which are also important players in combating other phytopathogens [86]. The relations between plants and fungi extend to the case of mycoheterotrophic plants, which are parasitizing fungi [87]. Both dodders [88,89] and Orobanchaceae [90–92] were shown to be rich in endophytic fungal biodiversity, and also several *Fusarium* spp. isolates showed promising results in terms of pathogenicity against the parasitic plants [93,94]. Phytopathogenic fungi were also the most exploited means of biological control of parasitic plants, which will be further discussed below.

2.2. *Symptomatics*

Since now, parasitic plants have been reported to be infected by multiple phytopathogens. However, they rarely exhibit visual symptoms of such infections, which is especially true for *Cuscuta* spp. and some holoparasitic members of Orobanchaceae, due to their simple, leafless morphology [1]. In terms of viral infections, they rarely experience visible phenotypic symptoms or are even reported as symptomless in various experiments [54,58,59]. Although some very early reports claim that *Cuscuta* spp. can suppress viruses from transmission to other plants [59], there is no recent confirmation of such inactivation and dodders can maintain high viral titer for years [59]. However, there are some reports that when parasitizing virus-infected hosts, some parasitic plants may exhibit a deformed phenotype, as in the case of *Ph. aegyptiaca*, infecting CMV-infected tobacco plants [95]. In terms of phytoplasma, Marcone [68] successfully employed *Cuscuta* spp. as vectors of several phytoplasmas, but did not report any visual symptoms in the dodder vector. Unlike them, *Orobanche* spp. showed clear evidence of phytoplasma infection, such as stem flattening and witches' broom [71].

Fungal infection on *Cuscuta gronovii* Willd. was shown to cause observable symptoms, including the discoloration and shriveling of the stem, blighted portion, necrotic lesions, and tip necrosis [96]. *Cuscuta pentagona* Engelm. *Alternaria destruens* E.G. Simmons 1998 also causes necrotic spots and blights [97]. In *Orobanche* spp., different *Fusarium* isolates caused a variety of symptoms such as the inhibition of seed germination, wilting, and necrotic lesions on the stem and inflorescences [93,94]. Similar symptoms were also observed in *Striga* spp. as a result of *Fusarium* infection [98]. The seed germination of *Ph. aegyptiaca* was also found to be inhibited by *Pseudomonas* and *Bacillus* isolates [76]. Overall, symptomatics in parasitic plants seem to be much less pronounced and specific than in other plants, raising the necessity of molecular methods for the identification and screening of phytopathogens.

2.3. Detection Methods

General detection methods in plant pathology were recently summarized by Khakimov et al. [99] and could be roughly divided into traditional (e.g., visual inspection, microscopy, cultivation methods) and molecular (or modern, e.g., immunological, genetic, and mass-spectrometric). Visual observation highly relies on detectable symptoms, such as characteristic necrosis, chlorosis, etc., while microscopic methods may result in false identification, and also depend on the availability of certain pathogenic structures—for example, fungal spores or micellium [99]. Moreover, they are not always sufficient to provide the species identification of the pathogen and are somehow difficult to apply on parasitic plants, such as dodders and holoparasitic Orobancheacea, for the above-mentioned reasons. The most commonly employed modern methods for detection could be divided into immunological and molecular genetic methods. The immunological methods rely on the specific detection of an antigen of the pathogen by an antibody. Molecular methods employ PCR amplification and further sequencing of specific, taxon-discriminative DNA fragments. Phytopathogen detection in parasitic plants does not require any specific methods, differing from that established in the state-of-the-art. However, there are some peculiarities related to the scarce material of certain plant parts.

Plant virus detection is performed mainly by either commercialized ELISA kits [54] or quantitative RT-PCR with virus-specific primers, which also allow the determination of the viral titer [56,59]. Both approaches, however, allow the identification of particular viruses, selected in advance, which may underestimate the distribution of other viruses. Some more efficient methods, allowing the simultaneous detection of multiple, including unknown, viruses, as next generation sequencing (NGS)-based methods, loop-mediated isothermal amplification (LAMP), etc., were recently summarized by Mehetre et al. [100]. Being the most diverse group of phytopathogens, systematic viral detection would require a greater variety of methods. The detection of phytoplasmas commonly involves PCR amplification with phytoplasma-specific primers, of the DNA region, extending from the 5′ end of the 16S rRNA gene to the 5′ region of the 23S rRNA gene, and its amplicon may be either further analyzed by a Restriction Fragment Length Polymorphism (RFLP) analysis [68] or sequenced [101]. The typically low titers of phytoplasmas require the employment of nested PCR, or preferably quantitative real-time PCR, or LAMP.

The detection of bacterial and fungal pathogens is generally more straightforward than viruses and phytoplasmas. Most of them cause visible symptoms on parasitic plants (see above), are often visible by a common microscope, and could be cultivated on growth media in the absence of a host [102,103]. Moreover, the simultaneous NGS-based identification of thousands of bacterial and fungal taxa in a single sample, based on the sequencing of the variable regions of the 16S rRNA gene (for bacteria) and the nuclear ribosomal internal transcribed spacer (ITS) region (for fungi), is already a routine method [90].

3. Putative Routes of Transmission

3.1. Plant-to-Plant Transmission

As already discussed, the haustorial connection between parasitic plants and their hosts represents a hotspot of bi-directional macromolecular trafficking [46]. Along with interspecific plasmodesmata between the host and parasite phloem [56], it is also the major route by which parasitic plants could infect their hosts with different phytopathogens, especially viruses and phytoplasma [55,67,104]. This was not confirmed, but also might be predicted for bacteria, either actively or passively moving through plant vascular tissues [105]. The penetration nature of haustorium establishment is also important for phytopathogens, ensuring an efficient invasion route, and overcoming the plant cuticle. However, the main purpose of the haustorium is to extract molecules from the host, suggesting that parasitic plants may easily acquire phytopathogens from their hosts (Figure 1), and further distribute them to other hosts. Especially for dodders, their role in transferring a variety of signaling molecules between plants was already established [106]. More importantly, their generalism and simultaneous infection of multiple hosts from different species and families [9] also ensure the efficient spreading of pathogens between species.

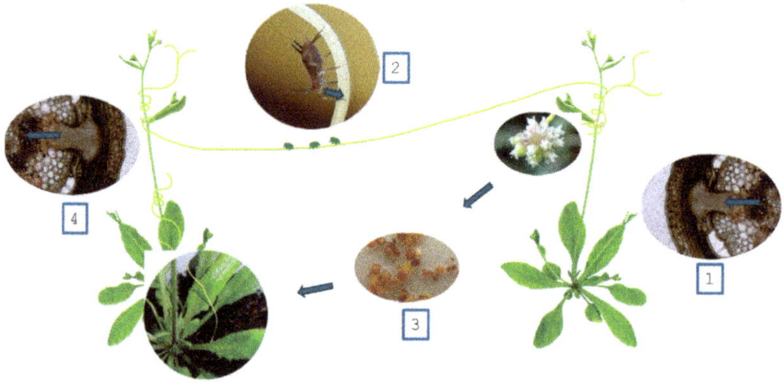

Figure 1. Putative transmission routes for phytopathogens, hypothesized for *Cuscuta* spp. as a vector. The parasitic plant acquires phytopathogens through the haustorial connection with an infected host plant (1) or from arthropods feeding on it (2). Subsequently, the phytopathogens are transmitted through the seed bank of *Cuscuta* (3) or directly (4) to a non-infected host plant.

However, the haustorial connection does not mean that all possible endophytes (incl. pathogens) would be efficiently distributed between the parasitic plant and the host. For example, the overlapping of endophytic fungi in *C. reflexa* and different hosts varied between 25% and 37% [88]. Parasitism of *C. campestris* affected the endophytic microbiome of *H. annuus*, but also with poor overlapping between the parasitic plant and the host plant [107]. Although different viruses can be acquired from the host, as in the case of *Ph. ramosa*, not all can successfully replicate in the parasite, and therefore not all can be further transmitted to other hosts [56]. There could be even specificity of virus acquisition and transmission among closely related parasitic plants—when a single virus (GLRaV-7) was detected in three different *Cuscuta* species, but only two were shown as vectors, and it was host-specific, e.g., one *Cuscuta* species may transfer the virus to one host, but not to another [59]. A similar specificity of phytoplasma transmission by *Cuscuta* spp. was shown by Marcone [68], where transmission rates differed greatly depending on both the particular phytoplasma and the dodder species, all of these suggesting that parasitic plants may be efficient, but very specific, phytopathogen vectors.

3.2. Arthropod-to-Plant Transmission

Hemipteran insects (Hemiptera: Linnaeus, 1758), topped by aphids, whiteflies, and leafhoppers, are considered the most important viral vectors, accounting for over 50% of the transmissions of known plant viruses [108]. Other arthropods such as mites, most notably eriophyid mites (Eriophyidae: Nalepa, 1898), are also known vectors of viral diseases [109,110] and nematodes are also significant vectors [111]. Similarly, hemipterans are also the most important vectors of phytoplasma [112,113]. Unlike these two groups, phytopathogenic bacteria and fungi are less commonly but not unlikely to be transmitted by arthropods [114–116].

As much as any other plant, parasitic plants are also associated with multiple arthropods that either feed or parasitize on them (Table 2). As clearly shown, both dodders and Orobanchaceae are subjected to feeding and parasitism by different groups of arthropods, potential vectors of phytopathogens. Moreover, it was also proven that molecules, such as mRNAs, could be transferred from the arthropods to the host plant [117]. Although it seems that the arthropod–parasitic plant–host plant route of phytopathogen transfer is not significant, as the insects may directly feed and infect the host plants, it must be noted that some arthropods are more or less specific to parasitic plants [118–120], or might not feed equally on the multiple hosts of the parasitic plant, or feed preferentially on the parasitic plant [121] as in the case of *Metcalfa pruinose* (Say, 1830), which fed exclusively on *C. campestris* when presented, but also attacked the host plant in the absence of the parasite. This will at least improve the chance of a phytopathogen to infect multiple plant species.

Table 2. Non-exhaustive list of arthropods, associated with different parasitic plants.

Group	Species	Parasitic Plant	Reference
Hemiptera: Aphididae	*Aphis fabae*	*Cuscuta lupuliformis*	[122]
		Cuscuta campestris	[123]
	Myzus persicae	*Cuscuta australis*	[117,124]
	Smynthurodes betae	*Phelipanche ramosa*	[125]
	Geoica utricularia	*Orobanche foetida*	[126]
Hemiptera: Flatidae	*Metcalfa pruinosa*	*Cuscuta campestris*	[121]
Hemiptera: Lygaeidae	*Oxycarenus hyalinipennis*	*Cuscuta campestris*	[123]
Diptera: Agromyzidae	*Melanagromyza cuscutae*	*Cuscuta* spp.	[118,127]
	Phytomyza orobanchia	*Orobanche* spp.	[119,128,129]
Coleoptera: Curculionidae	*Smicronyx* spp.	*Cuscuta* spp.	[120,130]
		Striga spp.	[131]

Although there is no experimental evidence of parasitic plants, serving as mediators of phytopathogens from arthropods to other plants, this potential route of transmission (Figure 1) seems plausible and requires more studies to assess the impact of both dodders and root parasites.

3.3. Seed Transmission

Most seeds of parasitic plants are characterized by long persistence in soil, where they can stay dormant for decades, waiting for the proper host to appear. This is mostly true for Orobanchaceae [132], which in combination with a large number of seeds, produced by a single plant, up to 200,000 in *Ph. ramosa* [133], gives a potentially enormous, long-term reservoir for various pathogens. *Cuscuta* spp. are also prominent seed producers, and although they do not require specific chemical compounds for germination, they are characterized by soil longevity and continuous germination over decades [134]. Such an enormous and persisting seed bank also represents a potential reservoir of numerous seed-borne pathogens from virtually every group [135], including fungi, bacteria, and viruses. Phytoplasmas are also frequently found in seeds of infected parents; the seed transmission of phytoplasmas is considered unlikely [112].

Besides some scarce reports [54], some of which are quite old [136], the potential of parasitic plant seeds to serve as a reservoir of plant viruses, transmitted to generations, is highly unexplored. The bacterial and fungal microbiota of root parasitic plants are much better studied and were shown to contain opportunistic, or obligate, phytopathogens as in *Ph. ramosa* [103,137] and *Cistanche phelypaea* (L.) Cout. [138]. Such pathogens are mostly soilborne and host plants are in contact with them anyway. However, the role of parasitic plants may be in the facilitation of pathogen penetration, simultaneously with the haustorium formation.

4. Phytopathogens as Biocontrol Agents of Parasitic Plants

Control of economically important parasitic plants is among the important aspects of contemporary agricultural practice, often complicated by their similarities to their hosts [12,13]. Being plants of their own, there is also a possibility that chemical control agents will affect their hosts equally, or even to a greater extent than the target parasites [139]. To overcome these limitations, several phytopathogens were studied and successfully applied as biocontrol agents, assuming they are highly specific to the parasite and affect neither host crop plants nor other plant species in proximity.

Both bacteria and fungi were extensively studied as potential biocontrol agents, and mostly isolated from natural sources, e.g., parasitic plants with visual symptoms of a disease. In *Cuscuta* spp., several species were found to be effective. One promising and already patented bioherbicide, specifically for dodder control, is *A. destruens* Strain 059 [97,140]. It was shown to affect a variety of dodder species, but it is not infecting other plant species, thus being a promising biocontrol agent. Some other promising results were shown for *Fusarium incarnatum* (Desm.) Sacc., *Alternaria dianthicola* Neerg., and *Curvularia pallescens* Boedijn. isolates from *C. gronovii* [96]. Of the bacterial pathogens, several *Bacillus* species were also shown to inhibit *Cuscuta* seed germination [141].

The inhibitory effect of bacterial isolates on Orobanchaceae members seemed to be much more extensive. These include the inhibition of radical elongation in *Ph. aegyptiaca* and *Orobanche cernua* Loefl. by two *Pseudomonas* and two *Bacillus* species [142], and in *Orobanche crenata* Forssk. and *Orobanche foetida* Poir. by two *Pseudomonas* species [143]. Besides their inhibitory activity on the parasitic plants, they also showed a positive effect on the growth of the host plants. The isolation of particular pathogenic strains, however, involves laborious screening of hundreds of strains. However, the most promising results for control of root parasitic plants were acquired with fungal pathogens of the *Alternaria* and *Fusarium* genera in both *Orobanche* spp. and *Striga* spp. The most exploited is probably *Fusarium oxysporum*, whose highly specific strains are efficient selective control agents for both the germination and growth of *Striga* [144–147], but also *Orobanche* spp. [148]. Other *Fusarium* species were also shown to be effective and selective pathogens on *Orobanche* [93,102]. There are hundreds of reports on the isolation and application of such fungal isolates as biocontrol agents on root parasitic plants, underlining the importance of such pathogens for food production security, especially in the poorest regions of Africa. One major advantage of such a bioherbicide is the possibility to propagate without any special equipment, so it is readily available to farmers in distant and poor regions [147]. However, most wild strains of putative phytopathogens are insufficiently virulent to be effective bioherbicides. The inhibitory effect of phytopathogens may be exerted by the direct infection of the plants, but also by the secretion of exometabolites, which specifically inhibit seed germination. Such exometabolites may be specific compounds [149], but also common metabolites such as amino acids, produced in excess [150]. The selection of an efficient hypervirulent biocontrol agent may be enhanced by the identification of such exometabolites and screening of multiple strains for the excessive secretion of such metabolites.

5. Conclusions and Future Perspectives

Although there is certain evidence that parasitic plants may serve as efficient vectors of phytopathogenic viruses, phytoplasmas, bacteria, and fungi, this aspect of their biology is highly understudied. This is because they often lack specific symptoms, but also because most of the efforts in studying parasitic plant biology were devoted to mechanisms of parasitism and methods of control, rather than phytopathogens. The available information is fragmented and does not allow a complete picture of the transmission routes, molecular mechanisms, and overall distribution of phytopathogens among the most economically damaging parasites of the *Cuscuta* genus and Orobanchaceae family. Besides the additional harm that transmitted phytopathogens add to the negative effect of parasitic plants on crop plants, they also represent a promising means of biological control, and more research on the overall endophytic diversity in both root and stem parasites is needed, to identify potential candidates for bioherbicides. Contemporary molecular methods, such as NGS-based DNA barcoding, offer the necessary tools for the high-throughput characterization of microbial diversity, associated with parasitic plants.

Author Contributions: L.Z. and S.S. conceptualized the review; B.M. prepared the figure and tables. All authors participated in the literature survey and writing of the manuscript. All authors have read and agreed to the published version of the manuscript.

Funding: This study is financed by the European Union-NextGenerationEU, through the National Recovery and Resilience Plan of the Republic of Bulgaria, project no. BG-RRP-2.004-0008 and grant KP-06-N31/10 and KP-06-COST/09 of the National Science Fund, Ministry of Education and Science, Bulgaria.

Institutional Review Board Statement: Not applicable.

Informed Consent Statement: Not applicable.

Data Availability Statement: Not applicable.

Conflicts of Interest: The authors declare no conflicts of interest.

References

1. Heide-Jørgensen, H. *Parasitic Flowering Plants;* Brill: Leiden, The Netherlands, 2008.
2. Nickrent, D.L. Parasitic angiosperms: How often and how many? *Taxon* **2020**, *69*, 5–27. [CrossRef]
3. Teixeira-Costa, L.; Davis, C.C. Life history, diversity, and distribution in parasitic flowering plants. *Plant Physiol.* **2021**, *187*, 32–51. [CrossRef]
4. Wicaksono, A.; Mursidawati, S.; Sukamto, L.A.; Teixeira da Silva, J.A. *Rafflesia* spp.: Propagation and conservation. *Planta* **2016**, *244*, 289–296. [CrossRef]
5. Park, I.; Song, J.-H.; Yang, S.; Kim, W.J.; Choi, G.; Moon, B.C. *Cuscuta* species identification based on the morphology of reproductive organs and complete chloroplast genome sequences. *Int. J. Mol. Sci.* **2019**, *20*, 2726. [CrossRef]
6. Fernández-Aparicio, M.; Yoneyama, K.; Rubiales, D. The role of strigolactones in host specificity of *Orobanche* and *Phelipanche* seed germination. *Seed Sci. Res.* **2011**, *21*, 55–61. [CrossRef]
7. Dor, E.; Plakhine, D.; Joel, D.M.; Larose, H.; Westwood, J.H.; Smirnov, E.; Ziadna, H.; Hershenhorn, J. A new race of sunflower broomrape (*Orobanche cumana*) with a wider host range due to changes in seed response to strigolactones. *Weed Sci.* **2020**, *68*, 134–142. [CrossRef]
8. Krasylenko, Y.A.; Janošíková, K.; Kukushkin, O.V. Juniper dwarf mistletoe (*Arceuthobium oxycedri*) in the Crimean Peninsula: Novel insights into its morphology, hosts, and distribution. *Botany* **2017**, *95*, 897–911. [CrossRef]
9. Barath, K.; Csiky, J. Host range and host choice of *Cuscuta* species in Hungary. *Acta Bot. Croat.* **2012**, *71*, 215–227. [CrossRef]
10. David, O.G.; Ayangbenro, A.S.; Odhiambo, J.J.; Babalola, O.O. *Striga hermonthica*: A highly destructive pathogen in maize production. *Environ. Chall.* **2022**, *8*, 100590. [CrossRef]
11. Dafaallah, A.B. Biology and physiology of witchweed (*Striga* spp.): A review. *Int. J. Acad. Multidiscip.* **2019**, *3*, 42–51.
12. Parker, C. Observations on the current status of *Orobanche* and *Striga* problems worldwide. *Pest Manag. Sci. Former. Pestic. Sci.* **2009**, *65*, 453–459. [CrossRef]
13. Parker, C. Parasitic weeds: A world challenge. *Weed Sci.* **2012**, *60*, 269–276. [CrossRef]
14. Jeschke, W.D.; Bäumel, P.; Räth, N.; Czygan, F.-C.; Proksch, P. Modelling of the flows and partitioning of carbon and nitrogen in the holoparasite *Cuscuta reflexa* Roxb. and its host *Lupinus albus* L. II. Flows between host and parasite and within the parasitized host. *J. Exp. Bot.* **1994**, *45*, 801–812. [CrossRef]
15. Bardgett, R.D.; Smith, R.S.; Shiel, R.S.; Peacock, S.; Simkin, J.M.; Quirk, H.; Hobbs, P.J. Parasitic plants indirectly regulate below-ground properties in grassland ecosystems. *Nature* **2006**, *439*, 969–972. [CrossRef]
16. Cameron, D.D.; Geniez, J.-M.; Seel, W.E.; Irving, L.J. Suppression of host photosynthesis by the parasitic plant *Rhinanthus minor*. *Ann. Bot.* **2008**, *101*, 573–578. [CrossRef]
17. Gurney, A.L.; Press, M.C.; Ransom, J.K. The parasitic angiosperm *Striga hermonthica* can reduce photosynthesis of its sorghum and maize hosts in the field. *J. Exp. Bot.* **1995**, *46*, 1817–1823. [CrossRef]
18. Shen, H.; Hong, L.; Ye, W.; Cao, H.; Wang, Z. The influence of the holoparasitic plant *Cuscuta campestris* on the growth and photosynthesis of its host *Mikania micrantha*. *J. Exp. Bot.* **2007**, *58*, 2929–2937. [CrossRef]
19. Runo, S.; Kuria, E.K. Habits of a highly successful cereal killer, *Striga*. *PLoS Pathog.* **2018**, *14*, e1006731. [CrossRef]
20. Fujioka, H.; Samejima, H.; Mizutani, M.; Okamoto, M.; Sugimoto, Y. How does *Striga hermonthica* Bewitch its hosts? *Plant Signal. Behav.* **2019**, *14*, 1605810. [CrossRef]
21. Kokla, A.; Melnyk, C.W. Developing a thief: Haustoria formation in parasitic plants. *Dev. Biol.* **2018**, *442*, 53–59. [CrossRef]
22. Yoshida, S.; Cui, S.; Ichihashi, Y.; Shirasu, K. The haustorium, a specialized invasive organ in parasitic plants. *Annu. Rev. Plant Biol.* **2016**, *67*, 643–667. [CrossRef] [PubMed]
23. Ishida, J.K.; Wakatake, T.; Yoshida, S.; Takebayashi, Y.; Kasahara, H.; Wafula, E.; de Pamphilis, C.W.; Namba, S.; Shirasu, K. Local Auxin Biosynthesis Mediated by a YUCCA Flavin Monooxygenase Regulates Haustorium Development in the Parasitic Plant *Phtheirospermum japonicum*. *Plant Cell* **2016**, *28*, 1795–1814. [CrossRef]

24. Cui, S.; Wakatake, T.; Hashimoto, K.; Saucet, S.B.; Toyooka, K.; Yoshida, S.; Shirasu, K. Haustorial hairs are specialized root hairs that support parasitism in the facultative parasitic plant *Phtheirospermum japonicum*. *Plant Physiol.* **2016**, *170*, 1492–1503. [CrossRef]
25. Westwood, J.H.; Yoder, J.I.; Timko, M.P.; Depamphilis, C.W. The evolution of parasitism in plants. *Trends Plant Sci.* **2010**, *15*, 227–235. [CrossRef] [PubMed]
26. Kirschner, G.K.; Xiao, T.T.; Jamil, M.; Al-Babili, S.; Lube, V.; Blilou, I. A roadmap of haustorium morphogenesis in parasitic plants. *J. Exp. Bot.* **2023**, *74*, 7034–7044. [CrossRef]
27. Xiao, T.T.; Kirschner, G.K.; Kountche, B.A.; Jamil, M.; Savina, M.; Lube, V.; Mironova, V.; Al Babili, S.; Blilou, I. A PLETHORA/PIN-FORMED/auxin network mediates prehaustorium formation in the parasitic plant *Striga hermonthica*. *Plant Physiol.* **2022**, *189*, 2281–2297. [CrossRef]
28. Weber, H.C. Evolution of the secondary haustoria to a primary haustorium in the parasitic *Scrophulariaceae/Orobanchaceae*. *Plant Syst. Evol.* **1987**, *156*, 127–131. [CrossRef]
29. Tsuchiya, Y.; Yoshimura, M.; Sato, Y.; Kuwata, K.; Toh, S.; Holbrook-Smith, D.; Zhang, H.; McCourt, P.; Itami, K.; Kinoshita, T. Probing strigolactone receptors in *Striga hermonthica* with fluorescence. *Science* **2015**, *349*, 864–868. [CrossRef] [PubMed]
30. Toh, S.; Holbrook-Smith, D.; Stogios, P.J.; Onopriyenko, O.; Lumba, S.; Tsuchiya, Y.; Savchenko, A.; McCourt, P. Structure-function analysis identifies highly sensitive strigolactone receptors in *Striga*. *Science* **2015**, *350*, 203–207. [CrossRef]
31. Yoneyama, K.; Awad, A.A.; Xie, X.; Yoneyama, K.; Takeuchi, Y. Strigolactones as germination stimulants for root parasitic plants. *Plant Cell Physiol.* **2010**, *51*, 1095–1103. [CrossRef]
32. Furuhashi, K.; Iwase, K.; Furuhashi, T. Role of Light and Plant Hormones in Stem Parasitic Plant (*Cuscuta* and *Cassytha*) Twining and Haustoria Induction. *Photochem. Photobiol.* **2021**, *97*, 1054–1062. [CrossRef] [PubMed]
33. Johnson, B.I.; De Moraes, C.M.; Mescher, M.C. Manipulation of light spectral quality disrupts host location and attachment by parasitic plants in the genus *Cuscuta*. *J. Appl. Ecol.* **2016**, *53*, 794–803. [CrossRef]
34. Runyon, J.B.; Mescher, M.C.; De Moraes, C.M. Volatile chemical cues guide host location and host selection by parasitic plants. *Science* **2006**, *313*, 1964–1967. [CrossRef] [PubMed]
35. Goyet, V.; Wada, S.; Cui, S.; Wakatake, T.; Shirasu, K.; Montiel, G.; Simier, P.; Yoshida, S. Haustorium inducing factors for parasitic Orobanchaceae. *Front. Plant Sci.* **2019**, *10*, 1056. [CrossRef]
36. Fernández-Aparicio, M.; Masi, M.; Cimmino, A.; Evidente, A. Effects of benzoquinones on radicles of *Orobanche* and *Phelipanche* species. *Plants* **2021**, *10*, 746. [CrossRef] [PubMed]
37. Jhu, M.-Y.; Sinha, N.R. *Cuscuta* species: Model organisms for haustorium development in stem holoparasitic plants. *Front. Plant Sci.* **2022**, *13*, 1086384. [CrossRef]
38. Teixeira-Costa, L. A living bridge between two enemies: Haustorium structure and evolution across parasitic flowering plants. *Braz. J. Bot.* **2021**, *44*, 165–178. [CrossRef]
39. Okayasu, K.; Aoki, K.; Kurotani, K.-I.; Notaguchi, M. Tissue adhesion between distant plant species in parasitism and grafting. *Commun. Integr. Biol.* **2021**, *14*, 21–23. [CrossRef]
40. Gaut, B.S.; Miller, A.J.; Seymour, D.K. Living with two genomes: Grafting and its implications for plant genome-to-genome interactions, phenotypic variation, and evolution. *Annu. Rev. Genet.* **2019**, *53*, 195–215. [CrossRef]
41. Kim, G.; Westwood, J.H. Macromolecule exchange in *Cuscuta*–host plant interactions. *Curr. Opin. Plant Biol.* **2015**, *26*, 20–25. [CrossRef]
42. Mower, J.P.; Stefanović, S.; Hao, W.; Gummow, J.S.; Jain, K.; Ahmed, D.; Palmer, J.D. Horizontal acquisition of multiple mitochondrial genes from a parasitic plant followed by gene conversion with host mitochondrial genes. *BMC Biol.* **2010**, *8*, 150. [CrossRef] [PubMed]
43. Park, S.-Y.; Shimizu, K.; Brown, J.; Aoki, K.; Westwood, J.H. Mobile host mRNAs are translated to protein in the associated parasitic plant *Cuscuta campestris*. *Plants* **2021**, *11*, 93. [CrossRef] [PubMed]
44. Westwood, J.H.; Kim, G. RNA mobility in parasitic plant–host interactions. *RNA Biol.* **2017**, *14*, 450–455. [CrossRef] [PubMed]
45. Haupt, S.; Oparka, K.J.; Sauer, N.; Neumann, S. Macromolecular trafficking between *Nicotiana tabacum* and the holoparasite *Cuscuta reflexa*. *J. Exp. Bot.* **2001**, *52*, 173–177. [CrossRef] [PubMed]
46. Aly, R. Trafficking of molecules between parasitic plants and their hosts. *Weed Res.* **2013**, *53*, 231–241. [CrossRef]
47. LeBlanc, M.; Kim, G.; Westwood, J.H. RNA trafficking in parasitic plant systems. *Front. Plant Sci.* **2012**, *3*, 203. [CrossRef] [PubMed]
48. Förste, F.; Mantouvalou, I.; Kanngießer, B.; Stosnach, H.; Lachner, L.A.M.; Fischer, K.; Krause, K. Selective mineral transport barriers at *Cuscuta*-host infection sites. *Physiol. Plant.* **2020**, *168*, 934–947. [CrossRef] [PubMed]
49. Sastry, K.S.; Zitter, T.A. Management of virus and viroid diseases of crops in the tropics. In *Plant Virus and Viroid Diseases in the Tropics: Volume 2: Epidemiology and Management*; Springer: Dordrecht, The Netherlands, 2014; pp. 149–480.
50. Bhat, A.I.; Rao, G.P. Isolation and Diagnosis of Virus Through Indicator Hosts. In *Characterization of Plant Viruses*; Springer Protocols Handbooks; Humana: New York, NY, USA, 2020; pp. 23–27.
51. Scholthof, K.B.G.; Adkins, S.; Czosnek, H.; Palukaitis, P.; Jacquot, E.; Hohn, T.; Hohn, B.; Saunders, K.; Candresse, T.; Ahlquist, P. Top 10 plant viruses in molecular plant pathology. *Mol. Plant Pathol.* **2011**, *12*, 938–954. [CrossRef]
52. Rybicki, E.P. A Top Ten list for economically important plant viruses. *Arch. Virol.* **2015**, *160*, 17–20. [CrossRef]
53. Hosford, R.M. Transmission of plant viruses by dodder. *Bot. Rev.* **1967**, *33*, 387–406. [CrossRef]

54. Teofanova, D.; Lozanova, Y.; Lambovska, K.; Pachedjieva, K.; Tosheva, A.; Odjakova, M.; Zagorchev, L. *Cuscuta* spp. populations as potential reservoirs and vectors of four plant viruses. *Phytoparasitica* **2022**, *50*, 555–566. [CrossRef]
55. Birschwilks, M.; Haupt, S.; Hofius, D.; Neumann, S. Transfer of phloem-mobile substances from the host plants to the holoparasite *Cuscuta* sp. *J. Exp. Bot.* **2006**, *57*, 911–921. [CrossRef] [PubMed]
56. Gal-On, A.; Naglis, A.; Leibman, D.; Ziadna, H.; Kathiravan, K.; Papayiannis, L.; Holdengreber, V.; Guenoune-Gelbert, D.; Lapidot, M.; Aly, R. Broomrape can acquire viruses from its hosts. *Phytopathology* **2009**, *99*, 1321–1329. [CrossRef] [PubMed]
57. Choi, D.; Shin, C.; Shirasu, K.; Hahn, Y. Two novel poty-like viruses identified from the transcriptome data of purple witchweed (*Striga hermonthica*). *Acta Virol.* **2021**, *65*, 365. [CrossRef] [PubMed]
58. Dikova, B. Establishment of Tobacco rattle virus (TRV) in weeds and *Cuscuta*. *Biotechnol. Biotechnol. Equip.* **2006**, *20*, 42–48. [CrossRef]
59. Mikona, C.; Jelkmann, W. Replication of Grapevine leafroll-associated virus-7 (GLRaV-7) by *Cuscuta* species and its transmission to herbaceous plants. *Plant Dis.* **2010**, *94*, 471–476. [CrossRef] [PubMed]
60. Jelkmann, W.; Hergenhahn, F.; Berwarth, C. Transmission of Little cherry virus-1 (LChV-1) by *Cuscuta europea* to herbaceous host plants. In Proceedings of the 21st International Conference on Virus and other Graft Transmissible Diseases of Fruit Crops, Neustadt, Germany, 5–10 July 2009.
61. Vachev, T.; Ivanova, D.; Minkov, I.; Tsagris, M.; Gozmanova, M. Trafficking of the Potato spindle tuber viroid between tomato and *Orobanche ramosa*. *Virology* **2010**, *399*, 187–193. [CrossRef] [PubMed]
62. Ivanova, D.; Vachev, T.; Baev, V.; Minkov, I.; Gozmanova, M. Identification of Potato Spindle Tuber Viroid Small RNA in *Orobanche ramosa* by Microarray. *Biotechnol. Biotechnol. Equip.* **2010**, *24*, 144–146. [CrossRef]
63. Lee, I.-M.; Davis, R.E.; Gundersen-Rindal, D.E. Phytoplasma: Phytopathogenic mollicutes. *Annu. Rev. Microbiol.* **2000**, *54*, 221–255. [CrossRef]
64. Bertaccini, A.; Duduk, B. Phytoplasma and phytoplasma diseases: A review of recent research. *Phytopathol. Mediterr.* **2009**, *48*, 355–378.
65. IRPCM Phytoplasma/Spiroplasma Working Team–Phytoplasma Taxonomy Group. '*Candidatus* Phytoplasma', a taxon for the wall-less, non-helical prokaryotes that colonize plant phloem and insects. *Int. J. Syst. Evol. Microbiol.* **2004**, *54*, 1243–1255. [CrossRef] [PubMed]
66. Kumari, S.; Nagendran, K.; Rai, A.B.; Singh, B.; Rao, G.P.; Bertaccini, A. Global status of phytoplasma diseases in vegetable crops. *Front. Microbiol.* **2019**, *10*, 441347. [CrossRef] [PubMed]
67. Přibylová, J.; Špak, J. Dodder transmission of phytoplasmas. *Phytoplasma Methods Protoc.* **2013**, *938*, 41–46.
68. Marcone, C.; Hergenhahn, F.; Ragozzino, A.; Seemüller, E. Dodder transmission of pear decline, European stone fruit yellows, rubus stunt, picris echioides yellows and cotton phyllody phytoplasmas to periwinkle. *J. Phytopathol.* **1999**, *147*, 187–192. [CrossRef]
69. Āzdemir, N.; Saygili, H.; Sahin, F.; Karsavuran, Y.; Bayrak, O.; Oral, B. Host range and genetic characterization of a phytoplasma causing tomato stolbur disease in Turkey. *Acta Hortic.* **2007**, *808*, 255–262. [CrossRef]
70. Shokri, M.; Jafary, H.; Moghadam, M.A. Molecular detection and survey of tomato big bud phytoplasma in Zanjan province. *Genet. Eng. Biosaf. J.* **2023**, *12*, 123–130.
71. Sarab, R.T.; Bakhsh, M.S.; Motlagh, M.A. First report of a phytoplasma associated with *Orobanche* spp. in Iran. In Proceedings of the 22nd Iranian Plant Protection Congress, Karaj, Iran, 27–30 August 2016.
72. Borkar, S.G.; Yumlembam, R.A. *Bacterial Diseases of Crop Plants*; CRC Press: Boca Raton, FL, USA, 2016.
73. Mansfield, J.; Genin, S.; Magori, S.; Citovsky, V.; Sriariyanum, M.; Ronald, P.; Dow, M.; Verdier, V.; Beer, S.V.; Machado, M.A. Top 10 plant pathogenic bacteria in molecular plant pathology. *Mol. Plant Pathol.* **2012**, *13*, 614–629. [CrossRef] [PubMed]
74. Eichenlaub, R.; Gartemann, K.-H.; Burger, A. Clavibacter michiganensis, a group of Gram-positive phytopathogenic bacteria. In *Plant-Associated Bacteria*; Springer: Dordrecht, The Netherlands, 2006; pp. 385–421.
75. Evseev, P.; Lukianova, A.; Tarakanov, R.; Tokmakova, A.; Shneider, M.; Ignatov, A.; Miroshnikov, K. *Curtobacterium* spp. and *Curtobacterium flaccumfaciens*: Phylogeny, genomics-based taxonomy, pathogenicity, and diagnostics. *Curr. Issues Mol. Biol.* **2022**, *44*, 889–927. [CrossRef] [PubMed]
76. Iasur Kruh, L.; Lahav, T.; Abu-Nassar, J.; Achdari, G.; Salami, R.; Freilich, S.; Aly, R. Host-parasite-bacteria triangle: The microbiome of the parasitic weed *Phelipanche aegyptiaca* and tomato-*Solanum lycopersicum* (Mill.) as a host. *Front. Plant Sci.* **2017**, *8*, 269. [CrossRef]
77. Ali, A.; Haider, M.S.; Muhammad, A. Biological control of fruit lesions caused by *Xanthomonas campestris* pathovars from *Cuscuta pedicellata* Ledeb. in vitro. *J. Pure Appl. Microbiol.* **2013**, *7*, 3149–3153.
78. Islam, R.; Rahman, M.S.; Rahman, S.M. GC-MS analysis and antibacterial activity of *Cuscuta reflexa* against bacterial pathogens. *Asian Pac. J. Trop. Dis.* **2015**, *5*, 399–403. [CrossRef]
79. Saadoun, I.M.; Hameed, K.M.; Al-Momani, F.; Ababneh, Q. Effect of three *Orobanche* spp. extracts on some local phytopathogens, Agrobacterium and Erwinia. *Turk. J. Biol.* **2008**, *32*, 113–117.
80. Ahmad, A.; Tandon, S.; Xuan, T.D.; Nooreen, Z. A Review on Phytoconstituents and Biological activities of *Cuscuta* species. *Biomed. Pharmacother.* **2017**, *92*, 772–795. [CrossRef] [PubMed]
81. Genovese, C.; D'Angeli, F.; Attanasio, F.; Caserta, G.; Scarpaci, K.S.; Nicolosi, D. Phytochemical composition and biological activities of *Orobanche crenata* Forssk.: A review. *Nat. Prod. Res.* **2021**, *35*, 4579–4595. [CrossRef] [PubMed]

82. Augustine, R.; Maikasuwa, B.; Etonihu, A. Phytochemical Screening, Thin Layer Chromatography and Antibacterial Activity of the Leaf Extracts of *Striga hermonthica*. *Chem. Res. J.* **2020**, *5*, 192.
83. Moore, D.; Robson, G.D.; Trinci, A.P. *21st Century Guidebook to Fungi*; Cambridge University Press: Cambridge, UK, 2020.
84. Dean, R.; Van Kan, J.A.; Pretorius, Z.A.; Hammond-Kosack, K.E.; Di Pietro, A.; Spanu, P.D.; Rudd, J.J.; Dickman, M.; Kahmann, R.; Ellis, J. The Top 10 fungal pathogens in molecular plant pathology. *Mol. Plant Pathol.* **2012**, *13*, 414–430. [CrossRef] [PubMed]
85. Gosling, P.; Hodge, A.; Goodlass, G.; Bending, G. Arbuscular mycorrhizal fungi and organic farming. *Agric. Ecosyst. Environ.* **2006**, *113*, 17–35. [CrossRef]
86. Akram, S.; Ahmed, A.; He, P.; He, P.; Liu, Y.; Wu, Y.; Munir, S.; He, Y. Uniting the role of endophytic fungi against plant pathogens and their interaction. *J. Fungi* **2023**, *9*, 72. [CrossRef]
87. Leake, J.R. The biology of myco-heterotrophic ('saprophytic') plants. *New Phytol.* **1994**, *127*, 171–216. [CrossRef]
88. Suryanarayanan, T.; Senthilarasu, G.; Muruganandam, V. Endophytic fungi from *Cuscuta reflexa* and its host plants. *Fungal Divers.* **2000**, *4*, 117–123.
89. Khare, E.; Vishwakarma, A.; Maurya, V.; Kaistha, S.D. Endophytic fungi from parasitic-plant *Cuscuta*, and their potential for producing L-asparaginase of pharmaceutical significance. *Environ. Sustain.* **2024**, *7*, 93–101. [CrossRef]
90. Ruraż, K.; Przemieniecki, S.W.; Piwowarczyk, R. Interspecies and temporal dynamics of bacterial and fungal microbiomes of pistil stigmas in flowers in holoparasitic plants of the *Orobanche* series *Alsaticae* (Orobanchaceae). *Sci. Rep.* **2023**, *13*, 6749. [CrossRef] [PubMed]
91. Gafar, N.Y.; Hassan, M.M.; Ahmed, M.M.; Osman, A.G.; Abdelgani, M.E.; Babiker, A.E. In vitro study of endophytic bacteria, carbohydrates and their combination on early developmental stages of *Striga hermonthica* (Del.) Benth. *Adv. Environ. Biol.* **2016**, *10*, 66–74.
92. Lombard, L.; van Doorn, R.; Groenewald, J.; Tessema, T.; Kuramae, E.; Etolo, D.; Raaijmakers, J.; Crous, P. *Fusarium* diversity associated with the Sorghum-*Striga* interaction in Ethiopia. *Fungal Syst. Evol.* **2022**, *10*, 177–215. [CrossRef] [PubMed]
93. Dor, E.; Hershenhorn, J.; Andolfi, A.; Cimmino, A.; Evidente, A. *Fusarium* verticillioides as a new pathogen of the parasitic weed *Orobanche* spp. *Phytoparasitica* **2009**, *37*, 361–370. [CrossRef]
94. Gibot-Leclerc, S.; Guinchard, L.; Edel-Hermann, V.; Dessaint, F.; Cartry, D.; Reibel, C.; Gautheron, N.; Bernaud, E.; Steinberg, C. Screening for potential mycoherbicides within the endophyte community of *Phelipanche ramosa* parasitizing tobacco. *FEMS Microbiol. Ecol.* **2022**, *98*, fiac024. [CrossRef] [PubMed]
95. Ibdah, M.; Dubey, N.K.; Eizenberg, H.; Dabour, Z.; Abu-Nassar, J.; Gal-On, A.; Aly, R. Cucumber Mosaic Virus as a carotenoid inhibitor reducing *Phelipanche aegyptiaca* infection in tobacco plants. *Plant Signal. Behav.* **2014**, *9*, e972146. [CrossRef]
96. Bangar, V.; Joshi, M.; Valvi, H. Isolation and pathogenicity of pathogen causing diseases on dodder (*Cuscuta gronovii*). *IJCS* **2019**, *7*, 1769–1773.
97. Cook, J.C.; Charudattan, R.; Zimmerman, T.W.; Rosskopf, E.N.; Stall, W.M.; MacDonald, G.E. Effects of *Alternaria destruens*, glyphosate, and ammonium sulfate individually and integrated for control of dodder (*Cuscuta pentagona*). *Weed Technol.* **2009**, *23*, 550–555. [CrossRef]
98. Idris, A.E.; Abouzeid, M.A.; Boari, A.; Vurro, M.; Evidente, A. Identification of phytotoxic metabolites of a new *Fusarium* sp. inhibiting germination of *Striga hermonthica* seeds. *Phytopathol. Mediterr.* **2003**, *42*, 65–70.
99. Khakimov, A.; Salakhutdinov, I.; Omolikov, A.; Utaganov, S. Traditional and current-prospective methods of agricultural plant diseases detection: A review. *IOP Conf. Ser. Earth Environ. Sci.* **2022**, *951*, 012002. [CrossRef]
100. Mehetre, G.T.; Leo, V.V.; Singh, G.; Sorokan, A.; Maksimov, I.; Yadav, M.K.; Upadhyaya, K.; Hashem, A.; Alsaleh, A.N.; Dawoud, T.M. Current Developments and Challenges in Plant Viral Diagnostics: A Systematic Review. *Viruses* **2021**, *13*, 412. [CrossRef] [PubMed]
101. Nair, S.; Manimekalai, R. Phytoplasma diseases of plants: Molecular diagnostics and way forward. *World J. Microbiol. Biotechnol.* **2021**, *37*, 102. [CrossRef] [PubMed]
102. Boari, A.; Vurro, M. Evaluation of *Fusarium* spp. and other fungi as biological control agents of broomrape (*Orobanche ramosa*). *Biol. Control* **2004**, *30*, 212–219. [CrossRef]
103. Durlik, K.; Żarnowiec, P.; Piwowarczyk, R.; Kaca, W. Culturable endophytic bacteria from *Phelipanche ramosa* (Orobanchaceae) seeds. *Seed Sci. Res.* **2021**, *31*, 69–75. [CrossRef]
104. Kamińska, M.; Korbin, M. Graft and dodder transmission of phytoplasma affecting lily to experimental hosts. *Acta Physiol. Plant.* **1999**, *21*, 21–26. [CrossRef]
105. van der Wolf, J.; De Boer, S.H. Phytopathogenic bacteria. In *Principles of Plant-Microbe Interactions: Microbes for Sustainable Agriculture*; Springer: Cham, Switzerland, 2014; pp. 65–77.
106. Hettenhausen, C.; Li, J.; Zhuang, H.; Sun, H.; Xu, Y.; Qi, J.; Zhang, J.; Lei, Y.; Qin, Y.; Sun, G. Stem parasitic plant *Cuscuta australis* (dodder) transfers herbivory-induced signals among plants. *Proc. Natl. Acad. Sci. USA* **2017**, *114*, E6703–E6709. [CrossRef]
107. Avila, A.T.; Van Laar, T.A.; Constable, J.V.; Waselkov, K. 16S rRNA Gene Diversity of Bacterial Endophytes in Parasitic *Cuscuta campestris* and Its *Helianthus annuus* Host. *Microbiol. Resour. Announc.* **2020**, *9*, e00968-20. [CrossRef] [PubMed]
108. Hogenhout, S.A.; Ammar, E.-D.; Whitfield, A.E.; Redinbaugh, M.G. Insect vector interactions with persistently transmitted viruses. *Annu. Rev. Phytopathol.* **2008**, *46*, 327–359. [CrossRef]
109. Sarwar, M. Mite (Acari Acarina) vectors involved in transmission of plant viruses. In *Applied Plant Virology*; Elsevier: Amsterdam, The Netherlands, 2020; pp. 257–273.

110. de Lillo, E.; Freitas-Astúa, J.; Watanabe Kitajima, E.; Ramos-González, P.L.; Simoni, S.; Tassi, A.D.; Valenzano, D. Phytophagous mites transmitting plant viruses: Update and perspectives. *Entomol. Gen.* **2021**, *41*, 439. [CrossRef]
111. Singh, S.; Awasthi, L.; Jangre, A.; Nirmalkar, V.K. Transmission of plant viruses through soil-inhabiting nematode vectors. In *Applied Plant Virology*; Elsevier: Amsterdam, The Netherlands, 2020; pp. 291–300.
112. Weintraub, P.G.; Beanland, L. Insect vectors of phytoplasmas. *Annu. Rev. Entomol.* **2006**, *51*, 91–111. [CrossRef] [PubMed]
113. Alma, A.; Lessio, F.; Nickel, H. Insects as phytoplasma vectors: Ecological and epidemiological aspects. In *Phytoplasmas: Plant Pathogenic Bacteria-II: Transmission and Management of Phytoplasma-Associated Diseases*; Springer: Singapore, 2019; pp. 1–25.
114. Bini, A.P.; Serra, M.C.D.; Pastore, I.F.; Brondi, C.V.; Camargo, L.E.A.; Monteiro-Vitorello, C.B.; van Sluys, M.-A.; Rossi, G.D.; Creste, S. Transmission of *Xanthomonas albilineans* by the spittlebug, *Mahanarva fimbriolata* (Hemiptera: Cercopidae), in Brazil: First report of an insect vector for the causal agent of sugarcane leaf scald. *J. Insect Sci.* **2023**, *23*, 28. [CrossRef] [PubMed]
115. Kisera Kwach, J.; Nyakomitta, P.S. Diversity of Insects Associated with Banana in Banana Xanthomonas Wilt Epidemic Areas of Western Kenya. *Asian J. Agric. Hortic. Res.* **2022**, *9*, 1–12. [CrossRef]
116. Esquivel, J.F.; Bell, A.A. Acquisition and Transmission of *Fusarium oxysporum* f. sp. *vasinfectum* VCG 0114 (Race 4) by Stink Bugs. *Plant Dis.* **2021**, *105*, 3082–3086. [PubMed]
117. Song, J.; Bian, J.; Xue, N.; Xu, Y.; Wu, J. Inter-species mRNA transfer among green peach aphids, dodder parasites, and cucumber host plants. *Plant Divers.* **2022**, *44*, 1–10. [CrossRef] [PubMed]
118. Piwowarczyk, R.; Mielczarek, Ł.; Panek-Wójcicka, M.; Ruraż, K. First report of *Melanagromyza cuscutae* (Diptera: Agromyzidae) from Poland. *Fla. Entomol.* **2020**, *103*, 124–126. [CrossRef]
119. Bayram, Y.; Cikman, E. Efficiency of *Pytomyza orobanchia* Kaltenbach (Diptera: Agromyzidae) on *Orobanche crenata* Forsk. (Orobanchaceae) in Lentil Fields at Diyarbakır and Mardin Provinces, Turkey. *Egypt. J. Biol. Pest Control* **2016**, *26*, 365–371.
120. Aistova, E.; Bezborodov, V. Weevils belonging to the Genus *Smicronyx* Schönherr, 1843 (Coleoptera, Curculionidae) affecting dodders (*Cuscuta* Linnaeus, 1753) in the Russian Far East. *Russ. J. Biol. Invasions* **2017**, *8*, 184–188. [CrossRef]
121. Zagorchev, L.; Albanova, I.; Lozanova, Y.; Odjakova, M.; Teofanova, D. Chitinase Profile of Arabidopsis, Subjected to the Combined Stress of Soil Salinity, *Cuscuta campestris* Parasitism and Herbivores. *Proc. Bulg. Acad. Sci.* **2022**, *75*, 835–844. [CrossRef]
122. Cagáň, L.; Komáromyová, E.; Barta, M.; Tóth, P. *Cuscuta lupuliformis* Krocker (Cuscutaceae)—A new host for bean aphid, *Aphis fabae* Scopoli (Homoptera, Aphididae). *Acta Fytotech. Zootech.* **2001**, *4*, 84.
123. Azami-Sardooei, Z.; Shahreyarinejad, S.; Rouzkhosh, M.; Fekrat, F. The first report on feeding of *Oxycarenus hyalinipennis* and *Aphis fabae* on dodder *Cuscuta campestris* in Iran. *J. Crop Prot.* **2018**, *7*, 121–124.
124. Zhuang, H.; Li, J.; Song, J.; Hettenhausen, C.; Schuman, M.C.; Sun, G.; Zhang, C.; Li, J.; Song, D.; Wu, J. Aphid (*Myzus persicae*) feeding on the parasitic plant dodder (*Cuscuta australis*) activates defense responses in both the parasite and soybean host. *New Phytol.* **2018**, *218*, 1586–1596. [CrossRef]
125. Piwowarczyk, R.; Guzikowski, S.; Depa, Ł.; Kaszyca, N. First report of *Smynthurodes betae* (Hemiptera: Aphididae) on *Phelipanche ramosa* (Orobanchaceae). *Fla. Entomol.* **2018**, *101*, 339–341. [CrossRef]
126. Boukhris-Bouhachem, S.; Youssef, S.B.; Kharrat, M. First report of *Geoica utricularia* (Hemiptera: Aphididae) population on parasitic broomrape *Orobanche foetida*. *Fla. Entomol.* **2011**, *94*, 343–344. [CrossRef]
127. Tóth, P.; Černý, M.; Cagáň, L. First records of *Melanagromyza cuscutae* Hering, 1958 (Diptera: Agromyzidae) from Slovakia and its new host plant. *Entomol. Fenn.* **2004**, *15*, 48–52. [CrossRef]
128. Klein, O.; Kroschel, J. Biological control of *Orobanche* spp. with *Phytomyza orobanchia*, a review. *Biocontrol* **2002**, *47*, 245–277. [CrossRef]
129. Al-Eryan, M.; Altahtawy, M.; El-Sherief, H.; Abu-Shall, A. Efficacy of *Phytomyza orobanchia* Kalt. in reduction of *Orobanche crenata* Forsk. seed yield under semi-field conditions. *Egypt. J. Biol. Pest Control* **2004**, *14*, 237–242.
130. Zhekova, E.; Petkova, D.; Ivanova, I. *Smicronyx smreczynskii* F. Solari, 1952 (Insecta: Curculionidae): Possibilities for biological control of two *Cuscuta* species (Cuscutaceae) in district of Ruse. *Acta Zool. Bulg.* **2014**, *66*, 431–432.
131. Otoidobiga, L.C.; Vincent, C.; Stewart, R.K. Relationship between *Smicronyx* spp. population and galling of *Striga hermonthica* (Del.) Benth. *Int. J. Trop. Insect Sci.* **1998**, *18*, 197–203. [CrossRef]
132. Rubiales, D.; Fernández-Aparicio, M.; Wegmann, K.; Joel, D. Revisiting strategies for reducing the seedbank of *Orobanche* and *Phelipanche* spp. *Weed Res.* **2009**, *49*, 23–33. [CrossRef]
133. Prider, J. The reproductive biology of the introduced root holoparasite *Orobanche ramosa* subsp. *mutelii* (Orobanchaceae) in South Australia. *Aust. J. Bot.* **2015**, *63*, 426–434.
134. Goldwasser, Y.; Miryamchik, H.; Rubin, B.; Eizenberg, H. Field dodder (*Cuscuta campestris*)—A new model describing temperature-dependent seed germination. *Weed Sci.* **2016**, *64*, 53–60. [CrossRef]
135. Chalam, V.C.; Deepika, D.; Abhishek, G.; Maurya, A. Major seed-borne diseases of agricultural crops: International trade of agricultural products and role of quarantine. In *Seed-Borne Diseases of Agricultural Crops: Detection, Diagnosis & Management*; Springer: Singapore, 2020; pp. 25–61.
136. Bennett, C. Latent virus of dodder and its effect on sugar beet and other plants. *Phytopathology* **1944**, *34*, 77–91.
137. Huet, S.; Pouvreau, J.-B.; Delage, E.; Delgrange, S.; Marais, C.; Bahut, M.; Delavault, P.; Simier, P.; Poulin, L. Populations of the parasitic plant *Phelipanche ramosa* influence their seed microbiota. *Front. Plant Sci.* **2020**, *11*, 1075. [CrossRef] [PubMed]

138. Petrosyan, K.; Thijs, S.; Piwowarczyk, R.; Ruraż, K.; Kaca, W.; Vangronsveld, J. Diversity and potential plant growth promoting capacity of seed endophytic bacteria of the holoparasite *Cistanche phelypaea* (Orobanchaceae). *Sci. Rep.* **2023**, *13*, 11835. [CrossRef] [PubMed]
139. Nadler-Hassar, T.; Shaner, D.L.; Nissen, S.; Westra, P.; Rubin, B. Are herbicide-resistant crops the answer to controlling *Cuscuta*? *Pest Manag. Sci. Former. Pestic. Sci.* **2009**, *65*, 811–816. [CrossRef] [PubMed]
140. Bewick, T.; Porter, J.; Ostrowski, R. Smolder™: A bioherbicide for suppression of dodder (*Cuscuta* spp.). In Proceedings of the Annual Meeting-Southern Weed Science Society, Tulsa, OK, USA, 24–26 January 2000.
141. Hadizadeh, F.; Mehrvarz, S.S.; Mirpour, M.S. Effect of *Bacillus* spp. on seed germination of selected species of the genus *Cuscuta* (Convolvulaceae). *Mod. Phytomorphol.* **2014**, *6*, 97.
142. Barghouthi, S.; Salman, M. Bacterial inhibition of *Orobanche aegyptiaca* and *Orobanche cernua* radical elongation. *Biocontrol Sci. Technol.* **2010**, *20*, 423–435. [CrossRef]
143. Zermane, N.; Souissi, T.; Kroschel, J.; Sikora, R. Biocontrol of broomrape (*Orobanche crenata* Forsk. and *Orobanche foetida* Poir.) by *Pseudomonas fluorescens* isolate Bf7-9 from the faba bean rhizosphere. *Biocontrol Sci. Technol.* **2007**, *17*, 483–497. [CrossRef]
144. Elzein, A.; Kroschel, J. *Fusarium oxysporum* Foxy 2 shows potential to control both *Striga hermonthica* and *S. Asiat. Weed Res.* **2004**, *44*, 433–438. [CrossRef]
145. Ndambi, B.; Cadisch, G.; Elzein, A.; Heller, A. Colonization and control of *Striga hermonthica* by *Fusarium oxysporum* f. sp. *strigae*, a mycoherbicide component: An anatomical study. *Biol. Control* **2011**, *58*, 149–159.
146. Oula, D.A.; Nyongesah, J.M.; Odhiambo, G.; Wagai, S. The effectiveness of local strains of *Fusarium oxysporium f. Sp. Strigae* to control *Striga hermonthica* on local maize in western Kenya. *Food Sci. Nutr.* **2020**, *8*, 4352–4360. [CrossRef]
147. Baker, C.S.; Sands, D.C.; Nzioki, H.S. The Toothpick Project: Commercialization of a virulence-selected fungal bioherbicide for *Striga hermonthica* (witchweed) biocontrol in Kenya. *Pest Manag. Sci.* **2024**, *80*, 65–71. [CrossRef] [PubMed]
148. Müller-Stöver, D.; Kohlschmid, E.; Sauerborn, J. A novel strain of *Fusarium oxysporum* from Germany and its potential for biocontrol of *Orobanche ramosa*. *Weed Res.* **2009**, *49*, 175–182. [CrossRef]
149. Anteyi, W.O.; Klaiber, I.; Rasche, F. Diacetoxyscirpenol, a *Fusarium* exometabolite, prevents efficiently the incidence of the parasitic weed *Striga hermonthica*. *BMC Plant Biol.* **2022**, *22*, 84. [CrossRef]
150. Sands, D.C.; Pilgeram, A.L. Methods for selecting hypervirulent biocontrol agents of weeds: Why and how. *Pest Manag. Sci.* **2009**, *65*, 581–587. [CrossRef]

pathogens

MDPI

Article

The 16SrXII-P Phytoplasma GOE Is Separated from Other Stolbur Phytoplasmas by Key Genomic Features

Rafael Toth [1,*], Bruno Huettel [2], Mark Varrelmann [3] and Michael Kube [1]

[1] Department of Integrative Infection Biology Crops-Livestock, University of Hohenheim, 70599 Stuttgart, Germany; michael.kube@uni-hohenheim.de
[2] Max Planck-Genome-Center Cologne, 50829 Cologne, Germany; huettel@mpipz.mpg.de
[3] Institute of Sugar Beet Research (IfZ), 37079 Göttingen, Germany; varrelmann@ifz-goettingen.de
* Correspondence: rafael.toth@uni-hohenheim.de

Abstract: The syndrome "bassess richesses" is a vector-borne disease of sugar beet in Germany. The gammaproteobacterium '*Candidatus* Arsenophonus phytopathogenicus' causes reduced sugar content and biomass, growth abnormalities, and yellowing. Co-infection with the 16SrXII-P stolbur phytoplasmas often leads to more severe symptoms and a risk of complete economic loss. This yellowing agent of the Mollicutes class had not been described before, so its differences from other stolbur phytoplasmas remained unanswered. The genome of strain GOE was sequenced, providing a resource to analyze its characteristics. Phylogenetic position was revised, genome organization was compared, and functional reconstructions of metabolic and virulence factors were performed. Average nucleotide identity analysis indicates that GOE represents a new '*Ca.* Phytoplasma' species. Our results show that GOE is also distinct from other stolbur phytoplasmas in terms of smaller genome size and G+C content. Its reductive evolution is reflected in conserved membrane protein repertoire and minimal metabolism. The encoding of a riboflavin kinase indicates a lost pathway of phytoplasmas outside the groups 16SrXII and 16SrXIII. GOE shows a complete *tra5* transposon harboring orthologs of SAP11, SAP54, and SAP05 effectors indicating an original phytoplasma pathogenicity island. Our results deepen the understanding of phytoplasma evolution and reaffirm the heterogeneity of stolbur phytoplasmas.

Keywords: SBR; sugar beet; phylogeny; virulence factor; phytoplasma pathogenicity island

1. Introduction

In the European Union, Germany is one of the largest producers of sugar beet (*Beta vulgaris* subsp. *vulgaris*), with 21,730 farms producing >29 million tonnes from approximately 364,000 hectares (WVZ, 2024). Today, German sugar beet growers are challenged by an outbreak of a bacterial disease called syndrome "basses richesses" (SBR), which now affects about 75,000 ha of sugar beet in Germany and is still spreading (unpublished data). When first described in Burgundy (France) in the 1990s, SBR was blamed for halving sugar beet growers' incomes, underlining the importance of this disease until today [1]. SBR is associated with degeneration of the vascular system, yellowing and necrosis of old leaves, proliferation, formation of new lanceolate leaves, and a reduction of up to 5% of the absolute sugar content and up to 25% sugar beet biomass [1–3]. The causative agent of SBR is '*Candidatus* Arsenophonus phytopathogenicus', a γ-proteobacterium [4–6]. In addition, co-infection of sugar beets with phytoplasmas in a small percentage of SBR-diseased sugar beets was noted early [4,5]. In contrast to '*Ca.* A. phytopathogenicus', phytoplasmas are wall-less bacteria of the class Mollicutes. They are also obligate biotrophs and vector-transmitted. The γ-proteobacterium and phytoplasma are transmitted by *Pentastiridius leporinus* (Linné) to sugar beet during phloem sap-feeding from and to sugar beet. Both pathogens infect the phloem but histological effects differ between phytoplasma associated with cell necrosis and cell wall lignification, while '*Ca.* A. phytopathogenicus' is in addition associated with the deposition of phenolic compounds in the lumen of phloem cells [5]. Further differences include the increased proliferation and reduction in biomass associated with phytoplasma infection of sugar beet [3].

In recent decades, the population of *P. leporinus* has increased dramatically following the exploitation of sugar beet as a host plant [7]. Vector fitness is also favored by the prominent crop rotation of sugar beet and winter wheat, which allows optimal development of pathogen-loaded larvae [8]. This could be an effect of short fallow treatments, taking into account a needed root system for sucking and the polyphagous nature of the larvae that can develop on different plant species [9,10]. This system, recognized early as problematic in France [3,8], is also blamed for the rapid spread of *P. leporinus* from 2008 until today in southwestern Germany and Switzerland [11–13]. In addition, optimal conditions, including high temperatures, favored the emergence of a second generation of *P. leporinus* in summer. Long vector presence in season and high prevalence increased the risk of infection of other important crops as it has been shown, e.g., for '*Ca.* A. phytopathogenicus' infection of potato [14]. Of particular importance is the rapid increase of infection by both pathogens [9,15]. Little is known about the phytoplasma strains/groups involved in the infection of sugar beet in the first decades in France and Germany. Today, a new phytoplasma subgroup is driving the epidemic in Germany, discovered by investigating high sugar beet losses in the Elbe River Valley [15] and later also identified in Poland [16]. This phytoplasma strain showed a novel 16S rRNA gene sequence restriction profile that differed from other phytoplasmas [17–19] and was classified as a 16SrXII-P subgroup [15]. This new subgroup is part of the 16SrXII/stolbur group, which includes a wide range of vectors and hosts and species such as '*Candidatus* Phytoplasma solani' (associated with bois noir disease of grapevine, potato stolbur [20], tomato stolbur [21], maize redness [22–24], and lavender decline [25,26]), '*Candidatus* Phytoplasma australiense' (also causing grapevine diseases) [27], '*Candidatus* Phytoplasma japonicum' (agent of Japanese Hydrangea phyllody) [28], and '*Candidatus* Phytoplasma fragariae' (strawberry yellows) [29]. Sugar beet with double infection of the '*Ca.* A. phytopathogenicus' and 16SrXII-P phytoplasma showed in combination with heat- and drought-stress rubbery taproot symptoms (Figure 1),which had previously only been described for '*Ca.* P. solani' 16SrXII-A subgroup infections in Serbia [30]. However, the spread of the 16SrXII-P phytoplasma in the following years rapidly became dominant in several other regions, including southern Germany [31].

The first description of the 16SrXII-P phytoplasmas indicated that this subgroup is a '*Ca.* P. solani' relative, but the number of available taxonomic markers was low [15]. The lack of genome information, especially complete phytoplasma genomes, is still striking in terms of impact. In 2024, 36 complete phytoplasma genomes were available (www.ncbi.nlm.nih.gov/datasets/genome/, accessed on 7 December 2024). They vary in size from 499–974 kb (www.ncbi.nlm.nih.gov/datasets/genome/?taxon=33926, accessed on 7 December 2024). The small phytoplasma genomes are characterized by reductive evolution [32] and encode a minimal metabolic repertoire resulting from colonization of nutrient-rich environments [33], as well as group-specific genes involved in vector and host interaction [34].

Figure 1. Rubbery taproot symptom on sugar beet in Elbe River Valley infected by '*Ca.* A. phytopathogenicus' and 16SrXII-P phytoplasma.

Until recently, only few genomic data were available for the stolbur group. Draft genomes were available for five strains of '*Ca.* P. solani'—strain 284/09 (FO393427), 231/09 (FO393428), SA-1 (MPBG00000000), STOL (JBFPNQ000000000), and ST19 (JBFSHS000000000)—and two strains for '*Ca.* P. australiense'—Tabriz.2 (JAINCS000000000) and Tabriz.4 (JAPFFB000000000)—as well as two complete genomes for the '*Ca.* P. australiense' strains PAa (AM422018) and NZSB11 (CP002548). In 2024, the complete genome of strain GOE (CP155828) from the 16SrXII-P stolbur subgroup was determined [35] and three complete genomes of the '*Ca.* P. solani' strains c1 (CP103788), c4 (CP103787), and c5 (CP103786) from *Convolvulus arvensis* (bindweed) and one from *Urtica dioica* (stinging nettle) namely strain o3 (CP103785) in Italy have recently been deposited in GenBank. This expanded database enabled us to perform a comprehensive analysis of strain GOE. Here, we revisited the distinct position of this 16SrXII-P strain in the stolbur group, performed a functional genome reconstruction with particular emphasis on host-dependent metabolism, and provided important insights into the virulence-associated phytoplasma mobilome.

2. Materials and Methods

2.1. Genomic Data

Comparative genome analyses within this work were based on all complete genome sequences of the 16SrXII stolbur group (NCBI:txid85632) including the complete genomes of the 16SrXII-P phytoplasma strain GOE (CP155828), the '*Ca.* P. solani' strains c1 (CP103788), c4 (CP103787), c5 (CP103786), and o3 (CP103785) (NCBI:txid69896), and the '*Ca.* P. australiense' (NCBI:txid59748) strains PAa (AM422018) and NZSB11 (CP002548) (NCBI:txid59748). All data were retrieved from NCBI (www.ncbi.nlm.nih.gov, accessed on 12 September 2024). Genomic benchmarks for comparison were assessed with the Artemis Genome Browser v18.2 [36] using the genetic code 11. If not otherwise stated, default settings were used.

2.2. Phylogenetic Comparison

Whole-genome phylogeny was assessed with average nucleotide identity (ANI) [37] computed with FastANI v1.3 [38]. In addition, sequence synteny analysis was conducted via Mauve v2.4.0 [39]. Whole-genome phylogeny analyses were performed with default settings.

To confirm the phylogeny on the whole-genome level, marker gene analysis was performed for phylogenetic assignment of strain GOE using the phytoplasma markers 16S rRNA and *tuf*. As a first comparison, the complete 16S rRNA gene sequences of GOE were used to infer identities with all other complete genomes of the stolbur group as well as with the reference strains STOL (AB639069) and 916/22 [15], obtained from BLAST analysis. Further, similar sequences for comparison were identified using the nested PCR amplicon of the primers R16F2n/R2 [40] and the fTufAy/rTufAy [41] amplicon from strain GOE and subjected as a query for a Basic Local Alignment Searching Tool (BLAST) analysis against the nucleotide collection (nt) databases from NCBI (accessed on 22 November 2024) [42]. Maximum target size was set to 500 and all filters and masks were disabled. BLAST outputs were inspected manually to select nucleotide sequences belonging to phytoplasmas of the stolbur group. In total, 254 nucleotide sequences were extracted for 16S rRNA and 46 for *tuf*, which were used for multiple sequence alignments, and which were calculated using the MUSCLE algorithm [43] within the Molecular Evolutionary Genetics Analysis (MEGA) software v.10.2.6 [44]. The evolutionary history was inferred via the maximum likelihood method within MEGA on default settings. Each tree was calculated with 1000 bootstraps to ensure statistical significance for cluster assignment. Sequences of '*Ca.* P. asteris' strain M8 [32] were used as an outgroup taxon to root the calculated trees, and 16S rRNA taxonomy was additionally analyzed with iPhyClassifier [45] and used for 16SrXII subgroup assignment via in silico restriction fragment length polymorphism (RFLP) analysis. Default settings have been used except where indicated.

2.3. Ortholog Prediction

For the comparison of shared and unique features of the complete 16SrXII stolbur phytoplasma genomes, orthogroups of the deduced amino acid sequences were predicted with OrthoFinder v2.5.5 [46] and visualized with an upset plot using Intervene v0.6.5 [47] with default parameters. The number of shared and unique deduced amino acid sequences of GOE were extracted from the OrthoFinder results and summarized in a pie chart.

2.4. Functional Reconstruction

For the analysis of encoded metabolic pathways and membrane transport systems, InterproScan v5.68 build 100.0 [48], KEGG database release 109.1 [49], and MetaCyc database v28.0 [50] were utilized. The deduced amino acid sequence of riboflavin kinase of GOE was subjected to a BLASTP comparison for the identification of orthologs in other phytoplasmas which were then used to infer maximum likelihood phylogeny on the amino acid level using MEGA as described in Section 2.2. RibF of *Alteracholeplasma palmae* J233^T [51] was used as an outgroup taxon to root the calculated tree.

Secretome analysis was performed with Phobius v1.01 [52] to identify deduced amino acid sequences encoding either a signal peptide or at least one transmembrane domain or both. Predicted effectors of the strain were confirmed via SignalP v6.0 [53]. Structural alignments of the deduced amino acid sequences of the immunodominant membrane protein (Imp) were constructed with T-Coffee [54]. Structure prediction of Imp was performed via Alphafold v2.0 [55] within the Galaxy Europe server (accessed on 25 June 2024). In addition, maximum likelihood phylogeny was assessed for Imp on nucleotide and amino acid levels within MEGA as described in Section 2.2. Insertion-element (IS-element)-associated regions were predicted via ISEScan v1.7.2.3 [56], further analyzed via palindrome within European Molecular Biology Open Software Suite v6.6.0 to predict repeat regions, and visualized with clinker v0.0.29 [57] and Artemis v18.2 [36].

The genome-wide organization of strain GOE was visualized with DNA Plotter within Artemis v18.2 [36]. Unless otherwise stated, default settings were used.

3. Results

3.1. Genomic Benchmarks of Complete Stolbur Phytoplasma Genomes

Strain GOE has a circular chromosome (Figure 2), like the other complete stolbur phytoplasmas. The chromosome size of the genomes ranged from ~704 kb of GOE to ~973 kb of '*Ca.* P. solani' o3 (Table 1). Both also represented the outer limits of the analyzed G+C contents with 26.17% and 28.58%, respectively (Table 1).

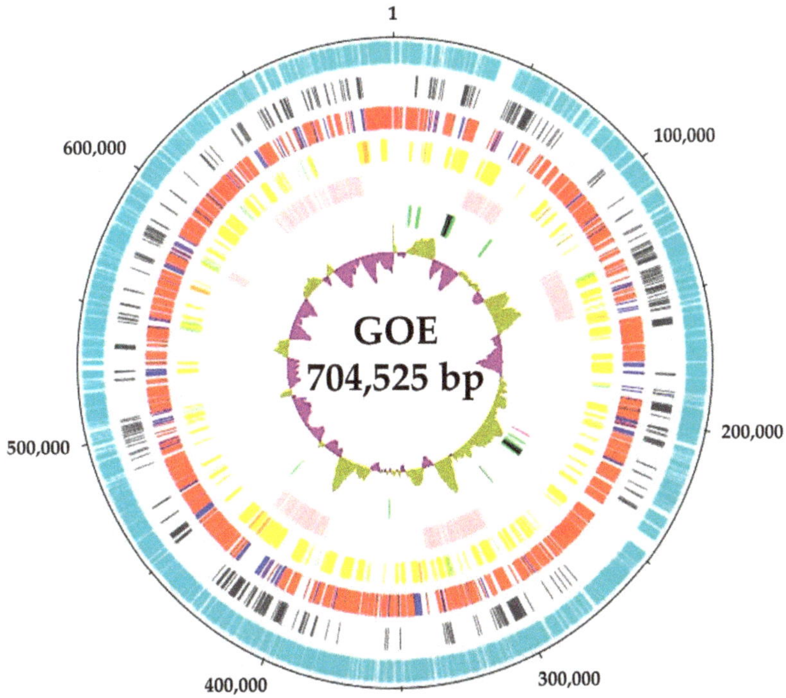

Figure 2. Genomic organization of the circular chromosome of strain GOE. Circular patterns (from outside): (1) (outer ring), scale in base pairs of the chromosome; chromosome regions including (2) (cyan), predicted protein-coding sequences; (3) (grey), hypothetical proteins; (4) (multi-colored), predicted protein-coding sequences: shared with all investigated genomes of the stolbur group (red), unique protein-coding sequences (blue); (5) (multi-colored), predicted membrane proteins: with only a signal peptide (pale green), only transmembrane domains (yellow), and both signal peptide and transmembrane domains (orange); (6) (pale pink) potential mobile unit (PMU)-like regions; (7) (multi-colored), RNAs: signal recognition particle RNA and RNase P RNA component class B (magenta), transfer RNAs (green), and rRNA operons (black), transfer-messenger RNAs (light blue); (8) G+C skew (olive and pink).

With a total of 663 CDSs, GOE has the lowest number of CDSs with a moderate coding density of 0.941 genes per kb, while '*Ca.* P. australiense' NZSb11 encodes the largest number with 1100 CDSs and the highest coding density with 1.146 genes per kb. All analyzed members of the stolbur group encode two complete rRNA operons typical for phytoplasmas [33], whereas the number of encoded tRNAs differs within the analyzed taxa, with 32 tRNAs for 'Ca. P. solani' and 35 tRNAs for 'Ca. P. australiense' (Table 1). All analyzed genomes harbor *ssrA* coding for a tmRNA [58] as well as the ncRNA genes *ffs* and *rnpB*. No plasmids were identified for GOE and the other complete genomes of the taxon 'Ca. P. solani', whereas for 'Ca. P. australiense', plasmids were found, of which one is assigned to NZSb11 [59,60]. All analyzed genomes were reconstructed from plant material, whereas GOE originated from insect tissue. In summary, the complete genome of the 16SrXII-P phytoplasma GOE represents the smallest genome of the stolbur phytoplasmas analyzed and is the first complete stolbur phytoplasma genome reconstructed from an insect host.

Table 1. Genomic features of strain GOE in comparison to the other complete 16SrXII phytoplasma genomes.

16SrXII-subgroup Taxon	16SrXII-P	16SrXII-A 'Ca. P. solani'				16SrXII-B/-C 'Ca. P. australiense'	
Strain	GOE	c1	c4	c5	o3	PAa	NZSb11
Chromosome							
Accession	CP155828	CP103788	CP103787	CP103786	CP103785	AM422018	CP002548
Length (bp)	704,525	751,320	751,188	824,084	973,640	879,324	959,779
GC content (%)	26.17	28.37	28.37	28.07	28.58	27	27
No. of CDSs (protein coding)	663	719	724	807	1000	684	1100
Coding density (genes/kb)	0.941	0.956	0.963	0.979	1.027	0.777	1.146
No. of rRNA operons	2	2	2	2	2	2	2
No. of tRNAs	32	32	32	32	32	35	35
No. of tmRNAs	1	1	1	1	1	1	1
No. of ncRNAs	2	2	2	2	2	2	2
Plasmids							
Accession							DQ318777
Length (bp)							3635
No. of CDSs (protein coding)							4
Host for reconstruction	*Pentastiridius leporinus*	Bindweed	Bindweed	Bindweed	Stinging nettle	Cotton	Strawberry
References	[35]	[61]				[62]	[59,63]

3.2. Phylogenetic Assessment

3.2.1. Average Nucleotide Identity and Sequence Synteny

Since the actual guidelines for the taxonomic assessment of the genus 'Ca. Phytoplasma' suggest the use of average nucleotide identity (ANI) [64], we computed pairwise ANI for the analyzed stolbur phytoplasma chromosomes. ANIs for GOE compared to the other genomes ranged from 82 to 83% (Table 2). All other members of the taxon 'Ca. P. solani', including strains c1, c4, c5, and o3 showed identities higher than 98% when compared to each other, indicating species affiliation, while PAa and NZsb11 of the taxon 'Ca. P. australiense' ANIs ranged from 80% to 82%. Our ANI analysis is consistent with a recent sequence marker analysis that suggests that stolbur phytoplasmas of the 16SrXII-P subgroup should be considered as 'Ca. P. solani'-related species [15]. Taken together, the ANI of GOE with a maximal value of 83% is clearly below 95%, which, according to the current taxonomic guidelines, indicates a new phytoplasma species [64].

Table 2. Average nucleotide identities of stolbur phytoplasma genome sequences.

Cluster	16SrXII-P	Solani				Australiense	
Strain	**GOE**	**c1**	**c4**	**c5**	**o3**	**PAa**	**NZSb11**
GOE	-	**83.04**	**83.05**	**82.87**	**82.85**	**82.00**	**82.23**
c1	**83.04**	-	99.99	99.15	98.42	**79.71**	**79.99**
c4	**83.05**	99.98	-	99.15	98.40	**79.73**	**80.04**
c5	**82.87**	99.15	99.15	-	98.53	**79.94**	**80.02**
o3	**82.85**	98.42	98.40	98.53	-	**80.12**	**79.57**
PAa	**82.00**	**79.71**	**79.73**	**79.94**	**80.12**	-	98.64
NZSb11	**82.23**	**79.99**	**80.04**	**80.02**	**79.57**	98.64	-

Bold values indicate crossing the species affiliation threshold of 95 percent [64]. ANI clusters are highlighted as follows: 16SrXII-P (green), Solani (blue), and Australiense (yellow).

Furthermore, sequence synteny analysis with Mauve resulted in the formation of three clusters (Figure 3). Sequence synteny also showed the separation of strain GOE from the other clusters representative of the taxa '*Ca.* P. solani' and '*Ca.* P. australiense'. Therefore, sequence synteny analysis supports cluster formation from ANI analysis.

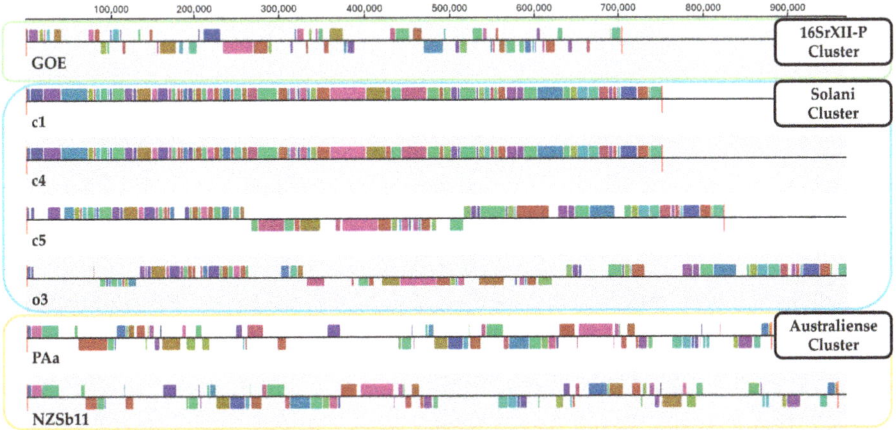

Figure 3. The sequence synteny of stolbur phytoplasma strains was determined using Mauve. Clusters obtained from ANI analysis were highlighted with green (16SrXII-P), blue (Solani), and yellow (Australiense) boxes surrounding the outer periphery. Sequence synteny is indicated by inner blocks of identical colors.

3.2.2. Marker Gene Analysis

BLAST analysis of the two complete 16S rRNA gene sequences showed that strain GOE shows 100% identity with the amplicon of the '*Ca.* P. solani'-related strain 916/22 of the 16SrXII-P subgroup [15]. '*Ca.* P. solani' strains c1, c5, and o3 showed an identity of >99% and c4 (98.75%) to the reference strain STOL (AB639069) of the 16SrXII-A subgroup [64,65]. The complete 16S rRNA gene sequence of GOE shows 98.95% identity to STOL and ranges from 97.70% to 99.02% with o3 (99.02% and 98.89%), c1 (98.95%), c5 (98.95% and 98.89%), and c4 (97.70%). The identities for the '*Ca.* P. australiense' strains range from 97.96% to 98.24% with PAa (98.24% and 98.05%) and NZSb11 (98.23 and 97.96%).

The maximum likelihood phylogeny of the R16F2n/R2 amplicon of the 16S rRNA showed the formation of nine clusters, whereby the strains of the analyzed genomes were divided into three clusters (Figure 4A). Strain GOE was assigned to a cluster with representative sequences for the 16SrXII-P subgroup. The strains c1, c4, c5, and o3 were assigned to a cluster mainly represented by sequences of '*Ca.* P. solani', while the '*Ca.* P. australiense' strains PAa and NZSb11 were assigned together (Table S1). Phylogenetic analysis of *tuf* analysis formed in total five clusters and confirmed the assignment of the analyzed strains to the same three clusters (Figure 4B). Additionally, the *tuf* analysis shows that GOE clusters besides other German strains also with the Polish 16SrXII-P strain PL359/23 (PP731991) [16].

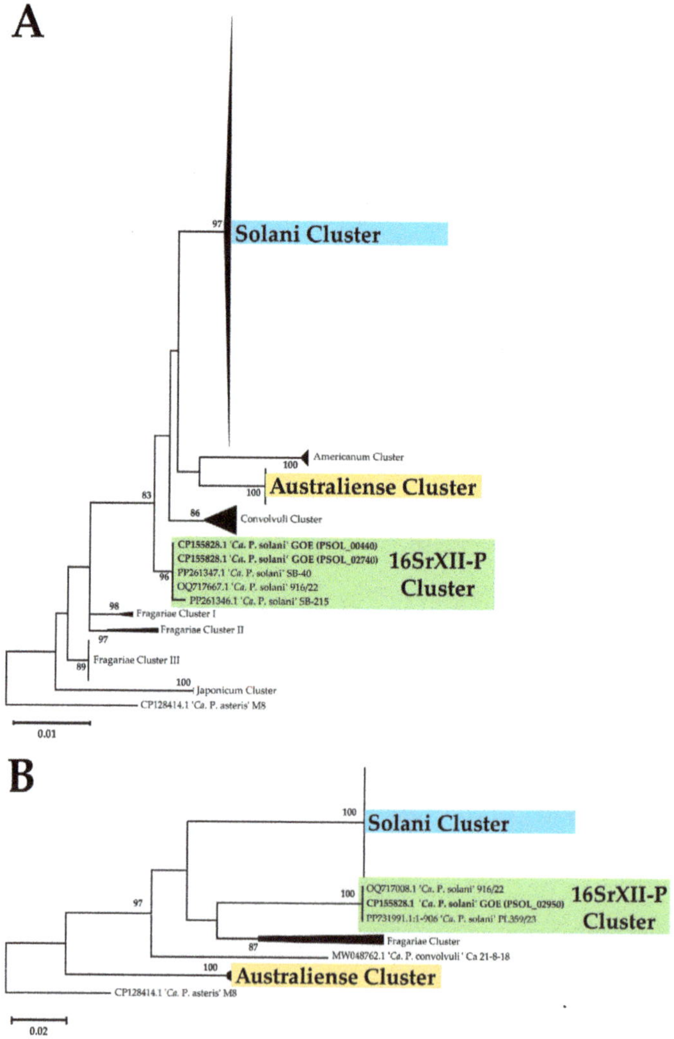

Figure 4. Maximum likelihood phylogeny based on the phytoplasma marker genes 16S rRNA (**A**) and *tuf* (**B**). Strain names with corresponding GenBank accessions are either given or listed in Table S1 in representative clusters. Only bootstrap support values of 70 or above are displayed, with data obtained from 1000 replicates. Scale bars indicate substitutions per site. Each cluster represents a distinct grouping of related sequences (Table S1). Clusters consistent with the whole-genome phylogeny are highlighted in the same color for clarity and comparison.

In silico RFLP analysis of the 16S rRNA of all stolbur genomes analyzed using the restriction enzyme MseI (Tru1I), which has been used to differentiate the 16SrXII-P subgroup [15], showed different patterns for the '*Ca.* P. solani' 16SrXII-A subgroup but not for the '*Ca.* P. australiense' 16SrXII-B/-C subgroups (Figure 5). The results of the virtual RFLP analysis for the analyzed stolbur strains across all 17 key restriction enzymes showed that GOE shared consistent patterns with the 16SrXII-P reference strain 916/22 (Figure S1A). RFLP patterns showed also that strains c1, c5, and o3 belong to 16SrXII-A with an 100% identity to the reference strain STOL (Figure S1B). Strain c4 differed by an identity of 97% and showed a different RFLP pattern by the key enzyme HinfI, indicating a different subgroup (Figure S1B). However, as BLAST analysis using the full-length 16S rRNA showed that strain c4 had the closest relationship to the 16SrXII-A subgroup strains c1, c5, and STOL, c4 appears to be a 16SrXII-A stolbur phytoplasma close to the subgroup boundary. The 16S rRNA sequences of PAa and NZSb11 showed interoperon heterogeneity when analyzed with the key enzyme BfaI, with one copy assigned to the 16SrXII-B and 16SrXII-C subgroups, respectively (Figure S1C) [66,67]. The key enzymes BfaI and HaeIII were identified to distinguish the 16SrXII-P phytoplasmas from the 16SrXII-B/-C subgroups by RFLP patterns. Taken together, RFLP analysis reaffirmed cluster assignment from whole-genome and marker-gene phylogeny.

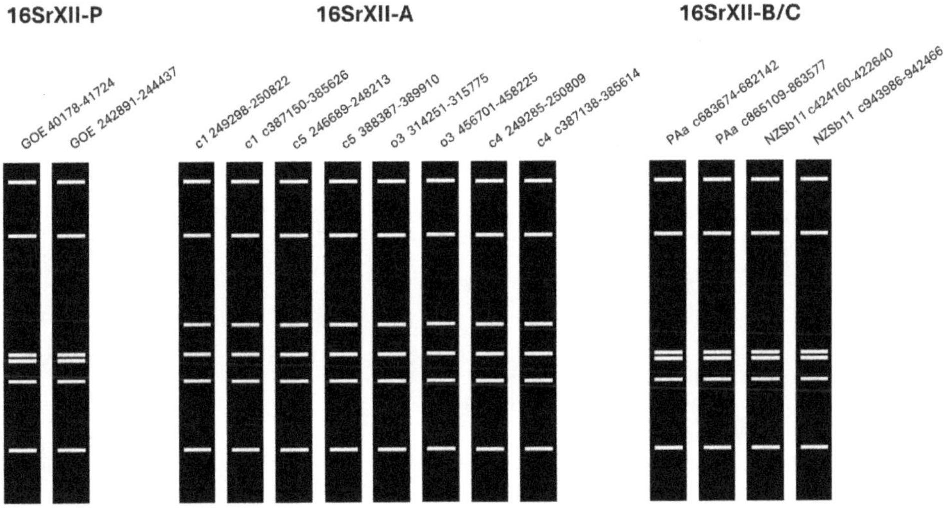

Figure 5. RFLP patterns from virtual gel images of 16S rRNA sequences from all analyzed stolbur genomes corresponding to the R16F2n/R2 amplicons obtained from in silico digestions with the key restriction enzyme MseI (Tru1I) in iPhyClassifier [45]. Numbers next to the strain names indicate the position in the genome and the letter "c" specifies a position on the reverse strand.

The phylogenetic analyses on the whole-genome and single-gene level confirm the separation of 16SrXII-P phytoplasmas from other stolbur group members and underline the heterogeneity of phytoplasmas within the 16SrXII stolbur group.

3.3. *Functional Comparison and Reconstruction*

3.3.1. Pan-Genome Analysis

Orthogroup prediction of the analyzed stolbur phytoplasmas was performed with a data set of 5697 deduced amino acid sequences. Out of these, 5314 (93%) deduced amino acid sequences were assigned to 805 orthogroups. Overall, 365 orthogroups were predicted that were shared by all genomes analyzed (Figure 6A). This core set comprises 411 (~62%) deduced amino acid sequences of GOE that were assigned to proteins involved in functions that contribute to the maintenance of the phytoplasma cell. In total, GOE shares 560 (84%) of its deduced amino acid sequences, whereas 103 (16%) are unique and represent accessory components of the pan-genome including both species-specific and unassigned sequences (Figure 6B). This unique part comprises 100 deduced amino acid sequences (97%), which were annotated as hypothetical proteins, with five of these being secreted and 31 associated with the cytoplasmic membrane (Table S2). Therefore, GOE differs with features associated with pathogen–host interaction from other stolbur phytoplasmas.

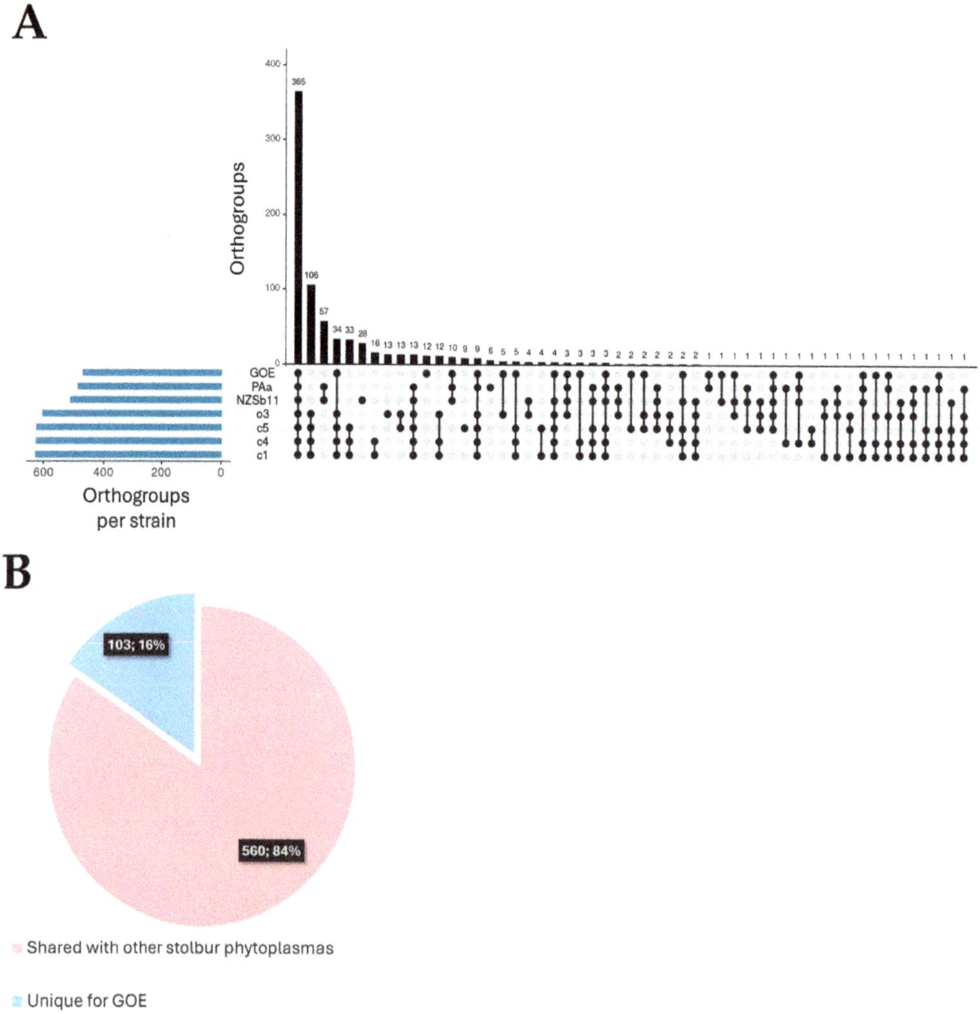

Figure 6. (**A**) Upset plot of predicted orthogroups. Black bars show the orthogroup number per intersection of respective strains (connected dots). The blue bar chart indicates the number of orthogroups per strain. (**B**) Pie chart indicating numbers of shared (pale red) and unique deduced protein sequences (blue) of strain GOE.

3.3.2. Metabolic Pathways

The functional reconstruction showed that the chromosome of strain GOE encodes the reduced metabolism typical for phytoplasmas [32,33], including three complete metabolic modules (Figure 7), which were shared by all analyzed stolbur genomes. Two of these are involved in carbohydrate metabolism, namely the core module of glycolysis/glucogenesis and acetogenesis. Glucose-6-phosphate can be catabolized to pyruvate, followed by oxidation and acetogenesis. In addition, the conserved malate–acetate pathway of phytoplasmas starting from the oxidative decarboxylation of malate to pyruvate is encoded [33,68]. In addition to carbohydrate metabolism, the glycolysis core module is also linked to glycerophospholipid metabolism via phosphatidylethanolamine biosynthesis, which is thought to contribute to the maintenance of the phytoplasma cell membrane [33]. Interestingly, all the stolbur phytoplasma genomes analyed so far possess the gene *ribF* which encodes a riboflavin kinase that is suggested to be involved in FMN/FAD synthesis. FMN/FAD are important cofactors for enzymes involved in redox reactions. Only a few studies have reported the presence of a putative *ribF* in phytoplasmas outside the 16SrXII stolbur group [33]. Our analysis of GOE's *ribF* revealed that members of the 16SrXIII group (Mexican periwinkle virescence group) also harbor a riboflavin kinase besides the stolbur group (Figure S2). Therefore, our analysis suggests that riboflavin kinase is a metabolic feature indicating a residual feature of the reductive evolution of phytoplasmas.

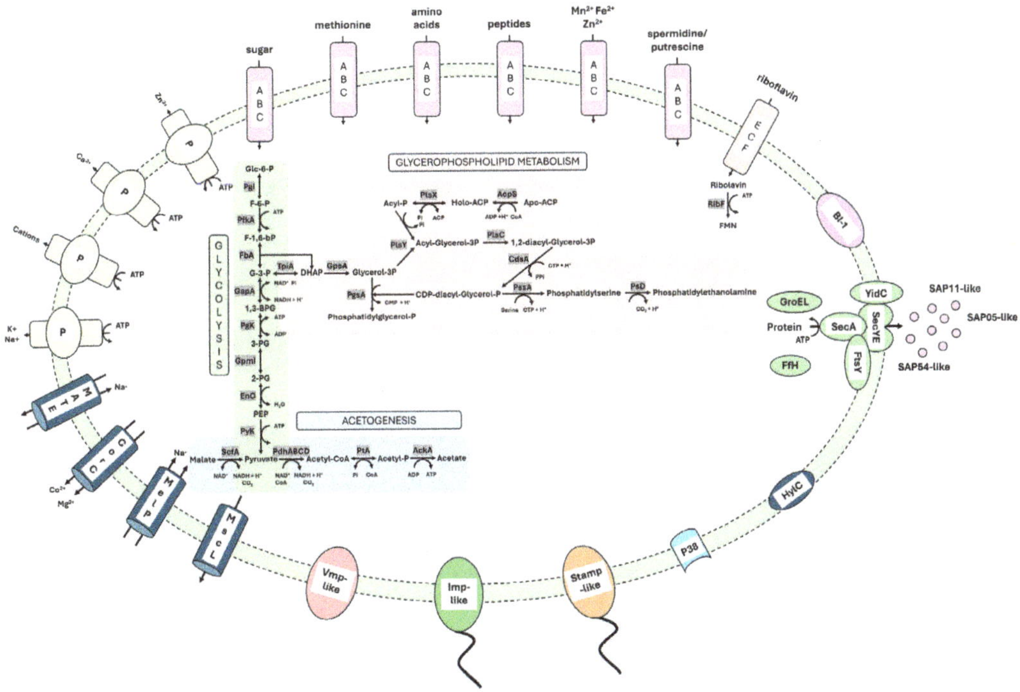

Figure 7. Schematic of complete metabolic pathways and representative membrane proteins involved in metabolism and virulence of the 16SrXII-P phytoplasma strain GOE.

3.3.3. Transporter

Due to their reduced metabolism, phytoplasmas rely on the exchange of essential metabolites from their host environment via several transporters [33]. A total of 34 genes encoding functional subunits of ATP-binding cassette (ABC) transporters were identified in

GOE, including six genes of the energy coupling factor (ECF) transporter family involved in the uptake of amino acids, peptides, bivalent cations, polyamides, sugars, and riboflavin (Figure 7).

Besides the ABC transporters, all analyzed stolbur phytoplasma genomes encode four P-type ATPases for the export of zinc, calcium, sodium, potassium, and unselective cation transport. In addition, all genomes encode two multidrug and toxic compound extrusion (MATE) transporters, a cobalt and magnesium exporter CorC, and the large conductance mechanosensitive channel, as well as a malate/sodium symporter (MelP).

3.3.4. Membrane-Associated Interaction

Phytoplasmas encode so-called immunodominant membrane proteins (IDPs) which are critical for their successful transmission and colonization of plant and insect hosts. Three main types of phytoplasma IDPs have been described: AMP, IMP, and IdPA [69]. Phytoplasmas of the 16SrXII stolbur group encode an AMP-type protein called stolbur antigenic membrane protein (STAMP), which is suggested to interact with actin filaments of their insect vectors [70]. Further, IMP is known for the interaction of phytoplasmas with plant actin [71]. All analyzed stolbur group members encode *stamp*, flanked by the genes *groES*, *groEL*, and *nadE*. *Imp* was also found to be ubiquitous in the analyzed stolbur genomes, flanked by the genes *rnc*, *dnaD*, and *pyrG*. Our analysis therefore supports that stolbur phytoplasmas share the common conserved genomic organization of phytoplasma IDPs as described previously [72].

As GOE was reconstructed from the insect vector *P. leporinus*, further in silico analysis of its deduced IMP-like protein sequence (PSOL_03490) was carried out for functionality and structure to investigate its suitability for transfer into the plant host (Figure 8).

The functional alignment of the IMP amino acid sequences revealed that the IMP-like protein of GOE shows conservation starting from the N-terminus (Figure 8A). All analyzed sequences possess a cytoplasmic, a transmembrane, and a non-cytoplasmic domain. The latter is involved in the interaction with the plant host and shows the highest sequence variation. In addition, all the sequences analyzed showed a similar tertiary structure (Figure 8B). Furthermore, the maximum likelihood phylogeny of Imp showed on nucleotide and amino acid levels the same clustering as described for the marker genes (Figure S3). These results indicate that IMP-like proteins of stolbur phytoplasmas differ mainly in their non-cytoplasmic interaction domain present in the host environment, which may be explained by their adaptation to different host plants and insect vectors.

The strain GOE encodes a gene for a protein similar to the adhesin P38 (PSOL_06560), which was conserved in all other compared stolbur phytoplasmas. P38 was initially described in onion yellows phytoplasmas and has been demonstrated to interact with both insect and plant hosts [73]. Within all analyzed genomes *p38* is flanked by *pyk*, encoding pyruvate kinase, and *folA*, which encodes a dihydrofolate reductase.

Further, GOE encodes a similar protein to the variable membrane protein 1 (PSOL_05990). Variable membrane proteins are highly divergent and are, like IDPs, suggested to be involved in the pathogen–host interactions of phytoplasmas [74]. All analyzed stolbur genomes encode Vmp1. In GOE, *vmp1* is flanked by the genes *uvrA*, encoding an exinuclease ABC subunit A, and *ribF*, encoding riboflavin kinase, while in c1, c4, c5, and o3, *vmp1* is flanked by *uvrA* and *ligA*, encoding a DNA ligase. The Vmp1 gene of strains PAa and NZSb11 is flanked by *ribF* and *engC*, which encodes a GTPase involved in the assistance of the maturation of ribosomes.

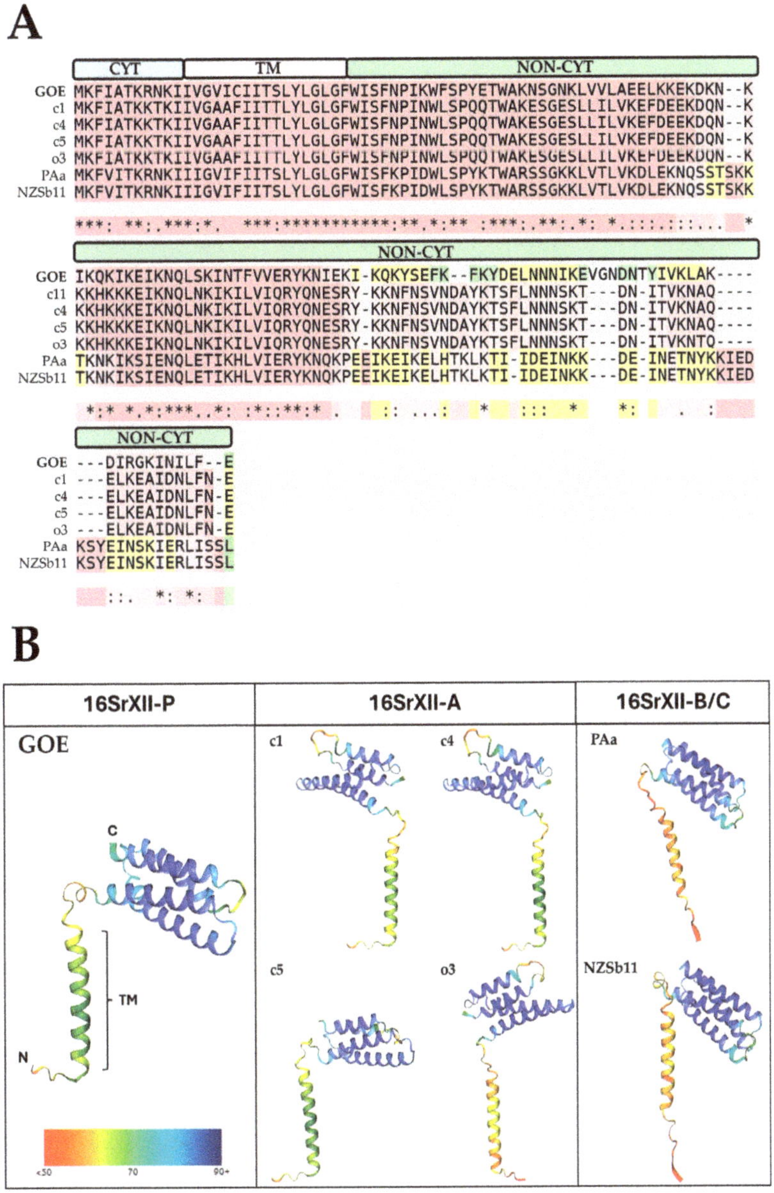

Figure 8. (**A**) Functional alignment of Imp using T-Coffee highlighting amino acid sequence conservation and chemical similarity. The peptide domain localization is shown above the alignment: cytoplasmic CYT (blue), transmembrane TM (pale pink), and non-cytoplasmic NON-CYT (green). Color code for amino acid conservation: Highly conserved (red), conserved but slightly variable (pale red), less conserved (yellow), and most variable (green). Asterisks (*) denote fully conserved residues, colons (:) represent strong chemical similarity, dots (.) indicate moderate similarity, and spaces show low or no conservation. (**B**) Structure prediction of Imp via AlphaFold. N and C indicate the N- and C-terminus of the protein, respectively. The colored bar indicates the percentage of prediction probability of the deduced protein structures.

Another potential virulence factor identified in the membrane of strain GOE is the bax inhibitor 1 (BI-1) (PSOL_2960). BI-1 proteins are involved in the suppression of programmed cell death [75], but the exact mechanism of phytoplasma BI-1 is unknown. All analyzed stolbur genomes encode a gene for BI-1 flanked by the genes *tuf* and *rsmG*. This is consistent with other studies that have analyzed the coding sequence of BI-1 in phytoplasma genomes and support the hypothesis that BI-1 is an evolutionarily conserved virulence factor that adapts to the host environment, also for stolbur phytoplasmas [32,76,77].

Moreover, all analyzed stolbur phytoplasmas encode *hylC*, which encodes a hemolysin III-like protein that is reported as a conserved candidate virulence factor and marker in phytoplasmas [78]. The gene *hylC* is flanked by *ackA* coding for an acetate kinase within all genomes analyzed. In GOE, besides *ackA*, the gene *nfnB* encoding a nitro reductase is flanking *hylC* while in c1, c4, c5, and o3, *hylC* is also flanked by *hisS* encoding a histidine tRNA ligase in all analyzed stolbur phytoplasmas. PAa and NZsb11 are additionally flanked by *mscL* encoding the large conductance mechanosensitive channel.

Taken together, the genome of GOE and those of the analyzed members of the stolbur group encode most of the previously described membrane-associated, conserved virulence factors that have been described or suggested to be involved in the interaction of phytoplasmas with their hosts.

3.3.5. Secretome Analysis

Secretome reconstruction showed that all analyzed stolbur phytoplasmas encode the common set of *sec* genes including *secA*, *secE*, *secY*, and *yidC* involved in the Sec-dependent secretion pathway representing the sole described secretion system in phytoplasmas yet [79]. In addition, of all analyzed genomes of strain GOE, the '*Ca.* P. solani' strains c1, c4, c5, and o3 encode the complete gene set (*ffs*, *ffh*, *ftsY*) for the signal recognition particle (SRP) pathway. The SRP pathway drives the co-translational integration of membrane proteins by targeting ribosomes to the Sec-dependent membrane complex of the secretion pathway [80]. In contrast, only PAa encodes a functional Ffh within the analyzed '*Ca.* P. australiense' genomes, while in NZSb11 *ffh* is disrupted, and it remains unclear whether an alternative protein fulfils the role of Ffh, as in the case of GroEL, which is considered to substitute for the function of the chaperone SecB in the Sec-dependent secretion system [33]. However, as little is known about functional redundancies within the SRP and Sec-dependent secretion pathways, this may indicate functional differences within the stolbur group in terms of protein translocation.

In total, 196 (~30%) of the deduced amino acid sequences of strain GOE possess at least one transmembrane (TM) domain, of which eleven additionally harbor a signal peptide domain (SP+TM), whereas 31 (~5%) proteins were identified that have only a signal peptide domain (SP) representing putative effector proteins (Figure 9). The comparison of these results with those of the other analyzed stolbur phytoplasma genomes revealed that GOE harbors the lowest number of deduced amino acid sequences that are either present in the phytoplasma membrane or secreted into the host environment. In contrast, strain o3 displays the highest number of deduced amino acid sequences. Therefore, these results may indicate that with increasing genome size, the number of proteins associated with pathogen–host interaction in stolbur phytoplasmas is also increasing. This is following a previous genome study of phytoplasmas of the 16SrI asteris group [32].

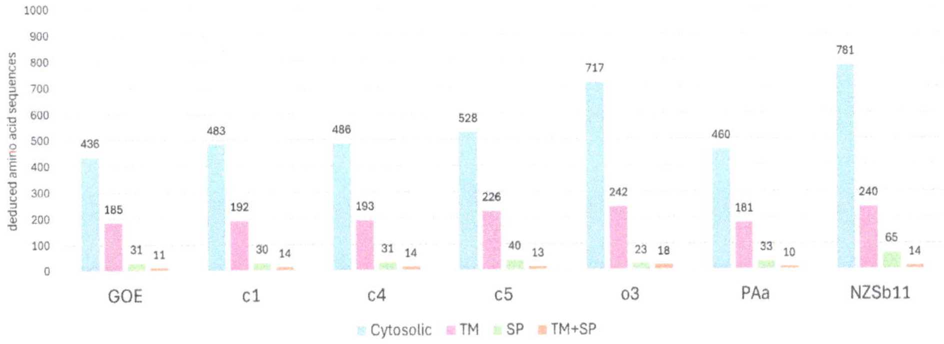

Figure 9. Number of deduced amino acid sequences that have been identified as either encoding a signal peptide (SP), transmembrane domain(s) (TM), or both (SP+TM) per strain.

3.3.6. Important Effector Proteins

We identified proteins similar to the experimentally verified effectors of the secreted aster yellows phytoplasma witches' broom proteins (SAPs) SAP11 (PSOL_01490) [81–83], SAP54 (PSOL_01470) [84], and SAP05 (PSOL_01730) [85] from '*Ca.* P. asteris' strains. Strains c1 and c4 encoded also all three SAP family effectors analyzed, while the annotations of strains c5 and o3 describe SAP54 only. No homologs for the three analyzed SAPs were identified within the '*Ca.* P. australiense' strains PAa and NZSb11 as described previously [62,86]. In comparison, the SAP11-like protein of GOE is enlarged in its C-terminal region compared to the identical SAP11-like proteins of c1 and c4 (111 aa) and shows an identity of 78.12%. In contrast to the SAP11-like protein, the SAP54-like protein (116 aa) of GOE showed a high sequence similarity with an identity of 89.74% to the identical SAP54-like proteins of strains c1 and c4, respectively, and 89.66% to the SAP54-like protein of strain c5 (116 aa), whereas the SAP54-like protein of o3 (97 aa) showed only a 39.78% identity. For SAP05, the homologs of strains GOE, c4, and c1 showed the same length of 322 amino acids and shared a sequence identity of 99.07%. Overall, the genome of strain GOE encodes all the well-characterized phytoplasma effectors of the SAP family, which have been described in many phytoplasmas and are located on transposable elements [81]. Furthermore, our analysis confirms the findings of a previous study that the coding of the SAP family effectors varies within the stolbur group [86].

Interestingly, all three SAP-like proteins analyzed are located on a transposon with a length of approximately 31 kb (Figure 10A), while for the '*Ca.* P. solani' strains c1, c4, c5, and o3 the annotated SAP effectors are separately encoded on different locations in the chromosome but were also flanked by potential mobile unit (PMU)-associated genes. The complete transposon of GOE has an IS3 family at each end. Each IS element is represented by the gene *tra5*, which encodes a transposase flanked by an inverted repeat of 11 bp and a direct repeat of 3 bp per site (5′-(ACA)TTTTAAAAAnnnnnnTTTTTTTAAAA(ACA)-3′), located on the reverse strand, but which are not unique within the GOE chromosome sequence. These repeat regions have the typical size for IS elements but are shorter than those described for PMUs in the genomes of '*Ca.* P. asteris' AYWB or '*Ca.* P. solani' SA-1 [81,86,87]. The transposon harbors genes that are associated with PMUs and also replication such as *dnaG*, *dnaB*, *tmk*, *rad50*, *himA*, *ssb*, as well as *sigF* [81]. In addition, the transposon showed a similar GC content (27.85%) to the GOE chromosome overall with 26.17%. In silico circulation of the transposon suggests the formation of replicative intermediates indicated by a putative origin of replication (Figure 10B). This had been described for an extrachromosomal PMU-associated replicon of '*Candidatus* Phytoplasma asteris' [88]. Taking into account the virulence factors encoded on the transposon, it must be considered as a complete phytoplasma pathogenicity island (PPAI).

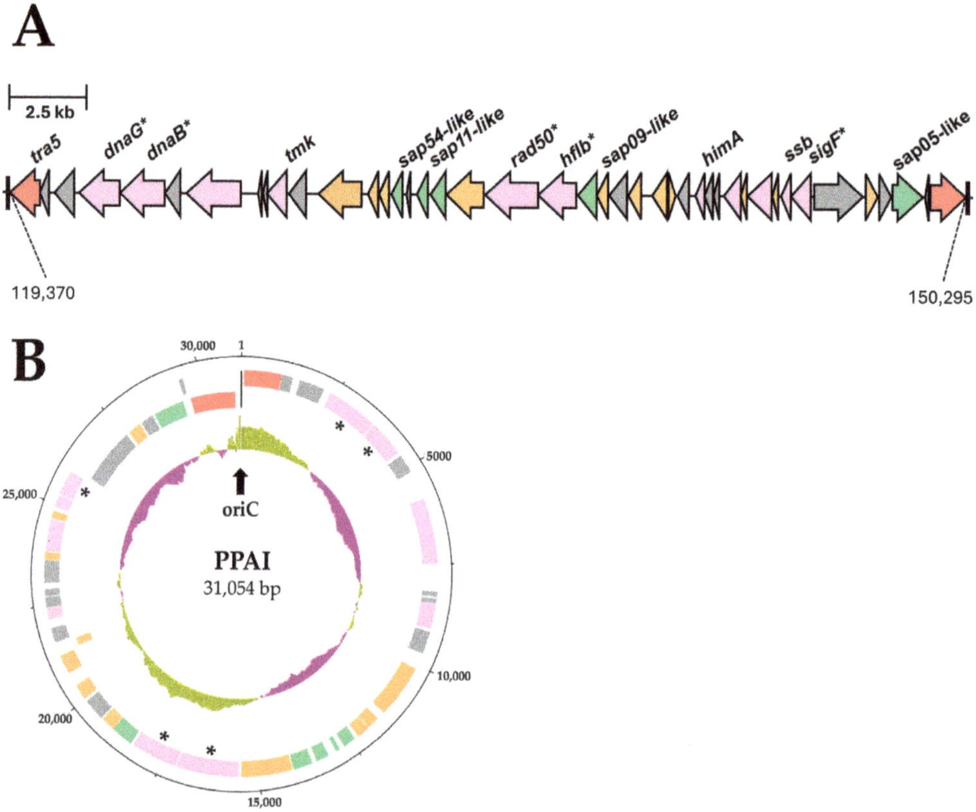

Figure 10. Phytoplasma pathogenicity island (PPAI) within the GOE chromosome (**A**) and its proposed extrachromosomal circular form (**B**). In the linear form (**A**), left-pointing arrows on the reverse strand and right-pointing arrows on the forward strand indicate coding orientation. Numbers connected by dashed lines indicate their position within the GOE genome. In the circular form (**B**), the following patterns are displayed from the outside in (1) outer ring representing the scale in base pairs, (2) predicted protein-coding sequences on the forward strand, (3) predicted protein-coding sequences on the reverse strand, and (4) GC skew analysis, where olive indicates above-average GC skew and violet indicates below-average GC skew. Genes are color-coded as follows: red—IS3 family transposases, purple—PMU-associated genes, orange—putative membrane proteins, green—putative secreted proteins. Asterisks indicate pseudogenes. Repeat regions are indicated by black bars. The black arrow marks the proposed *oriC* region.

4. Discussion

The 16SXII-P subgroup dominates the current phytoplasma infection in sugar beet in Germany and has also been detected in Poland [15,16]. However, the *P. leporinus*-driven epidemic is still ongoing [7,14,15,31] and it remains unclear to what extent the 16SrXII-P phytoplasma will spread throughout Germany and into other neighboring countries.

The comparative analysis of the 16SrXII-P subgroup from *P. leporinus* [35] highlighted the separated phylogenetic position and genomic content. This is in accordance with the first description of the 16SrXII-P phytoplasma 916/22 highlighting the distance to '*Ca.* P. solani' but new species description was hampered by access to a limited number of genes and due to an 16S rDNA identity of 98.67% to the STOL reference [15], thereby missing the taxon threshold of 98.65% identity [64]. This also applies to the two complete and identical 16S rRNA genes of GOE showing 98.95% identity. Other gene markers can be used in addition with assigned thresholds comprising *tuf* (97.5%), *secY* (95%), *secA* (97.5), *rplV-rpsC* (97.5%), or *groEL* (97.6%) [64] and had been clearly fulfilled by amplicon sequence analysis of *tuf*, *secY*, and *rp*-locus for 916/22 standing in opposite to the 16S rRNA result [15]. However, an ANI value of <95% is considered as overvoting the criterium, if 16S rDNA identity is >98.65% [64]. While no genome of the reference strain STOL is available, a maximal ANI of 83.04% (strain c1) is reached in the analyzed complete genomes (Table 2). In contrast, ANI values of c1, c4, c5, and o3 range from 99.99% (c1, c4) to a minimum of 98.42% in '*Ca.* P. solani', indicating one taxon for this criterium. Separation of the 16SrXII-P subgroup is also obvious in comparison to '*Ca.* P. australiense', reaching a maximum of 82.23%. It is also notable that the '*Ca.* P. solani' strains fulfil taxon separation from GOE (max. 83.05%) and '*Ca.* P. australiense' (max. 80.12%).

Our functional reconstruction and comparison showed that the 16SrXII-P phytoplasma GOE and the other stolbur phytoplasmas analyzed encode the minimal metabolism with the core module of glycolysis coupled with acetogenesis as the energy-yielding instance as well as the glycerophospholipid metabolism for the preservation of the cytoplasmic membrane, as described previously for the genomes of the 16SrI asteris group [32]. This minimal coding of metabolic pathways can still differ between phytoplasma groups, as for example the 16SrX group member '*Ca.* P. mali' lacks the energy-producing part of glycolysis from D-glyceraldehyde-3-phosphate to pyruvate [33,89], while phytoplasmas of the 16SrV group encode the carboxylic acid metabolism from malate to lactate [90]. However, neither of these features could be confirmed for the analyzed genomes of the stolbur group. What is noteworthy is that our analysis revealed the coding of a riboflavin kinase, which has been suggested to be involved in FMN/FAD synthesis as an important co-factor and appears to be a lost pathway outside of the 16SrXII stolbur group and the closely related 16SrXIII Mexican periwinkle virescence group. Apart from the synthesis of FMN/FAD, no information is available on the role of riboflavin in phytoplasmas in the context of uptake from the host environment. It has been shown that riboflavin levels are linked to the induction of plant resistance against phytopathogenic bacteria such as *Pseudomonas syringae* pv. tomato in *Arabidopsis thaliana* [91]. One may speculate that the uptake and utilization of riboflavin is involved in the immunomodulation of the host's defense system. This would then be another brick in the series of host manipulations. As phytoplasmas are limited to the plant phloem, they secrete effector proteins to manipulate their hosts via the Sec-dependent secretion pathway as the major functional secretion system of phytoplasmas [79]. The secreted aster yellows witches' broom proteins (SAPs) [81] represent with their well-examined effectors SAP11 [82], SAP54 [84], and SAP05 [85] an important group of virulence factors of phytoplasmas that cause growth abnormalities and interfere with the host hormone system [92]. We identified homologs of the phytoplasma effectors SAP11, SAP54, and SAP05 within the genome sequence of GOE, c1, and c4, whereas for c5 and o3 only a homolog for SAP54 was identified, and for the '*Ca.* P. australiense' strains PAa and NZSb11, no homologs have been identified. Such a variable coding of these effectors has been shown also in other phytoplasma groups like the 16SrI asteris group [32,93]. It has been demonstrated that the SAP11 homolog of '*Ca.* P. solani' alters plant morphology by destabilization of TEOSINTE BRANCHED 1-CYCLOIDEA-PROLIFERATING CELL FACTOR (TCP) transcription factors (TFs) when expressed in *A. thaliana*. Plants showed reduced biomass, size, proliferation, crinkled leaves, as well as deformations in the root system [83]. Further, SAP05 is known to induce massive proliferation in *A. thaliana* but with a different mechanism by degradation of the SQUAMOSA promoter binding protein-like genes and GATA motif-specific TFs independent from ubiquitination [85]. SAP54-like proteins degrade MADs-box transcriptional factors leading to the development of leaf-like flowers and therefore sterile plants [84]. No information by the infection on the seed production is available to date. Besides phytoplasma host plants, SAP11 and SAP54 are also suggested to have an influence on the behavior of the insect vectors of phytoplasmas in terms of attractiveness and reproduction [82,94,95], which needs clarification for the *P. leporinus* vector system. The presence of crucial phytoplasma effectors fits the observed symptoms associated to infection by '*Ca.* A. phytopathogenicus' and/or 16SrXII-P phytoplasma in sugar beet in the field. Nevertheless, the exact mode of action of the SAP family proteins and how they interact with the transcription factors of affected plants and insect vectors remains elusive.

The analysis of the coding of effector proteins in phytoplasma genomes revealed their location on regions that are flanked by *tra5* genes of the IS3 family that form putative transposable elements which were named potential mobile units (PMUs) that are present as multiple fragmented copies in the phytoplasma genome [81]. They were also suggested to have the potential to excise from the chromosome, as they could also be detected as extrachromosomal replicons in high copy numbers in the host plant and insect [88]. Comparative analyses of PMUs across different phytoplasma species led to the hypothesis that PMUs can be exchanged by horizontal gene transfer and that effector proteins evolve faster than other genes not located on a PMU, explaining the species- or strain-specific encoding of effector proteins and leading to lineage-specific adaptation to different hosts [96–98]. We identified to our knowledge the first complete transposon harboring orthologs of the effector proteins SAP11, SAP54, and SAP05, representing a transposable phytoplasma pathogenicity island (PPAI) with a size of ~31 kb, suggesting the ability to excise as a replicon (Figure 10). In contrast, the circular PMU1 of the '*Ca.* P. asteris' strain AY-WB is with a total size of ~19 kb shorter and lacks the coding of homologs to the well-examined SAPs. Furthermore, in contrast to GOE, the inverted repeats of PMU1 from AY-WB show a size of 327–328 bp, which is atypical for IS3 family transposases, but both indicate an *oriC* in these regions when GC skew was analyzed [88]. Considering that this transposon has a similar GC content to the entire GOE chromosome, this suggests that such transposons entered phytoplasmas early on and evolved with them. It should also be considered that the transposon PPAI may be frequently available in the phloem of the host plant or in the insect vector as demonstrated for PMU1 [88] and can be taken up horizontally from other phytoplasmas and bacteria in the phloem or as suggested for other phytoplasmas [97]. It has been described that selfish DNA in prokaryotes can be integrated into eukaryotic plant genomes like IS elements [99]. It is therefore possible to speculate whether the transposon reported in the GOE genome is still functional and can be also integrated into the plant host. If so, such pathogenicity islands could explain the effective host adaptation of phytoplasmas, as long-term integration of effectors into and expression by the host could have led to an effective colonization strategy that overcomes the plant defence system. Furthermore, such a host-mediated effector expression would be of great breeding importance for phytoplasma-infected crops since the development of symptoms triggered by the effector expression independent from an infection would bias the selection of tolerant varieties.

5. Conclusions

This comparative genomic study of phytoplasma strain GOE provides deep insights into a representative of the important sugar beet stolbur pathogen subgroup 16SrXII-P, which is transmitted by the cixiid vector *P. leporinus* causing the current epidemic in Germany. Phylogenetic genome and single-gene analyses underline the differentiation from other stolbur taxa, while genomic benchmarks and functional reconstruction emphasize the general reductive evolution of phytoplasmas as highly adapted bacteria for vector and host infection. This includes the discovery of a complete pathogenicity island formed by a transposon containing the major effector proteins of phytoplasmas, providing new insights into evolutionary mechanisms and a starting point for further studies to analyze spread, but also the manipulation of organogenesis in sugar beet by encoded effectors.

Supplementary Materials: The following supporting information can be downloaded at https://www.mdpi.com/article/10.3390/pathogens14020180/s1, Figure S1: Virtual gel images of the RFLP analysis of the analysed stolbur genomes and representative reference strains obtained from the digestion with all 17 key restriction enzymes. The gels are grouped according to their 16SrXII subgroup affiliation as follows: (A) 16SrXII-P, (B) 16SrXII-A, and (C) 16SrXII-B/-C. Corresponding accessions and positions in the genome sequences are indicated at the bottom of each virtual gel image; Figure S2: Maximum likelihood phylogeny based on the gene *ribF*. Names of strains with their respective GenBank accession are given. 16Sr group/subgroup affiliation is highlighted next to the clusters. Only bootstrap support values of 70 or above are displayed, with data obtained from 1,000 replicates. Scale bars indicate substitutions per site; Figure S3: Maximum likelihood phylogeny of the analysed stolbur genomes based on the gene *imp* (A) and the deduced amino acid sequences of Imp (B). Names of strains with their respective GenBank accession are given. Only bootstrap support values of 70 or above are displayed, with data obtained from 1,000 replicates. Scale bars indicate substitutions per site; Table S1: Cluster composition of the maximum likelihood phylogeny; Table S2: Unique deduced amino acid sequences of 16SrXII-P phytoplasma strain GOE.

Author Contributions: Conceptualization, M.K.; methodology, R.T., B.H., M.V. and M.K.; software, R.T., B.H. and M.K.; validation, R.T., B.H. and M.K.; formal analysis, R.T. and M.K.; investigation, R.T. and M.K.; resources, M.V., B.H. and M.K.; data curation, R.T., B.H. and M.K.; writing—original draft preparation, R.T. and M.K.; writing—review and editing, R.T., B.H., M.V. and M.K.; visualization, R.T.; supervision, M.K.; project administration, M.V. and M.K.; funding acquisition, M.V. and M.K. All authors have read and agreed to the published version of the manuscript.

Funding: This work is funded by the Federal Ministry of Economic Affairs and Climate Protection (BMWK) funding no. 22,943N.

Data Availability Statement: The datasets generated during and/or analyzed during the current study can be find in the main text and the Supplementary Materials.

Acknowledgments: We would like to thank the "Gemeinschaft zur Förderung von Pflanzeninnovation e. V." (GFPi) and the "Industrielle Gemeinschaftsforschung" (IGF) for coordinating and supporting our work in the project "Differentiation of pathogens and course of infection in SBR-associated bacterioses in sugar beet to derive resistance testing methods to ensure yield stability".

Conflicts of Interest: The authors declare no conflicts of interest.

References

1. Richard-Molard, M.; Garessus, S.; Malatesta, G.; Valentin, P.; Fonné, G.; Gerst, M.; Grousson, C. Le syndrôme des "basses richesses": Investigations au champ et tentatives d'identification de l'agent pathogène et du vecteur. In Proceedings of the 58th Congres de L'Institut International de Recherches Betteravieres, Bruxelles, France, 19 June 1995; pp. 19–22.
2. Muchembled, C.; Garressus, S.; Ecalle, F.; Boudon-Padieu, E.; Gatineau, F. Low sugar content syndrome. In Proceedings of the Fifth International Conference on Pests in Agriculture, Montpellier, France, 7–9 December 1999; pp. 529–536.
3. Bressan, A.; Sémétey, O.; Nusillard, B.; Clair, D.; Boudon-Padieu, E. Insect Vectors (Hemiptera: Cixiidae) and Pathogens Associated with the Disease Syndrome "Basses Richesses" of Sugar Beet in France. *Plant Dis.* **2008**, *92*, 113–119. [CrossRef] [PubMed]
4. Gatineau, F.; Larrue, J.; Clair, D.; Lorton, F.; Richard-Molard, M.; Boudon-Padieu, E. A New Natural Planthopper Vector of Stolbur Phytoplasma in the Genus *Pentastiridius* (Hemiptera: Cixiidae). *Eur. J. Plant Pathol.* **2001**, *107*, 263–271. [CrossRef]
5. Gatineau, F.; Jacob, N.; Vautrin, S.; Larrue, J.; Lherminier, J.; Richard-Molard, M.; Boudon-Padieu, E. Association with the Syndrome "Basses Richesses" of Sugar Beet of a Phytoplasma and a Bacterium-Like Organism Transmitted by a *Pentastiridius* sp. *Phytopathology* **2002**, *92*, 384–392. [CrossRef]

6. Sémétey, O.; Bressan, A.; Richard-Molard, M.; Boudon-Padieu, E. Monitoring of proteobacteria and phytoplasma in sugar beets naturally or experimentally affected by the disease Syndrome "Basses richesses". *Eur. J. Plant Pathol.* **2007**, *117*, 187–196. [CrossRef]
7. Behrmann, S.; Schwind, M.; Schieler, M.; Vilcinskas, A.; Martinez, O.; Lee, K.-Z.; Lang, C. Spread of bacterial and virus yellowing diseases of sugar beet in South and Central Germany from 2017–2020. *Sugar Ind.* **2021**, *146*, 476–485. [CrossRef]
8. Bressan, A.; Moral García, F.J.; Boudon-Padieu, E. The prevalence of '*Candidatus* Arsenophonus phytopathogenicus' infecting the planthopper *Pentastiridius leporinus* (Hemiptera: Cixiidae) increase nonlinearly with the population abundance in sugar beet fields. *Environ. Entomol.* **2011**, *40*, 1345–1352. [CrossRef]
9. Behrmann, S.C.; Witczak, N.; Lang, C.; Schieler, M.; Dettweiler, A.; Kleinhenz, B.; Schwind, M.; Vilcinskas, A.; Lee, K.-Z. Biology and Rearing of an Emerging Sugar Beet Pest: The Planthopper *Pentastiridius leporinus*. *Insects* **2022**, *13*, 656. [CrossRef]
10. Pfitzer, R.; Rostás, M.; Häußermann, P.; Häuser, T.; Rinklef, A.; Detring, J.; Schrameyer, K.; Voegele, R.T.; Maier, J.; Varrelmann, M. Effects of succession crops and soil tillage on suppressing the syndrome "basses richesses" vector *Pentastiridius leporinus* in sugar beet. *Pest Manag. Sci.* **2024**, *80*, 3379–3388. [CrossRef]
11. Schröder, M.; Rissler, D.; Schrameyer, K. Syndrome des"Basses Richesses" (SBR)—Erstmaliges Auftreten an Zuckerrübe in Deutschland. *J. Kulturpflanz.* **2012**, *64*, 396–397.
12. Pfitzer, R.; Schrameyer, K.; Voegele, R.T.; Maier, J.; Lang, C.; Varrelmann, M. Ursachen und Auswirkungen des Auftretens von Syndrome des "Basses Richesses" in deutschen Zuckerrübenanbaugebieten. *Sugar Ind.* **2020**, *145*, 234–244.
13. Mahillon, M.; Groux, R.; Bussereau, F.; Brodard, J.; Debonneville, C.; Demal, S.; Kellenberger, I.; Peter, M.; Steinger, T.; Schumpp, O. Virus Yellows and Syndrome "Basses Richesses" in Western Switzerland: A Dramatic 2020 Season Calls for Urgent Control Measures. *Pathogens* **2022**, *11*, 885. [CrossRef] [PubMed]
14. Behrmann, S.C.; Rinklef, A.; Lang, C.; Vilcinskas, A.; Lee, K.-Z. Potato (*Solanum tuberosum*) as a New Host for *Pentastiridius leporinus* (Hemiptera: Cixiidae) and '*Candidatus* Arsenophonus phytopathogenicus'. *Insects* **2023**, *14*, 281. [CrossRef] [PubMed]
15. Duduk, B.; Ćurčić, Ž.; Stepanović, J.; Böhm, J.W.; Kosovac, A.; Rekanović, E.; Kube, M. Prevalence of a '*Candidatus* Phytoplasma solani'-Related Strain Designated as New 16SrXII-P Subgroup over '*Candidatus* Arsenophonus phytopathogenicus' in Sugar Beet in Eastern Germany. *Plant Dis.* **2023**, *107*, 3792–3800. [CrossRef] [PubMed]
16. Duduk, B.; Stepanović, J.; Fránová, J.; Zwolińska, A.; Rekanović, E.; Stepanović, M.; Vučković, N.; Duduk, N.; Vico, I. Geographical variations, prevalence, and molecular dynamics of fastidious phloem-limited pathogens infecting sugar beet across Central Europe. *PLoS ONE* **2024**, *19*, e0306136. [CrossRef]
17. Lee, I.-M.; Gundersen-Rindal, D.E.; Davis, R.E.; Bartoszyk, I.M. Revised Classification Scheme of Phytoplasmas based on RFLP Analyses of 16S rRNA and Ribosomal Protein Gene Sequences. *Int. J. Syst. Evol. Microbiol.* **1998**, *48*, 1153–1169. [CrossRef]
18. Wei, W.; Lee, I.-M.; Davis, R.E.; Suo, X.; Zhao, Y. Automated RFLP pattern comparison and similarity coefficient calculation for rapid delineation of new and distinct phytoplasma 16Sr subgroup lineages. *Int. J. Syst. Evol. Microbiol.* **2008**, *58*, 2368–2377. [CrossRef]
19. Zhao, Y.; Wei, W.; Lee, I.-M.; Shao, J.; Suo, X.; Davis, R.E. Construction of an interactive online phytoplasma classification tool, iPhyClassifier, and its application in analysis of the peach X-disease phytoplasma group (16SrIII). *Int. J. Syst. Evol. Microbiol.* **2009**, *59*, 2582–2593. [CrossRef]
20. Jović, J.; Ember, I.; Mitrović, M.; Cvrković, T.; Krstić, O.; Krnjajić, S.; Acs, Z.; Kölber, M.; Toševski, I. Molecular detection of potato stolbur phytoplasma in Serbia. *Bull. Insectology* **2011**, *64*, 83–84.
21. Favali, M.A.; Musetti, R.; Fossati, F.; Vighi, C. Association of stolbur phytoplasmas with diseased tomatoes in Italy. *EPPO Bull.* **2000**, *30*, 347–350. [CrossRef]
22. Duduk, B.; Bertaccini, A. Corn with Symptoms of Reddening: New Host of Stolbur Phytoplasma. *Plant Dis.* **2006**, *90*, 1313–1319. [CrossRef]
23. Jović, J.; Cvrković, T.; Mitrović, M.; Krnjajić, S.; Redinbaugh, M.G.; Pratt, R.C.; Gingery, R.E.; Hogenhout, S.A.; Toševski, I. Roles of stolbur phytoplasma and *Reptalus panzeri* (Cixiidae, Auchenorrhyncha) in the epidemiology of Maize redness in Serbia. *Eur. J. Plant Pathol.* **2007**, *118*, 85–89. [CrossRef]
24. Jović, J.; Cvrković, T.; Mitrović, M.; Krnjajić, S.; Petrović, A.; Redinbaugh, M.G.; Pratt, R.C.; Hogenhout, S.A.; Tosevski, I. Stolbur phytoplasma transmission to maize by *Reptalus panzeri* and the disease cycle of maize redness in Serbia. *Phytopathology* **2009**, *99*, 1053–1061. [CrossRef] [PubMed]
25. Boudon-Padieu, E.; Cousin, M.T. Yellow decline of *Lavandula hybrida* Rev and L. vera DC. *Int. J. Trop. Plant Dis.* **1999**, *17*, 1–34.
26. Sémétey, O.; Gaudin, J.; Danet, J.-L.; Salar, P.; Theil, S.; Fontaine, M.; Krausz, M.; Chaisse, E.; Eveillard, S.; Verdin, E.; et al. Lavender Decline in France Is Associated with Chronic Infection by Lavender-Specific Strains of '*Candidatus* Phytoplasma solani'. *Appl. Environ. Microbiol.* **2018**, *84*, e01507-18. [CrossRef]

27. Davis, R.E.; Dally, E.L.; Gundersen, D.E.; Lee, I.M.; Habili, N. '*Candidatus* Phytoplasma australiense', a new phytoplasma taxon associated with Australian grapevine yellows. *Int. J. Syst. Evol. Microbiol.* **1997**, *47*, 262–269. [CrossRef]
28. Sawayanagi, T.; Horikoshi, N.; Kanehira, T.; Shinohara, M.; Bertaccini, A.; Cousin, M.T.; Hiruki, C.; Namba, S. '*Candidatus* Phytoplasma japonicum', a new phytoplasma taxon associated with Japanese Hydrangea phyllody. *Int. J. Syst. Evol. Microbiol.* **1999**, *49 Pt 3*, 1275–1285. [CrossRef]
29. Valiunas, D.; Staniulis, J.; Davis, R.E. '*Candidatus* Phytoplasma fragariae', a novel phytoplasma taxon discovered in yellows diseased strawberry, Fragaria x ananassa. *Int. J. Syst. Evol. Microbiol.* **2006**, *56*, 277–281. [CrossRef]
30. Ćurčić, Ž.; Stepanović, J.; Zübert, C.; Taški-Ajduković, K.; Kosovac, A.; Rekanović, E.; Kube, M.; Duduk, B. Rubbery Taproot Disease of Sugar Beet in Serbia Associated with '*Candidatus* Phytoplasma solani'. *Plant Dis.* **2021**, *105*, 255–263. [CrossRef]
31. Eini, O.; Shoaei, Z.; Varrelmann, M. Molecular detection and multilocus sequence analysis of '*Candidatus* Phytoplasma solani'-related strains infecting potato and sugar beet plants in Southern Germany. Res. Sq., 2024; *submitted*. [CrossRef]
32. Toth, R.; Ilic, A.-M.; Huettel, B.; Duduk, B.; Kube, M. Divergence within the Taxon '*Candidatus* Phytoplasma asteris' Confirmed by Comparative Genome Analysis of Carrot Strains. *Microorganisms* **2024**, *12*, 1016. [CrossRef]
33. Kube, M.; Mitrovic, J.; Duduk, B.; Rabus, R.; Seemüller, E. Current view on phytoplasma genomes and encoded metabolism. *Sci. World J.* **2012**, *2012*, 185942. [CrossRef]
34. Bertaccini, A.; Oshima, K.; Maejima, K.; Namba, S. Phytoplasma Effectors and Pathogenicity Factors. In *Phytoplasmas: Plant Pathogenic Bacteria—III*; Springer: Singapore, 2019; pp. 17–34. ISBN 978-981-13-9632-8.
35. Toth, R.; Huettel, B.; Eini, O.; Varrelmann, M.; Kube, M. The complete genome sequence of the stolbur pathogen '*Candidatus* Phytoplasma solani' from *Pentastiridius leporinus*. *Microbiol. Resour. Announc.* **2024**, *14*, e0064024. [CrossRef] [PubMed]
36. Carver, T.; Harris, S.R.; Berriman, M.; Parkhill, J.; McQuillan, J.A. Artemis: An integrated platform for visualization and analysis of high-throughput sequence-based experimental data. *Bioinformatics* **2012**, *28*, 464–469. [CrossRef] [PubMed]
37. Goris, J.; Konstantinidis, K.T.; Klappenbach, J.A.; Coenye, T.; Vandamme, P.; Tiedje, J.M. DNA-DNA hybridization values and their relationship to whole-genome sequence similarities. *Int. J. Syst. Evol. Microbiol.* **2007**, *57*, 81–91. [CrossRef] [PubMed]
38. Jain, C.; Rodriguez-R, L.M.; Phillippy, A.M.; Konstantinidis, K.T.; Aluru, S. High throughput ANI analysis of 90K prokaryotic genomes reveals clear species boundaries. *Nat. Commun.* **2018**, *9*, 5114. [CrossRef]
39. Darling, A.C.E.; Mau, B.; Blattner, F.R.; Perna, N.T. Mauve: Multiple alignment of conserved genomic sequence with rearrangements. *Genome Res.* **2004**, *14*, 1394–1403. [CrossRef]
40. Gundersen, D.E.; Lee, I.-M. Ultrasensitive detection of phytoplasmas by nested-PCR assays using two universal primer pairs. *Phytopathol. Mediterr.* **1996**, *35*, 144–151.
41. Schneider, B.; Gibb, K.S. Sequence and RFLP analysis of the elongation factor Tu gene used in differentiation and classification of phytoplasmas. *Microbiology* **1997**, *143 Pt 10*, 3381–3389. [CrossRef]
42. Altschul, S.F.; Gish, W.; Miller, W.; Myers, E.W.; Lipman, D.J. Basic local alignment search tool. *J. Mol. Biol.* **1990**, *215*, 403–410. [CrossRef]
43. Edgar, R.C. MUSCLE: Multiple sequence alignment with high accuracy and high throughput. *Nucleic Acids Res.* **2004**, *32*, 1792–1797. [CrossRef]
44. Kumar, S.; Stecher, G.; Li, M.; Knyaz, C.; Tamura, K. MEGA X: Molecular Evolutionary Genetics Analysis across Computing Platforms. *Mol. Biol. Evol.* **2018**, *35*, 1547–1549. [CrossRef]
45. Zhao, Y.; Wei, W.; Lee, I.-M.; Shao, J.; Suo, X.; Davis, R.E. The iPhyClassifier, an interactive online tool for phytoplasma classification and taxonomic assignment. *Phytoplasma* **2013**, *938*, 329–338. [CrossRef]
46. Emms, D.M.; Kelly, S. OrthoFinder: Phylogenetic orthology inference for comparative genomics. *Genome Biol.* **2019**, *20*, 238. [CrossRef] [PubMed]
47. Khan, A.; Mathelier, A. Intervene: A tool for intersection and visualization of multiple gene or genomic region sets. *BMC Bioinform.* **2017**, *18*, 287. [CrossRef] [PubMed]
48. Zdobnov, E.M.; Apweiler, R. InterProScan—An integration platform for the signature-recognition methods in InterPro. *Bioinform.* **2001**, *17*, 847–848. [CrossRef] [PubMed]
49. Kanehisa, M.; Sato, Y.; Morishima, K. BlastKOALA and GhostKOALA: KEGG Tools for Functional Characterization of Genome and Metagenome Sequences. *J. Mol. Biol.* **2016**, *428*, 726–731. [CrossRef]
50. Caspi, R.; Billington, R.; Keseler, I.M.; Kothari, A.; Krummenacker, M.; Midford, P.E.; Ong, W.K.; Paley, S.; Subhraveti, P.; Karp, P.D. The MetaCyc database of metabolic pathways and enzymes—A 2019 update. *Nucleic Acids Res.* **2020**, *48*, D445–D453. [CrossRef]
51. Tully, J.G.; Whitcomb, R.F.; Rose, D.L.; Bové, J.M.; Carle, P.; Somerson, N.L.; Williamson, D.L.; Eden-Green, S. *Acholeplasma brassicae* sp. nov. and *Acholeplasma palmae* sp. nov., two non-sterol-requiring Mollicutes from plant surfaces. *Int. J. Syst. Evol. Microbiol.* **1994**, *44*, 680–684. [CrossRef]
52. Käll, L.; Krogh, A.; Sonnhammer, E.L.L. Advantages of combined transmembrane topology and signal peptide prediction—The Phobius web server. *Nucleic Acids Res.* **2007**, *35*, W429–W432. [CrossRef]

53. Teufel, F.; Almagro Armenteros, J.J.; Johansen, A.R.; Gíslason, M.H.; Pihl, S.I.; Tsirigos, K.D.; Winther, O.; Brunak, S.; von Heijne, G.; Nielsen, H. SignalP 6.0 predicts all five types of signal peptides using protein language models. *Nat. Biotechnol.* **2022**, *40*, 1023–1025. [CrossRef]
54. Notredame, C.; Higgins, D.G.; Heringa, J. T-Coffee: A novel method for fast and accurate multiple sequence alignment. *Int. J. Syst. Evol. Microbiol.* **2000**, *302*, 205–217. [CrossRef]
55. Jumper, J.; Evans, R.; Pritzel, A.; Green, T.; Figurnov, M.; Ronneberger, O.; Tunyasuvunakool, K.; Bates, R.; Žídek, A.; Potapenko, A.; et al. Highly accurate protein structure prediction with AlphaFold. *Nature* **2021**, *596*, 583–589. [CrossRef] [PubMed]
56. Xie, Z.; Tang, H. ISEScan: Automated identification of insertion sequence elements in prokaryotic genomes. *Bioinformatics* **2017**, *33*, 3340–3347. [CrossRef] [PubMed]
57. Gilchrist, C.L.M.; Chooi, Y.-H. clinker & clustermap.js: Automatic generation of gene cluster comparison figures. *Bioinformatics* **2021**, *37*, 2473–2475. [CrossRef] [PubMed]
58. Zübert, C.; Ilic, A.-M.; Duduk, B.; Kube, M. The Genome Reduction Excludes the Ribosomal Rescue System in Acholeplasmataceae. *Microb. Physiol.* **2022**, *32*, 45–56. [CrossRef]
59. Liefting, L.W.; Andersen, M.T.; Lough, T.J.; Beever, R.E. Comparative analysis of the plasmids from two isolates of '*Candidatus* Phytoplasma australiense'. *Plasmid* **2006**, *56*, 138–144. [CrossRef]
60. Tran-Nguyen, L.T.T.; Gibb, K.S. Extrachromosomal DNA isolated from tomato big bud and '*Candidatus* Phytoplasma australiense' phytoplasma strains. *Plasmid* **2006**, *56*, 153–166. [CrossRef]
61. Carminati, G. '*Candidatus* Phytoplasma solani': From Insect Transmission to Complete Genome Sequencing of Several Strains Inducing Different Symptoms in Tomato. Ph.D. Thesis, University of Udine, Udine, Italy, 13 March 2023.
62. Tran-Nguyen, L.T.T.; Kube, M.; Schneider, B.; Reinhardt, R.; Gibb, K.S. Comparative genome analysis of '*Candidatus* Phytoplasma australiense '(subgroup tuf-Australia I; rp-A) and '*Ca.* Phytoplasma asteris' strains OY-M and AY-WB. *J. Bacteriol.* **2008**, *190*, 3979–3991. [CrossRef]
63. Andersen, M.T.; Liefting, L.W.; Havukkala, I.; Beever, R.E. Comparison of the complete genome sequence of two closely related isolates of '*Candidatus* Phytoplasma australiense' reveals genome plasticity. *BMC Genom.* **2013**, *14*, 529. [CrossRef]
64. Bertaccini, A.; Arocha-Rosete, Y.; Contaldo, N.; Duduk, B.; Fiore, N.; Montano, H.G.; Kube, M.; Kuo, C.-H.; Martini, M.; Oshima, K.; et al. Revision of the '*Candidatus* Phytoplasma' species description guidelines. *Int. J. Syst. Evol. Microbiol.* **2022**, *72*, 5353. [CrossRef]
65. Seemüller, E.; Schneider, B.; Mäurer, R.; Ahrens, U.; Daire, X.; Kison, H.; Lorenz, K.H.; Firrao, G.; Avinent, L.; Sears, B.B. Phylogenetic classification of phytopathogenic mollicutes by sequence analysis of 16S ribosomal DNA. *Int. J. Syst. Bacteriol.* **1994**, *44*, 440–446. [CrossRef]
66. Liefting, L.W.; Andersen, M.T.; Beever, R.E.; Gardner, R.C.; Forster, R.L. Sequence heterogeneity in the two 16S rRNA genes of Phormium yellow leaf phytoplasma. *Appl. Environ. Microbiol.* **1996**, *62*, 3133–3139. [CrossRef] [PubMed]
67. Jomantiene, R.; Davis, R.E.; Valiunas, D.; Alminaite, A. New Group 16SrIII Phytoplasma Lineages in Lithuania Exhibit rRNA Interoperon Sequence Heterogeneity. *Eur. J. Plant Pathol.* **2002**, *108*, 507–517. [CrossRef]
68. Saigo, M.; Golic, A.; Alvarez, C.E.; Andreo, C.S.; Hogenhout, S.A.; Mussi, M.A.; Drincovich, M.F. Metabolic regulation of phytoplasma malic enzyme and phosphotransacetylase supports the use of malate as an energy source in these plant pathogens. *Microbiology* **2014**, *160*, 2794–2806. [CrossRef] [PubMed]
69. Kakizawa, S.; Oshima, K.; Namba, S. Diversity and functional importance of phytoplasma membrane proteins. *Trends Microbiol.* **2006**, *14*, 254–256. [CrossRef]
70. Fabre, A.; Danet, J.-L.; Foissac, X. The stolbur phytoplasma antigenic membrane protein gene stamp is submitted to diversifying positive selection. *Gene* **2011**, *472*, 37–41. [CrossRef]
71. Boonrod, K.; Munteanu, B.; Jarausch, B.; Jarausch, W.; Krczal, G. An immunodominant membrane protein (Imp) of '*Candidatus* Phytoplasma mali' binds to plant actin. *Mol. Plant Microbe Interact.* **2012**, *25*, 889–895. [CrossRef]
72. Konnerth, A.; Krczal, G.; Boonrod, K. Immunodominant membrane proteins of phytoplasmas. *Microbiology* **2016**, *162*, 1267–1273. [CrossRef]
73. Neriya, Y.; Maejima, K.; Nijo, T.; Tomomitsu, T.; Yusa, A.; Himeno, M.; Netsu, O.; Hamamoto, H.; Oshima, K.; Namba, S. Onion yellow phytoplasma P38 protein plays a role in adhesion to the hosts. *FEMS Microbiol. Lett.* **2014**, *361*, 115–122. [CrossRef]
74. Cimerman, A.; Pacifico, D.; Salar, P.; Marzachì, C.; Foissac, X. Striking diversity of *vmp1*, a variable gene encoding a putative membrane protein of the stolbur phytoplasma. *Appl. Environ. Microbiol.* **2009**, *75*, 2951–2957. [CrossRef]
75. Hückelhoven, R.; Dechert, C.; Kogel, K.-H. Overexpression of barley BAX inhibitor 1 induces breakdown of mlo-mediated penetration resistance to *Blumeria graminis*. *Proc. Natl. Acad. Sci. USA* **2003**, *100*, 5555–5560. [CrossRef]
76. Quaglino, F.; Kube, M.; Jawhari, M.; Abou-Jawdah, Y.; Siewert, C.; Choueiri, E.; Sobh, H.; Casati, P.; Tedeschi, R.; Lova, M.M.; et al. '*Candidatus* Phytoplasma phoenicium' associated with almond witches'-broom disease: From draft genome to genetic diversity among strain populations. *BMC Microbiol.* **2015**, *15*, 148. [CrossRef] [PubMed]

77. Gao, X.; Ren, Z.; Zhao, W.; Li, W. '*Candidatus* Phytoplasma ziziphi' encodes non-classically secreted proteins that suppress hypersensitive cell death response in *Nicotiana benthamiana*. *Phytopathol. Res.* **2023**, *5*, 11. [CrossRef]
78. Passera, A.; Zhao, Y.; Murolo, S.; Pierro, R.; Arsov, E.; Mori, N.; Moussa, A.; Silletti, M.R.; Casati, P.; Panattoni, A.; et al. Multilocus Genotyping Reveals New Molecular Markers for Differentiating Distinct Genetic Lineages among '*Candidatus* Phytoplasma solani' Strains Associated with Grapevine Bois Noir. *Pathogens* **2020**, *9*, 970. [CrossRef] [PubMed]
79. Kakizawa, S.; Oshima, K.; Nishigawa, H.; Jung, H.-Y.; Wei, W.; Suzuki, S.; Tanaka, M.; Miyata, S.-I.; Ugaki, M.; Namba, S. Secretion of immunodominant membrane protein from onion yellows phytoplasma through the Sec protein-translocation system in *Escherichia coli*. *Microbiology* **2004**, *150*, 135–142. [CrossRef]
80. Koch, H.-G.; Moser, M.; Müller, M. Signal recognition particle-dependent protein targeting, universal to all kingdoms of life. *Rev. Physiol. Biochem. Pharmacol.* **2003**, *146*, 55–94. [CrossRef]
81. Bai, X.; Zhang, J.; Ewing, A.; Miller, S.A.; Jancso Radek, A.; Shevchenko, D.V.; Tsukerman, K.; Walunas, T.; Lapidus, A.; Campbell, J.W.; et al. Living with Genome Instability: The Adaptation of Phytoplasmas to Diverse Environments of Their Insect and Plant Hosts. *J. Bacteriol.* **2006**, *188*, 3682–3696. [CrossRef]
82. Sugio, A.; Kingdom, H.N.; MacLean, A.M.; Grieve, V.M.; Hogenhout, S.A. Phytoplasma protein effector SAP11 enhances insect vector reproduction by manipulating plant development and defense hormone biosynthesis. *Proc. Natl. Acad. Sci. USA* **2011**, *108*, E1254–E1263. [CrossRef]
83. Drcelic, M.; Skiljaica, A.; Polak, B.; Bauer, N.; Seruga Music, M. '*Candidatus* Phytoplasma solani' Predicted Effector SAP11-like Alters Morphology of Transformed Arabidopsis Plants and Interacts with AtTCP2 and AtTCP4 Plant Transcription Factors. *Pathogens* **2024**, *13*, 893. [CrossRef]
84. MacLean, A.M.; Sugio, A.; Makarova, O.V.; Findlay, K.C.; Grieve, V.M.; Tóth, R.; Nicolaisen, M.; Hogenhout, S.A. Phytoplasma effector SAP54 induces indeterminate leaf-like flower development in Arabidopsis plants. *Plant Physiol.* **2011**, *157*, 831–841. [CrossRef]
85. Huang, W.; MacLean, A.M.; Sugio, A.; Maqbool, A.; Busscher, M.; Cho, S.-T.; Kamoun, S.; Kuo, C.-H.; Immink, R.G.H.; Hogenhout, S.A. Parasitic modulation of host development by ubiquitin-independent protein degradation. *Cell* **2021**, *184*, 5201–5214.e12. [CrossRef]
86. Music, M.S.; Samarzija, I.; Hogenhout, S.A.; Haryono, M.; Cho, S.-T.; Kuo, C.-H. The genome of '*Candidatus* Phytoplasma solani' strain SA-1 is highly dynamic and prone to adopting foreign sequences. *Syst. Appl. Microbiol.* **2019**, *42*, 117–127. [CrossRef] [PubMed]
87. Mahillon, J.; Chandler, M. Insertion sequences. *Microbiol. Mol. Biol. Rev.* **1998**, *62*, 725–774. [CrossRef]
88. Toruño, T.Y.; Music, M.S.; Simi, S.; Nicolaisen, M.; Hogenhout, S.A. Phytoplasma PMU1 exists as linear chromosomal and circular extrachromosomal elements and has enhanced expression in insect vectors compared with plant hosts. *Mol. Microbiol.* **2010**, *77*, 1406–1415. [CrossRef] [PubMed]
89. Kube, M.; Schneider, B.; Kuhl, H.; Dandekar, T.; Heitmann, K.; Migdoll, A.M.; Reinhardt, R.; Seemüller, E. The linear chromosome of the plant-pathogenic mycoplasma '*Candidatus* Phytoplasma mali'. *BMC Genom.* **2008**, *9*, 306. [CrossRef] [PubMed]
90. Böhm, J.W.; Duckeck, D.; Duduk, B.; Schneider, B.; Kube, M. Genome Comparison of '*Candidatus* Phytoplasma rubi' with Genomes of Other 16SrV Phytoplasmas Highlights Special Group Features. *Appl. Microbiol.* **2023**, *3*, 1083–1100. [CrossRef]
91. Dong, H.; Beer, S.V. Riboflavin induces disease resistance in plants by activating a novel signal transduction pathway. *Phytopathology* **2000**, *90*, 801–811. [CrossRef]
92. Sugio, A.; MacLean, A.M.; Kingdom, H.N.; Grieve, V.M.; Manimekalai, R.; Hogenhout, S.A. Diverse targets of phytoplasma effectors: From plant development to defense against insects. *Annu. Rev. Phytopathol.* **2011**, *49*, 175–195. [CrossRef]
93. Cho, S.-T.; Kung, H.-J.; Huang, W.; Hogenhout, S.A.; Kuo, C.-H. Species Boundaries and Molecular Markers for the Classification of 16SrI Phytoplasmas Inferred by Genome Analysis. *Front. Microbiol.* **2020**, *11*, 1531. [CrossRef]
94. MacLean, A.M.; Orlovskis, Z.; Kowitwanich, K.; Zdziarska, A.M.; Angenent, G.C.; Immink, R.G.H.; Hogenhout, S.A. Phytoplasma effector SAP54 hijacks plant reproduction by degrading MADS-box proteins and promotes insect colonization in a RAD23-dependent manner. *PLoS Biol.* **2014**, *12*, e1001835. [CrossRef]
95. Orlovskis, Z.; Singh, A.; Kliot, A.; Huang, W.; Hogenhout, S.A. Molecular Matchmakers: Phytoplasma Effector SAP54 Targets MADS-Box Factor SVP to Enhance Attraction of Fecund Female Vectors by Modulating Leaf Responses to Male Presence. *bioRxiv* **2024**. [CrossRef]
96. Tokuda, R.; Iwabuchi, N.; Kitazawa, Y.; Nijo, T.; Suzuki, M.; Maejima, K.; Oshima, K.; Namba, S.; Yamaji, Y. Potential mobile units drive the horizontal transfer of phytoplasma effector phyllogen genes. *Front. Genet.* **2023**, *14*, 1132432. [CrossRef] [PubMed]
97. Chung, W.-C.; Chen, L.-L.; Lo, W.-S.; Lin, C.-P.; Kuo, C.-H. Comparative analysis of the peanut witches'-broom phytoplasma genome reveals horizontal transfer of potential mobile units and effectors. *PLoS ONE* **2013**, *8*, e62770. [CrossRef] [PubMed]

98. Huang, C.-T.; Cho, S.-T.; Lin, Y.-C.; Tan, C.-M.; Chiu, Y.-C.; Yang, J.-Y.; Kuo, C.-H. Comparative Genome Analysis of '*Candidatus* Phytoplasma luffae' Reveals the Influential Roles of Potential Mobile Units in Phytoplasma Evolution. *Front. Microbiol.* **2022**, *13*, 773608. [CrossRef] [PubMed]
99. Kohl, S.; Bock, R. Transposition of a bacterial insertion sequence in chloroplasts. *Plant J.* **2009**, *58*, 423–436. [CrossRef]

pathogens

MDPI

Article

Involvement of MicroRNAs in the Hypersensitive Response of Capsicum Plants to the Capsicum Chlorosis Virus at Elevated Temperatures

Wei-An Tsai *, Christopher A. Brosnan, Neena Mitter and Ralf G. Dietzgen *

Centre for Horticultural Science, Queensland Alliance for Agriculture and Food Innovation, The University of Queensland, St. Lucia, QLD 4072, Australia
* Correspondence: w.tsai@uq.edu.au (W.-A.T.); r.dietzgen@uq.edu.au (R.G.D.); Tel.: +61-0413364957 (W.-A.T.); +61-0478504111 (R.G.D.)

Abstract: The orthotospovirus capsicum chlorosis virus (CaCV) is an important pathogen affecting capsicum plants. Elevated temperatures may affect disease progression and pose a potential challenge to capsicum production. To date, CaCV-resistant capsicum breeding lines have been established; however, the impact of an elevated temperature of 35 °C on this genetic resistance remains unexplored. Thus, this study aimed to investigate how high temperature (HT) influences the response of CaCV-resistant capsicum to the virus. Phenotypic analysis revealed a compromised resistance in capsicum plants grown at HT, with systemic necrotic spots appearing in 8 out of 14 CaCV-infected plants. Molecular analysis through next-generation sequencing identified 105 known and 83 novel microRNAs (miRNAs) in CaCV-resistant capsicum plants. Gene ontology revealed that phenylpropanoid and lignin metabolic processes, regulated by Can-miR408a and Can- miR397, are likely involved in elevated-temperature-mediated resistance-breaking responses. Additionally, real-time PCR validated an upregulation of Can-miR408a and Can-miR397 by CaCV infection at HT; however, only the *Laccase 4* transcript, targeted by Can-miR397, showed a tendency of negative correlation with this miRNA. Overall, this study provides the first molecular insights into how elevated temperature affects CaCV resistance in capsicum plants and reveals the potential role of miRNA in temperature-sensitive tospovirus resistance.

Keywords: capsicum chlorosis orthotospovirus; resistance-breaking; small RNA libraries; high-throughput sequencing

1. Introduction

Capsicum or pepper, belonging to the genus *Capsicum* and family Solanaceae, is a nutritionally important vegetable that originated in Central and South America, the Caribbean, and Mexico, with at least 27 reported species [1,2]. Capsicum production has increased due to strong demand and culinary use of its fruit [3,4]; however, its growth is frequently threatened by different abiotic and biotic stresses and can be affected by those stresses simultaneously [3–6]. Therefore, it is of increasing importance to study the impact of combined stresses on capsicum plants, such as virus infection and high temperature.

Capsicum chlorosis virus (CaCV) is a serious pathogen that infects not only capsicum but also chili, tomato, pineapple, and peanut crops [7–9]. The virus has been reported in Australia, China, Greece, India, Iran, Taiwan, Thailand, and the USA [10]. In southern China, CaCV is considered a significant disease in peanuts, with incidence reaching up to 20% and causing noticeable yield losses, particularly when plants are infected at early growing stages [10,11]. In India, the virus affects the production of chili peppers, with disease incidences over 20% reported [10,12]. In Australia, the emergence of CaCV has been reported in large commercial capsicum production areas, including Bundaberg and the northern dry tropics of Queensland [9]. The typical CaCV symptoms on capsicum plants include stunting, marginal and interveinal chlorosis and leaf deformation on emerging leaves, and concentric chlorotic or necrotic lesions on mature leaves [9,13]. Recently, CaCV-resistant advanced breeding lines (PI 290972 × *C. annuum* cv. Mazurka and cv. Warlock inbred lines) were selected by the Queensland Department of Agriculture and Fisheries (DAF) breeding program [13,14]. CaCV is taxonomically classified in the species *Capsicum chlorosis orthotospovirus* in the genus *Orthotospovirus*, family *Tospoviridae*, order *Bunyavirales* [15]. The genome of this orthotospovirus consists of three segments of single-stranded RNA (ssRNA). Among them, the large (L) segment is of negative polarity and encodes the RNA-dependent RNA polymerase (RdRp), while both medium (M) and small (S) segments are of ambisense coding polarity with viral movement protein (NSm) and glycoproteins (Gn/Gc) encoded on the M segment, and a viral silencing suppresser (NSs) and nucleocapsid protein (N) encoded on the S segment [13,16,17].

The progression of virus diseases in plants is significantly affected by temperature since both virus propagation and plant development are temperature-dependent [18]. Given current concerns about global warming, elevated temperatures have been recognized as an important climate-changing variable impacting plant–virus interactions [19–21]. To date, evidence shows that exposure to elevated temperatures suppresses effector-triggered immunity (ETI)-mediated plant resistance [22–24]. ETI is a robust plant defense response triggered by recognizing pathogen effector molecules through nucleotide-binding leucine-rich repeat (NLR) proteins [25]. Typical ETI responses are associated with calcium production, salicylic acid (SA) accumulation, and a burst of oxidative reactive oxygen species (ROS). Furthermore, these plant responses result in two typical manifestations of disease resistance, hypersensitive response (HR) and systemic acquired resistance (SAR) [25]. Examples of temperature-mediated breaking of disease resistance include the potato virus Y (PVY)–potato and tomato spotted wilt virus (TSWV)–capsicum pathosystems [24,26–29]. In these pathosystems, virus-resistant plant varieties, carrying resistance (R) genes, are unable to mount effective ETI at elevated temperatures. The capsicum *Tsw* gene and potato R gene, *Ny-DG*—which confer resistance to TSWV and PVY, respectively—are compromised at 32 °C or above 28 °C [26,29].

MicroRNAs (miRNAs), which are post-transcriptional regulators, are found in most eukaryotes [30]. In plants, miRNAs are conserved regulators of plant developmental processes and stress responses [31,32]. The 20–24 nucleotide (nt) miRNAs originate from miRNA genes (MIR). These genes are transcribed by RNA polymerase II (Pol II) into a stem–loop structure, named primary (pri)-miRNAs. Pri-miRNAs are processed by DICER-LIKE 1 (DCL1) and several assisting proteins to form the precursor (pre)-miRNAs and mature miRNA duplex. At the last step of this pathway, the guide strand of the miRNA duplex is loaded onto specific ARGONAUTE (AGO) proteins for target mRNA cleavage or translational inhibition [33]. MiRNAs are essential elements affecting the plant–virus arms race [34,35]. Several miRNAs have been reported to act as regulators to support plant antiviral resistance [36,37]. For example, the expression of miR6019/miR6020, which down-regulates expression of the NLR receptor N gene in tobacco plants, is reduced, allowing effective induction of NLR-mediated ETI during tobacco mosaic virus (TMV) infection [37]. During infection of tomato leaf curl New Delhi virus, Sly-miR159 appears to be down-regulated, which increases the expression of its target *SlMyB33* and the downstream R gene *SlSw5a* in resistant tomatoes. This modulation enhances the plant defense against the virus by triggering HR [38]. Conversely, some miRNAs act as regulators that are favorable for virus infections, like miR319 [39,40]. The expression of miR319 in rice plants is induced by rice ragged stunt virus infections. This leads to a suppression of jasmonic acid (JA)-mediated defenses, which results in an increased susceptibility of rice plants to virus disease [39].

Although the interplay between plant miRNAs and negative-sense RNA viruses has been explored [41], the role of miRNAs in capsicum resistance to CaCV at different temperatures remains unknown. Therefore, this study investigated the effects of high temperature on capsicum resistance to CaCV through phenotypic and miRNA analysis. Small RNA (sRNA) high-throughput sequencing (HTS) was used for a systemic comparative analysis of differentially expressed miRNAs in plants challenged with or without CaCV at ambient (25 °C) and elevated temperatures (35 °C/30 °C), elucidating miRNA-related underlying mechanisms involved in the capsicum–CaCV pathosystem.

2. Materials and Methods

2.1. Plants and Growth Conditions

CaCV-resistant capsicum plants from the third backcross generation of *Capsicum chinense* PI 290972 × commercial CaCV-susceptible *C. annuum* cultivars [14] were grown in a temperature-controlled glasshouse compartment at an ambient temperature (AT) of approximately 25 °C. Seedlings aged four weeks with two true leaves were mechanically inoculated with CaCV isolate QLD 3432 [17]. Inoculum was prepared by grinding fresh CaCV-infected symptomatic capsicum leaves of the susceptible cultivar Yolo Wonder in 10 mM phosphate buffer, pH 7.6, containing freshly added 20 mM sodium sulphite with a mortar and pestle. Mock treatment used buffer only. Inoculum and buffer were rub-inoculated onto the carborundum-dusted first two leaves and two cotyledons. Subsequently, half of each plant group was transferred to a growth cabinet at high temperature (HT) of 35 °C/30 °C (16 h day/8 h night), light intensity of 230 $\mu mol \cdot m^{-2} \cdot s^{-1}$, and relative humidity of 60%.

2.2. RNA Isolation, Small RNA Library Construction. and Sequencing

Capsicum plants, treated under four conditions, were used for sRNA library construction: mock-inoculated plants grown at AT ('AM' hereafter); CaCV-inoculated plants grown at AT ('AV' hereafter); mock-inoculated plants grown at HT ('HM' hereafter); and CaCV-inoculated plants grown at HT ('HV' hereafter). Systemic leaves from three individual plants, each grown in one of those four conditions, were collected at 10 dpi. For each biological replicate, three leaf disks from the third leaf from the top were sampled. Total RNA from sample lysates in lysis buffer (ISOLATE II RNA Plant Kit, Bioline, London, UK) was purified by Direct-zol RNA Miniprep kit (Zymo Research, Irvine, CA, USA) according to the manufacturer's instructions. On-column DNase I treatment was performed using the same kit. Total RNA (2–3 µg per sample) was sent to the Australian Genome Research Facility (AGRF, Melbourne, VIC, Australia) for sRNA library construction and HTS. The sRNA libraries were prepared using NEBNext® Small RNA Library Prep Set (New England Biolabs, Ipswich, MA, USA) for Illumina® following the manufacturer's instructions. Single-end 50 bp sequencing was performed on an Illumina NOVASEQ 6000, AGRF, Melbourne, VIC, Australia.

2.3. Small RNA Sequencing Data Pre-Processing and microRNA Identification

Raw reads generated by sRNA sequencing were pre-processed using the source code of iwa software (https://github.com/cma2015/iwa-miRNA/tree/master/Source_code/modulei, accessed on 21 January 2021) [42]. First, the FASTX-Toolkit v0.0.14 (http://hannonlab.cshl.edu/fastx_toolkit, accessed on 31 January 2021) was used to trim low-quality reads and adaptor (NEBnext: AGATCGGAAGAG). The threshold for low-quality read trimming was set as a minimum quality score of 20, with more than 80 percent of bases covered. Second, length_cutoff.sh script from iwa software was used to remove the reads with lengths <18 nt or >26 nt. Subsequently, all the length-limited reads were aligned against the *C. annuum* (CM334) reference genome v1.6 (http://peppergenome.snu.ac.kr/download.php, accessed on 14 January 2021) using Bowtie v1.2.2 with parameters -v1 –best –strata -m20 [43]. The mapped reads were used for quantification of capsicum miRNAs, and the reads that failed to map to the capsicum genome were used for CaCV viral siRNA (vsiRNA) quantification.

The MiRNA Compilation and MiRNA Selection modules in iwa software were used to identify the putative known and novel miRNAs [42]. Raw reads were uploaded into iwa software and were pre-processed by the same parameters mentioned above. In the MiRNA Compilation module, miRDeep-P2 was selected for miRNA prediction [44]. The

PmiREN (http://www.pmiren.com/) [45], sRNAanno (http://www.plantsrnas.org/) [46], and PsRNA (https://plantsmallrnagenes.science.psu.edu/) [47] databases were used for annotating capsicum miRNAs. In the MiRNA Selection module, high-throughput criteria and the one-class support vector machine (SVM) classifier were applied to determine if tested miRNA candidates are real miRNAs [48,49]. InteractiVenn was then used to visualize all identified miRNAs using Venn diagrams [50]. After miRNA identification, all novel and known miRNAs that were identified in libraries across at least two conditions were used as the reference for miRNA quantification. The mapper and quantifier modules in MiRDeep2 v0.0.7 [51] were then used to collapse genome-mapped reads and count the reads of the identified miRNAs.

2.4. *Analysis of Virus-Derived siRNAs*

The genome of CaCV-Qld-3432, containing 8913 nt in the L segment, 4846 nt in the M segment, and 3944 nt in the S segment (GenBank accession numbers KM589495, KM589494, KM589493) [17], was used for sRNA alignment. The profile of 21-nucleotide, 22-nucleotide, and 24-nucleotide vsiRNA normalized read counts was obtained using SCRAM aligner module (https://sfletc.github.io/scram/). Subsequently, the SCRAM plotting module was used to show the read coverage across CaCV reference genome segments [52].

2.5. *Differential Expression Analysis of miRNAs*

Differentially expressed (DE) miRNAs in the two treatments were analyzed using DEseq2 (Galaxy Version 2.11.40.6 + galaxy1) [53]. Here, four pairwise comparisons were performed: AV vs. AM; HV vs. HM; HM vs. AM; and HV vs. AV. Total read counts for each miRNA were normalized by median ratio normalization. The resulting *p*-values were corrected for multiple testing using Benjamini and Hochberg's false discovery rate (FDR) [54]. MiRNAs were judged to be DE if the FDR-adjusted *p*-value (Padj) was $\leq$0.01 and log2 fold-change was >1.0 or <−1.0. Heatmaps were created using bioinfokit python package (Version 2.0.8) to visualize significantly DE miRNAs [55].

2.6. *miRNA Target Prediction and Enrichment Analysis*

PsRNAtarget (2017 release, https://www.zhaolab.org/psRNATarget/home, accessed on 2 February 2021), with a strict expectation score $\leq$ 3, was used to predict miRNA targets [56]. Mature miRNA sequences were aligned against transcript sequences retrieved from the Plaza 4.0 dicot database. Subsequently, the Plaza 4.0 workbench was used to obtain gene ontology (GO) terms according to the molecular function, biological process, and cellular component with default parameters [57].

Single enrichment analysis (SEA) was performed using agriGO v2.0 (http://systemsbiology.cau.edu.cn/agriGOv2/, accessed on 3 February 2021) [58]. The GO terms of targets predicted from all identified miRNAs were input as a custom background list. Since separate analyses of up- and downregulated genes could identify pathways that are more relevant to phenotypic differences, the GO terms of those targets predicted from up- or downregulated miRNAs were input separately as a query list [59]. For SEA, Fisher's exact test with a minimum of 5 mapping entries per term was selected as the statistical test, and the Benjamini–Yekutieli FDR with a significance level of 0.05 was selected as the correction method [60].

2.7. *Evaluation of the Expression of miRNAs and Their Potential Target Genes*

The abundance of selected miRNAs was examined using linear specific (S)-poly (A)-tailed real-time RT-PCR, as described by Xie and collaborators [61–63]. Total RNA purification and DNase treatment were performed using the methods described in Section 2.2 above. DNase-treated RNAs were then polyadenylated using poly (A) polymerase (New England BioLabs, Ipswich, MA, USA) according to the manufacturer's instructions. For each miRNA, cDNA was synthesized from polyadenylated RNA (1 μg) by using Superscript IV reverse transcriptase (Invitrogen, Carlsbad, CA, USA) with S-poly (A)-tailed RT primers (Table S1). These RT primers (SRT) that target specific miRNAs were designed using sRNAprimerDB software version 1.0 (http://www.srnaprimerdb.com/, accessed in May 2021) [63]. After cDNA synthesis, real-time PCR was performed using a SensiFAST SYBR No-ROX kit (Bioline, London, UK) following the manufacturer's instructions with cycling conditions of 95 °C for 2 min, followed by 40 cycles of 95 °C for 5 s, 61 °C for 10 s, and 72 °C for 10 s. The miRNA-specific primers (SqPF) and universal reverse primer (universal_SqPR) were used for real-time PCR quantification of mature miRNAs (Table S1). For target gene quantification, cDNA was synthesized from DNase-treated RNA using the SensiFAST™ cDNA Synthesis Kit (Bioline, London, UK) following the manufacturer's instructions with reaction conditions of 25 °C for 10 min, 42 °C for 15 min, 85 °C for 5 min. Then, real-time PCR was performed using the method described above. Based on the transcript sequences obtained from our previous study [13], primers were designed using Geneious Prime and are listed in Table S1. U6 and Actin genes were used as internal controls for miRNAs and target genes quantification, respectively [64,65] (Table S1). Relative expression levels of mature miRNAs and target genes were calculated by the $2^{-\Delta\Delta CT}$ method [66].

Northern blot hybridization was used for the detection and quantification of mature Can-miR164b/c since a specific product of mature miRNA164b/c was unable to be amplified by using linear S-poly (A)-tailed real-time RT-PCR. Total RNA was separated on a 17% polyacrylamide gel [67]. Subsequentially, gel-separated RNA was transferred onto a Hybond-N+ nylon membrane (Roche) using a Bio-Rad mini trans-blot system. For U6 and Can-miR164b/c detection, DIG-labeled DNA probes of U6 (5′-TCATCCTTGCGCAGGGGCCA) and Can-miR164b/c (5′-TGCACGTGCCCTGCTTCTCCA) were generated using the DIG Oligonucleotide 3′-End Labelling Kit (Roche, Basel, Switzerland). The relative accumulation level of Can-miR164b/c was then calculated by densitometry of the Northern blot chemiluminescent images using iBright analysis software (Version 5.0.0). The densitometry value of the AM condition was set as reference in each blot. The normalized U6 values served as an internal control for the normalized values of Can-miR164b/c.

Statistical tests were performed using GraphPadPrism software (Version 9.3.1). Data were analyzed with unpaired Student's *t*-tests and considered significantly different if the two-tailed *p*-value was <0.05.

2.8. Cleavage Sites Validation of Target Genes of miRNAs

Modified 5′ RNA ligase-mediated amplification of cDNA ends (5′-RLM-RACE) was used to validate the cleavage sites of predicted miRNA target genes [68,69]. This modified procedure starts with ligating the 5′-RNA adapter (RLM_RNA_adapter, Table S1) to total RNAs using T4 RNA ligase (New England Biolabs, Ipswich, MA, USA). Then, the adapter-ligated RNAs were used to synthesize first-strand cDNA using Superscript IV reverse transcriptase (Invitrogen) with the oligo dT primer following the manufacturer's instructions. The cDNA was used as the template for touchdown PCR using Phusion High-Fidelity DNA Polymerase (New England Biolabs, Ipswich, MA, USA) with the primers RLM, gene-specific primers (RLM_GSP)—including RLM_AGO1b_GSP, RLM_LAC4_GSP, and RLM_CLAVATA_GSP (Table S1)—and GeneRacer_5P. The cycling program of touchdown PCR was set as the initial annealing temperature of 68 °C with 3 cycles, followed by a reduced annealing temperature of 66 °C with 3 cycles, 63 °C with 7 cycles, and 60 °C with 22 cycles. The products of the first amplification were used as the templates for the second PCR with nested primer pairs RLM_nGSP/GeneRacer_n5P (Table S1) that annealed internally to the first primer pairs. The products of the second PCR were then cloned into pGEM®-T Easy vector (Promega, Madison, WI, USA) and sequenced to analyze the cleavage sites.

3. Results

3.1. CaCV Symptom Development in Resistant Capsicum Plants at Elevated Temperatures

To evaluate the effect of elevated temperature on capsicum resistance to CaCV, symptoms were observed over a 10-day period in CaCV-infected capsicum plants grown at HT and AT. Four-week-old CaCV-resistant capsicum plants grown at AT were inoculated with CaCV prior to half of them being transferred to HT. At 5 dpi, all resistant plants challenged with CaCV showed HR-mediated necrotic spots on virus-inoculated leaves at both AT and HT (Figure 1a). At 10 dpi, 8 out of 14 (57%) CaCV-infected plants grown at HT showed distinct necrotic spots on upper systemic leaves, while none of the CaCV-infected plants grown at AT had symptoms on the systemic leaves (Figure 1b). All mock-inoculated CaCV-resistant plants grown at HT or AT remained symptomless and had a similar appearance at the two temperatures. These results indicate that the HR was successfully triggered by CaCV infection in the inoculated leaves of CaCV-resistant capsicum, which was independent of growth temperature. However, the HR appears to be unable to restrict CaCV infection in some plants grown at HT, allowing CaCV to move to systemic leaves and trigger HR there.

Figure 1. Effect of elevated temperature on capsicum chlorosis virus (CaCV) infection in CaCV-resistant capsicum plants: (**a**) at 5 days post inoculation (dpi), a hypersensitive response (red arrows) was apparent in CaCV-inoculated leaves of capsicum plants grown at high temperature (HT) of 35 °C and ambient temperature (AT) of 25 °C; (**b**) at 10 dpi, necrotic spots (indicated by red arrows) were apparent on systemic leaves of some capsicum plants grown at HT but not on any plants grown at AT. Three representative plants are shown for each treatment.

3.2. Small RNAs in CaCV-Infected and Virus-Free Capsicum Grown at Different Temperatures

To identify miRNAs that may be involved in capsicum resistance to CaCV at HT and/or AT, systemic leaves from three CaCV-infected and three mock-inoculated plants grown at either temperature were collected. For HV treatment, leaf disks were sampled from three plants with systemic necrotic spot symptoms. A total of twelve sRNA libraries, containing three libraries in each of the four conditions, were generated for Illumina sequencing and the data are summarized in Table S2. In total, 123,888,579, 95,340,112, 160,227,368, and 127,753,253 raw reads were obtained from the libraries of the AV, AM, HV, and HM treatments, respectively. After trimming adaptor sequences and low-quality reads, 121,643,325, 93,441,595, 157,681,773, and 125,749,412 cleaned reads, and 58,305,917, 45,565,384, 102,316,388, and 65,410,603 length-limited reads remained for AV, AM, HV, and HM treatments, respectively (Table S2). In the analysis of length distribution, 24-nucleotide sRNAs were the largest sRNA group among all length-limited reads in the libraries of AV, AM, and HM treatments (Figure 2a,b,d), whereas 21-nucleotide and 22-nucleotide sRNAs accounted for the highest percentage of all sRNAs in the HV libraries in addition to the 24-nucleotide sRNA peak (Figure 2c). To further analyze vsiRNAs and miRNAs, length-limited reads were aligned to the capsicum genome. An average of 70.0%, 70.2%, 56.4%, and 73.3% of reads for the AV, AM, HV, and HM treatments, respectively, mapped to the capsicum reference genome. The reads that failed to map to the capsicum genome in the AV and HV libraries were retrieved for vsiRNA analysis and indicated that the expression of CaCV-derived vsiRNAs was significantly higher in CaCV-infected capsicum grown at HT than in those plants grown at AT (Figure 3, compare panels a and b). Visualization of vsiRNA coverage in CaCV-infected capsicum grown at HT showed that vsiRNAs were more evenly dispersed across the entire L segment than across M and S segments (Figure 3a). Viral siRNAs were less abundant across the intergenic region of both M and S segments, indicating vsiRNA hotspots in the coding regions of M and S segments (Figure 3a). Moreover, the abundance of 21-nucleotide, 22-nucleotide, and 24-nucleotide vsiRNAs at the higher temperature was highest in several hotspots in the CaCV S segment in all three HV libraries (Figure 3a).

3.3. Identification of Known and Novel miRNAs

The sRNA datasets were grouped according to the four treatments (AV, AM, HV, and HM), and the tool miRDeep-P2 in iwa-miRNA software was used to predict miRNA clusters. The capsicum miRNAs were annotated using the miRNA annotations retrieved from PmiREN, sRNAanno, and PsRNA databases. After aggregating annotated and predicted miRNAs into a candidate miRNAs list, miRNAs that passed both high-throughput-based and machine learning-based criteria were selected. From the four grouped datasets, 105 known and 83 novel miRNAs were identified. Among all identified miRNAs, 89 known and 47 novel miRNAs were present in libraries across more than one treatment. In addition, 69 known and 21 novel miRNAs were identified in all libraries (Figure S1). The mature, star, and precursor sequences of identified miRNAs present in libraries across more than one treatment are listed in Table S3. The abundance of miRNAs in individual sRNA libraries was determined by aligning reads against the mature, star, and precursor sequences of those identified miRNAs. They were given the prefix 'Can-' for *Capsicum annuum*. Overall, Can-MIR159 and Can-MIR166 were the most abundant known miRNA families and Can-MIRN19 was the most abundant novel miRNA family in all libraries across all treatments.

3.4. Differentially Expressed miRNAs

The differential expression of miRNAs was calculated using DEseq2 in four pairwise comparisons: AV vs. AM; HV vs. HM; HM vs. AM; and HV vs. AV. Five DE miRNAs were identified at AT when comparing CaCV-infected and mock-inoculated treatments, which included one upregulated and four downregulated miRNAs (Figure 4a). In HV vs. HM and HM vs. AM comparisons, 34 and 48 DE miRNAs, respectively, were detected. Of those DE miRNAs, 18 upregulated and 16 downregulated miRNAs were identified in the HT CaCV-infected treatment versus mock treatment; 8 upregulated and 40 downregulated miRNAs were identified in mock treatments comparing HT and AT conditions (Figure 4b,c).

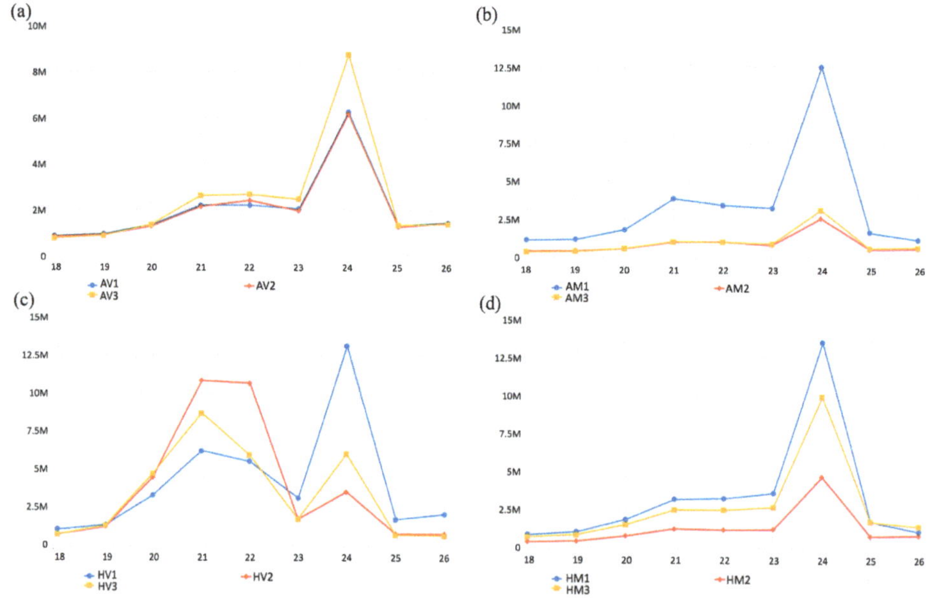

Figure 2. Small RNA length distribution of 18-nucleotide to 26-nucleotide (million (M) reads) in 12 small RNA libraries. Each diagram represents three independent libraries constructed from samples collected from (**a**) capsicum chlorosis virus (CaCV)-inoculated plants grown at ambient temperature (AV), (**b**) mock-inoculated plants grown at ambient temperature (AM), (**c**) CaCV-inoculated plants grown at higher temperature (HV), and (**d**) mock-inoculated plants grown at higher temperature (HM).

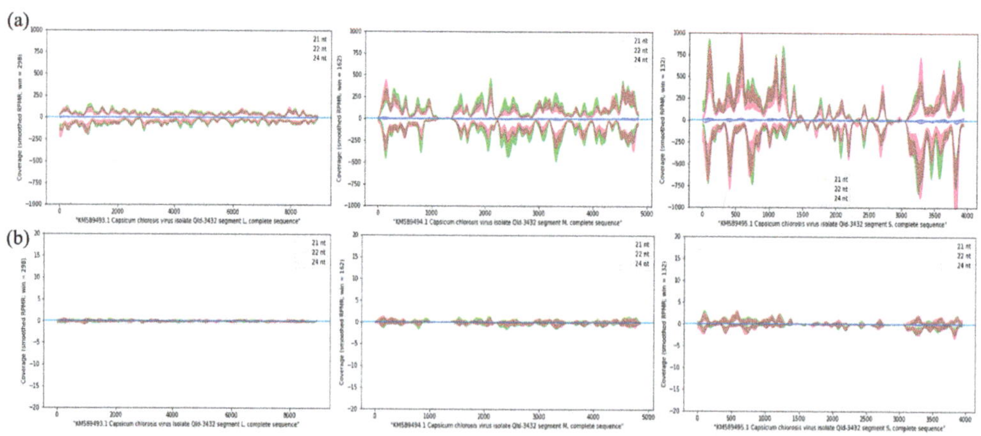

Figure 3. Profiles of capsicum chlorosis virus (CaCV)-derived viral siRNAs. The 21-, 22-, and 24-nucleotide vsiRNA coverage across the CaCV genome L, M, and S segments in CaCV-infected capsicum plants grown at (**a**) higher temperature (HV), and (**b**) ambient temperature (AV). Standard error of 3 biological replicates is presented as the smoothed plots. The reads-per-million (RPM) scale is set at ±1000 for three datasets of the HV treatment and at ±20 for three datasets of the AV treatment.

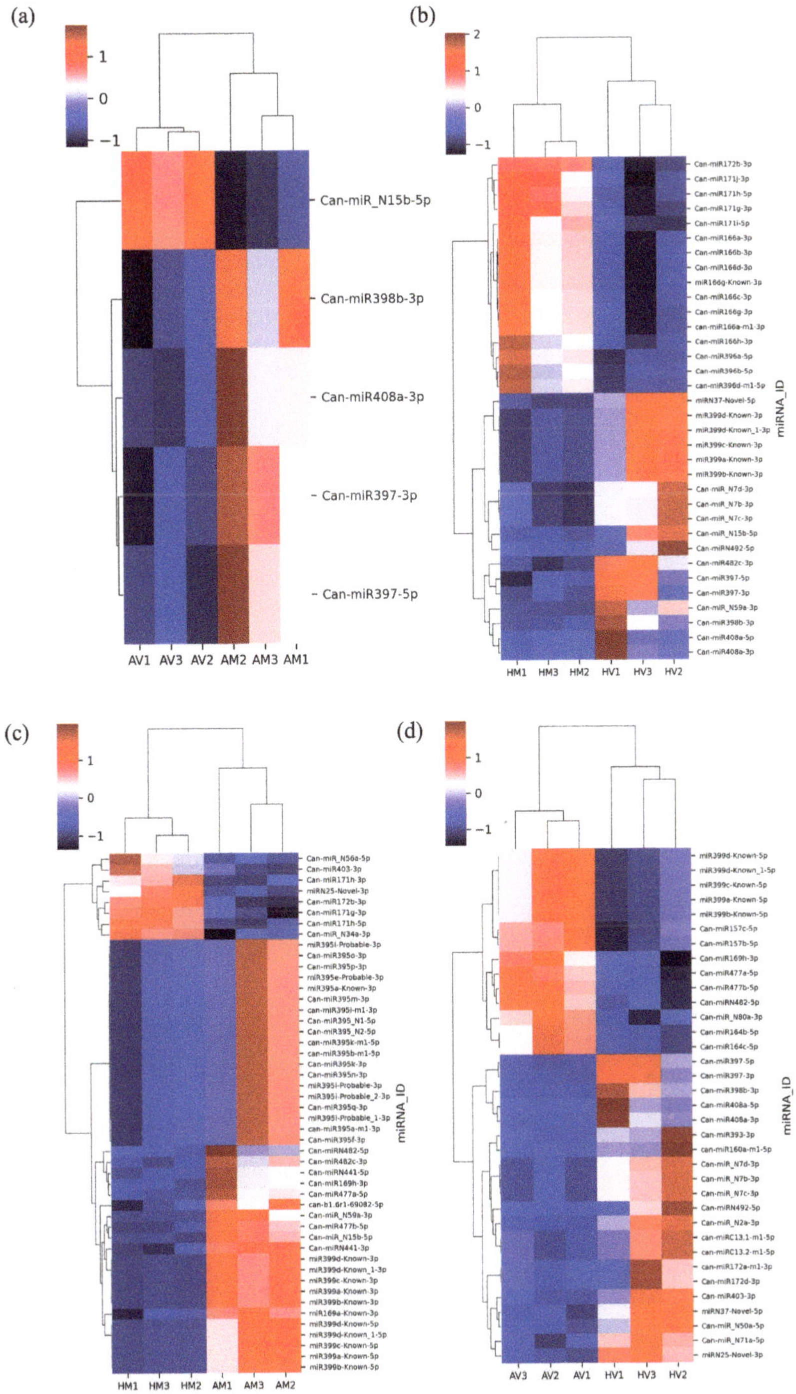

Figure 4. Heat maps showing differentially expressed capsicum miRNAs in four pairwise comparisons: (**a**) capsicum chlorosis virus (CaCV)-infected capsicum grown at ambient temperature (AT)

compared to mock-inoculated capsicum grown at AT (AV vs. AM); (**b**) CaCV-infected capsicum grown at higher temperature (HT) compared to mock-inoculated capsicum grown at HT (HV vs. HM); (**c**) mock-inoculated capsicum grown at HT compared to AT (HM vs. AM); (**d**) CaCV-infected capsicum grown at HT compared to AT (HV vs. AV); Columns represent independent biological replicates; rows represent different miRNAs. Clustering is based on Z-score hierarchical calculation. Red and blue indicate miRNAs with high and low expression, respectively, as shown on the color scale.

To identify miRNAs that are potentially involved in temperature-sensitive resistance to CaCV, we focused on the comparative expression levels of miRNAs in HV and AV conditions. A total of 35 DE miRNAs, including 21 upregulated and 14 downregulated miRNAs, were identified at HT compared to AT in CaCV-infected plants (Figure 4d). Among those, DE miRNAs, Can-MIR408, Can-MIR398, Can-MIR397, Can-MIR393, Can-MIRN492, and Can-MIRN7 were the most highly upregulated miRNA families with log2-fold changes above 2.0 in HV compared to AV conditions (Table 1). Interestingly, Can-miR408a-3p, Can-miR398b-3p, Can-miR397-3p, and Can-miR397-5p were upregulated when comparing the HV and HM treatments but were downregulated when comparing the AV and AM treatments (Table 1). For those downregulated miRNAs in the CaCV-infected treatment comparing HT and AT growth conditions, some miRNAs in the Can-MIR169, Can-MIR477, Can-MIRN482, and Can-MIR399 families were responsive to HT and showed a reduced expression in HM versus AM treatment. Notably, Can-miR157b-5p, Can-miR157c-5p, Can-miR164b-5p, Can-miR164c-5p, and Can-miR_N80a-3p were downregulated only in HV compared to AV treatment but not in HM compared to AM treatment (Table 1). These data indicate that increased or reduced expression of some miRNAs in Can-MIR408, Can-MIR398, Can-MIR397, Can-MIR157, Can-MIR164, and Can-MIR_N80a families may be associated with CaCV-induced changes between HT and AT treatments.

Table 1. Differentially expressed miRNAs in virus-infected capsicum versus mock-inoculated capsicum grown at high compared to ambient temperature.

miRNA ID	HV/AV_log2 (FC)	AV/AM_log2 (FC)	HV/HM_log2 (FC)	HM/AM_log2 (FC)
Can-miR408a-5p	4.94651183		2.15675264	
Can-miR408a-3p	4.52786751	−1.431929	1.87576962	
Can-miR398b-3p	4.3647467	−1.1693631	2.7875221	
Can-miR397-5p	3.41052681	−1.3643359	1.80648604	
Can-miR397-3p	3.09698569	−1.1034217	1.41436391	
Can-miRN492-5p	2.4800554		2.07163552	
Can-miR_N7b-3p	2.22044295		1.95841442	
Can-miR_N7c-3p	2.22044295		1.95841442	
Can-miR_N7d-3p	2.16627692		1.84961374	
Can-miR393-3p	2.00053338			
Can-miRN37-Novel-5p	1.97325814		1.2736888	
Can-miR_N2a-3p	1.66675404			
Can-miR172a-m1-3p	1.62135413			
Can-miR172d-3p	1.61286777			
Can-miR_N71a-5p	1.60761055			
Can-miR_N50a-5p	1.51467255			
Can-miR403-3p	1.29105902			1.04987498
Can-miRN25-Novel-3p	1.27226491			1.27136116
Can-miR160a-m1-5p	1.25709852			
Can-miRC13.1-m1-5p	1.19626839			
Can-miRC13.2-m1-5p	1.19626839			
Can-miR157c-5p	−1.0923326			
Can-miR157b-5p	−1.1162378			
Can-miR169h-3p	−1.1385036			−1.8141374
Can-miR_N80a-3p	−1.4595388			
Can-miR477a-5p	−1.6760649			−3.4314472

Table 1. *Cont.*

miRNA ID	HV/AV_log2 (FC)	AV/AM_log2 (FC)	HV/HM_log2 (FC)	HM/AM_log2 (FC)
Can-miR164b-5p	−1.7200209			
Can-miR164c-5p	−1.7417164			
Can-miR477b-5p	−1.7912909			−2.9432463
Can-miRN482-5p	−1.8617146			−2.2237655
Can-miR399a-Known-5p	−1.9889216			−5.5813609
Can-miR399b-Known-5p	−1.9889216			−5.5813609
Can-miR399c-Known-5p	−1.9889216			−5.5813609
Can-miR399d-Known_1-5p	−1.9889216			−5.5813609
Can-miR399d-Known-5p	−1.9889216			−5.5813609

HV/AV_log2 (FC)—log2-fold change when comparing miRNA expression in CaCV-infected capsicum grown at high temperature (HT) with infected plants grown at ambient temperature (AT); AV/AM_log2 (FC)—log2-fold change when comparing miRNA expression in CaCV-infected capsicum grown at AT to that in mock-inoculated capsicum grown at AT; HV/HM_log2 (FC)—log2-fold change when comparing miRNA expression in CaCV-infected capsicum grown at HT to that in mock-inoculated capsicum grown at HT; HM/AM_log2 (FC)—log2-fold change when comparing miRNA expression in mock-inoculated capsicum grown at HT and AT.

3.5. Functional Analysis of Predicted Target Genes of DE miRNAs in Response to CaCV under Ambient and High-Temperature Conditions

For a better understanding of the potential regulatory roles of DE miRNAs during CaCV infection and at elevated temperature, target genes of all identified miRNAs were predicted by psRNAtarget software with stringent criteria (expectation score ≤ 3) and were functionally annotated with GO terms. An overview of all DE miRNAs and their predicted mRNA targets in the four pairwise comparisons AV vs. AM, HV vs. HM, HM vs. AM, and HV vs. AV is shown in the Venn diagrams depicted in Figure S2. A large proportion of the upregulated miRNAs and their predicted targets in HV compared to AV overlapped with those upregulated miRNAs and their predicted targets in HV compared to HM treatments (Figure S2a,c), suggesting a significant virus effect. However, downregulated miRNAs and their predicted targets in HV compared to AV overlapped with those downregulated miRNAs and their predicted targets in HM compared to AM (Figure S2b,d), suggesting a significant temperature effect.

To identify pathways that are relevant to the observed phenotypic differences between HV and AV treatments, GO enrichment analyses were conducted for the targets that were predicted from up- or downregulated miRNAs. Besides analyzing targets of DE miRNAs in HV vs. AV treatments, targets of DE miRNAs in other pairwise comparisons (AV vs. AM, HV vs. HM, and HM vs. AM) were also included. All significantly enriched (Padj < 0.05) GO terms describing biological process (BP), molecular function (MF), and cellular component (CC) for the predicted targets of DE miRNAs are listed in Tables S4–S9. A total of 156 targets were predicted from 21 miRNAs that were upregulated in CaCV-infected plants at HT compared to AT. Among them, 143 miRNA targets could be functionally annotated. Subsequently, 8 and 12 GO terms listed in MF and BP were enriched through the SEA of those miRNA targets. The hierarchical relationships of the significantly enriched terms for BP and MF are presented in Figures 5 and 6, respectively. DE miRNAs and their corresponding target genes with enriched GO terms are shown in Table S4. Briefly, miRNA targets that were annotated with significantly enriched GO terms in BP and MF categories—including the phenylpropanoid metabolic process (GO:0009698), lignin catabolic process (GO:0046274), lignin metabolic process (GO:0009808), phenylpropanoid catabolic process (GO:0046271), hydroquinone: oxygen oxidoreductase activity (GO:0052716), and copper ion binding (GO:0005507)—were regulated by Can-miR408a-3p and Can-miR397-5p (Figures 5 and 6 and Table S4). In agreement with the relevance of DE miRNAs among the pairwise comparisons HV vs. AV, HV vs. HM, and AV vs. AM, the most highly enriched GO terms were also enriched through SEA of targets of downregulated miRNAs in AV compared to AM as well as targets of upregulated miRNAs in HV compared to HM (Figures S3 and S4). In addition, the GO-enriched term, cysteine-type peptidase activity (GO:0008234) for MF, was associated with a group of targets that were affected by upregulated miRNAs (Can-miR_N7b-3p, Can-miR_N7c-3p, Can-miR_N7d-3p, miRN37-Novel-5p, and Can-miR408a-5p) in HV compared to both AV and HM treatments (Figures 6 and S4 and Tables S4 and S6).

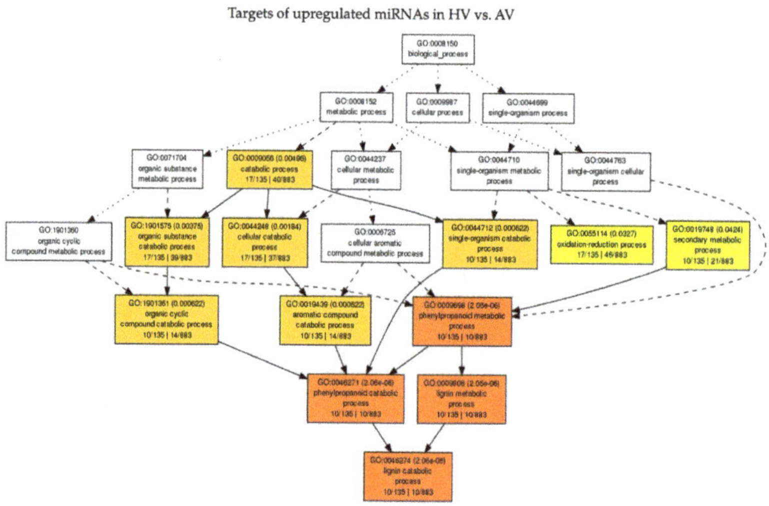

Figure 5. Acyclic graph showing hierarchical relationship of the enriched gene ontology (GO) terms associated with targets predicted from upregulated miRNAs in CaCV-infected capsicum grown at high temperature (HV) compared to ambient temperature (AV) in the biological process category. The color scale from yellow to red indicates an increasingly significant enrichment of GO terms.

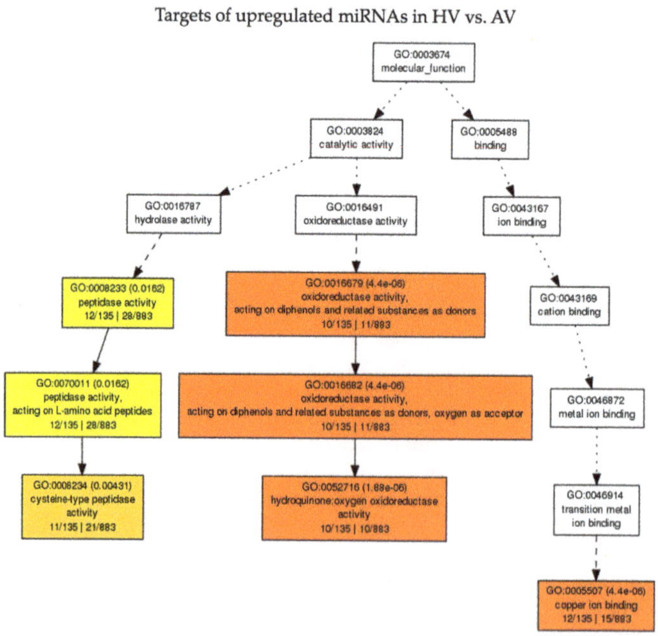

Figure 6. Acyclic graph showing hierarchical relationship of the enriched gene ontology (GO) terms associated with targets predicted from upregulated miRNAs in CaCV-infected capsicum grown at high

temperature (HV) compared to ambient temperature (AV) in the molecular function category. The color scale from yellow to red indicates an increasingly significant enrichment of GO terms.

For miRNAs that were downregulated at higher temperature in CaCV-infected plants, a total of 48 target mRNAs were functionally annotated. Among those targets, only two GO terms (GO:0003677 and GO:0003676) were enriched in the MF category. Moreover, these two enriched terms were associated with targets that were predicted from eight miRNAs, including Can-miR157b-5p, Can-miR157c-5p, Can-miR164b-5p, Can-miR164c-5p, Can-miR169h-3p, Can-miR477a-5p, Can-miR_N80a-3p, and Can-miRN482-5p (Table S5). Based on the functional prediction of miRNA targets and the DE miRNAs, Can-miR408a-3p, Can-miR397-5p, Can-MIR157, and Can-MIR164 families may represent important known miRNAs involved in temperature-dependent resistance-breaking.

3.6. Validation of the Expression Patterns of Selected miRNAs and Their Target Genes

To confirm the sRNA sequencing data, S- poly(A)-tailed real-time RT-PCR or Northern blot with signal intensity quantification was used to quantify the levels of selected miRNAs at 10 dpi. Based on the results of the differential expression analysis of miRNAs and functional analysis of targets, Can-miR408a-3p, Can-miR397-5p, and Can-miR164b/c-5p were selected based on their potential involvement in temperature-dependent resistance-breaking. In addition, Can-miR168-5p, which was upregulated with a log2-fold change of 1.0 and Padj of 0.017 in HV compared to the AV condition, was selected despite the cut-off threshold of Padj ≤ 0.01. This exception was based on the reported importance of miR168. Accumulation of miR168 compromises AGO1-mediated antiviral RNA silencing in several virus-infected plant species [70–72]. In addition, accumulation of miR168 may affect the expression of many target genes since AGO1 plays a central role in RNA silencing [73]. The results of the real-time RT-PCR and Northern blot (Figure S5) confirmed the increased expression levels of Can-miR408a-3p, Can-miR397-5p, and Can-miR168-5p as well as the decreased expression level of Can-miR164b/c-5p in HV compared to AV treatments, and showed that they were consistent with the expression profiles obtained by sRNA sequencing (Figure 7a). When HV was compared to HM, the expression trends of Can-miR408a-3p, Can-miR397-5p, and Can-miR164b/c-5p were similar between the results observed in real-time RT-PCR and sRNA sequencing (Figure 7b). However, expression trends of Can-miR408a-3p, Can-miR397-5p, and Can-miR168-5p in the AV and AM comparison were opposite between the results of real-time RT-PCR and sRNA sequencing (Figure 7c).

In addition to analyzing the expression of mature miRNAs, quantitative real-time RT-PCR was also used to investigate the effect of these miRNAs on their target genes in the HV condition. Targets were selected based on the target prediction data obtained from psRNAtarget and WPMIAS software [74] as well as the target validation data reported by Zhang and collaborators [75]. CAN.G1061.9 (*CLAVATA1-related*), CAN.G671.1 (*Laccase 5, LAC5*), and CAN.G1305.35 (*Plantacyanin*) were chosen for Can-miR408a-3p target analysis; CAN.G394.71 (*LAC2*), CAN.G351.2 (*LAC4*), and CAN.G355.8 (*LAC4*) were chosen for Can-miR397-5p target analysis; CAN.G637.6 (*AGO1b*) was chosen for Can-miR168-5p target analysis; NAM/ATAF1,2/CUC2 (*NAC*) transcription factors—CAN.G943.28 (*NAC1*) and CAN.G587.8 (*NAC5*)—were chosen for Can-miR164b/c-5p target analysis. Unexpectedly, expression patterns of the targets CAN.G671.1, CAN.G1305.35, CAN.G394.71, CAN.G351.2, CAN.G355.8, and CAN.G637.6 were not inverse to the expression of their corresponding miRNAs when comparing the HV and AV conditions (Figure 8c,d,f–h,m). Conversely, the expected inverse expression pattern was observed in Can-miR408a-3p/CAN.G1061.9, Can-miR164b/c-5p/CAN.G943.28, and Can-miR164b/c-5p/CAN.G587.8 pairs (Figure 8b,j,k). Notably, we found that the transcript levels of all three selected targets of Can-miR397-5p in mock-inoculated plants drastically increased at HT (Figure 8f–h). Moreover, the transcript level of one of the targets, CAN.G355.8, was downregulated, while that of the other two targets (Figure 8h), CAN.G394.71 and CAN.G351.2, remained unchanged in CaCV-infected plants as compared with mock-inoculated plants at HT (Figure 8f,g). These results suggest that the downregulation of CAN.G355.8 may be due to the negative regulation of upregulated Can-miR397-5p in HV compared to HM. Similarly, an inverse expression pattern between Can-miR168-5p and CAN.G637.6 was also observed in HV vs. HM (Figure 8m). Different from other targets that showed inverse or no correlation tendency with miRNAs, CAN.G1305.35 showed a positively correlated expression pattern with Can-miR408a-3p in three pairwise comparisons (AV vs. AM, HV vs. HM, and HV vs. AV) (Figure 8d). Overall, diverse regulatory effects of miRNAs on their targets were observed when plants were treated at different temperatures and biotic stress conditions.

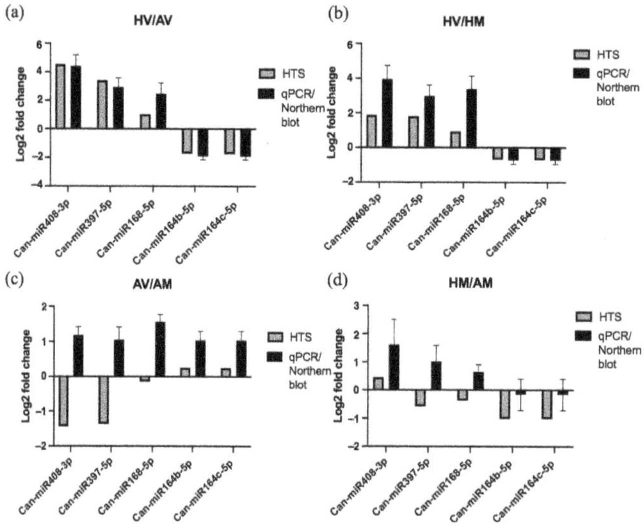

Figure 7. Linear specific (S)-poly (A)-tailed quantitative real-time RT-PCR (qPCR) of miR408, miR397, and miR168 or Northern blot hybridization confirmation of miR164b and miR164c expression patterns compared to those obtained by sRNA high-throughput sequencing (HTS). The expression levels (log2-fold change) of five miRNAs in four pairwise comparisons: (**a**) CaCV-infected plants at high temperature (HV) vs. CaCV-infected plants at ambient temperature (AV); (**b**) HV vs. mock-inoculated plants at high temperature (HM); (**c**) AV vs. mock-inoculated plants at ambient temperature (AM); and (**d**) HM vs. AM are displayed for HTS and qPCR or Northern blot. *U6* was used as internal reference for calibrating the expression of miRNAs. The error bars represent the mean (±standard error of the mean) of 4 biological replicates.

3.7. *Identification of miRNA-Mediated Cleavage of Target Genes*

To confirm putative miRNA targets, the specific cleavage sites were identified experimentally using 5′-RACE. Three targets, including CAN.G1061.9, CAN.G355.8, and CAN.G637.6, were chosen for the 5′-RACE verification based on the observed inverse expression patterns of miRNA/target pairs. CAN.G1061.9 was selected for its inverse expression pattern at HV vs. AV, while CAN.G355.8 and CAN.G637.6 were chosen for their inverse patterns at HV vs. HM. The other targets, CAN.G943.28 and CAN.G587.8, which showed opposite correlation with Can-miR164b/c-5p, were verified by degradome-based miRNAs-target analysis using WPMIAS (Table S10). Based on the results of 5′-RACE, the cleavage sites were all located in the regions that were predicted as miRNA-binding sites (Figure 9). The majority cleavage site in CAN.G1061.9 was mapped to the 11th position of the complementary sequence at the 5′-end of Can-miR408a-3p (Figure 9a). On the other hand, the main cleavage site in CAN.G355.8 and CAN.G637.6 was mapped to the 10th position of the complementary sequences at the 5′-end of Can-miR397-5p and Can-miR168-5p, respectively (Figure 9b,c). These results suggest that CAN.G1061.9, CAN.G355.8, and CAN.G637.6 can be cleaved by Can-miR408a-3p, Can-miR397-5p, and Can-miR168-5p, respectively, in capsicum.

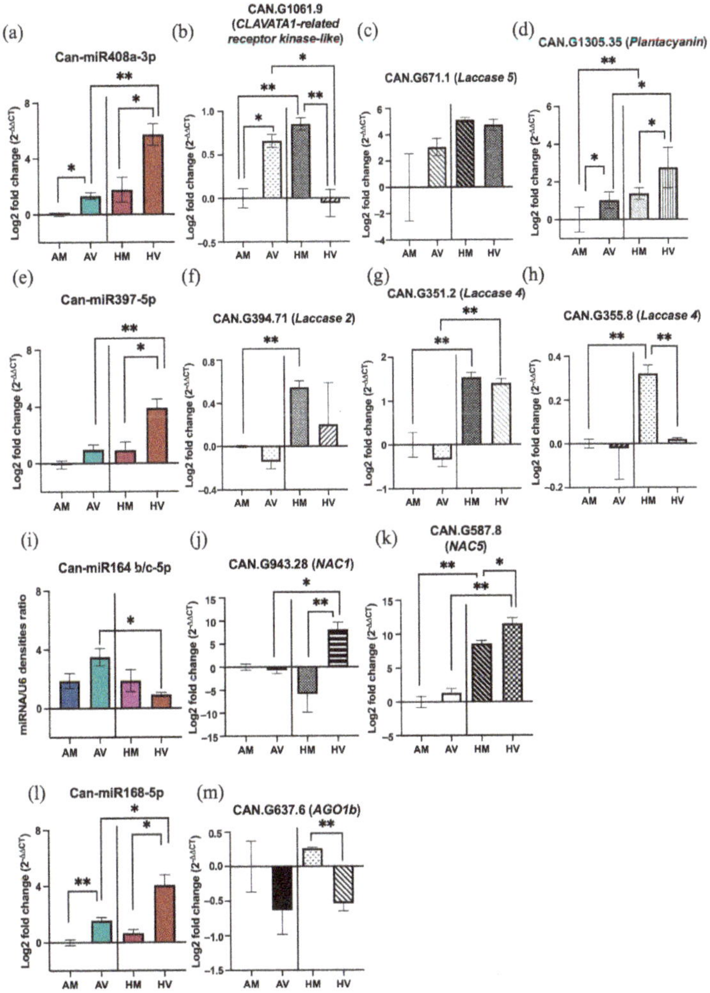

Figure 8. MicroRNA-mediated regulation involved in capsicum resistance response to CaCV at high temperature (HV) or ambient temperature (AV). Expression patterns of miRNAs (in color) and targets (in black and white) in four pairwise comparisons (AV vs. mock-inoculated plants at ambient temperature (AM); HV vs. mock-inoculated plants at high temperature (HM); HM vs. AM; and HV vs. AV) were analyzed by real-time RT-PCR with the log2-fold change 2–ΔΔCt method or by Northern blot. Real-time RT-PCR of (**a**) Can-miR408a-3p and its targets: (**b**) CAN.G1061.9, (**c**) CAN.G671.1, and (**d**) CAN.G1305.35. Real-time RT-PCR analysis of (**e**) Can-miR397-5p and its targets: (**f**) CAN.G394.71, (**g**) CAN.G351.2, and (**h**) CAN.G355.8. Northern blot analysis of (**i**) Can- miR164b/c-5p, and real-time PCR analysis of its targets: (**j**) CAN.G394.28 and (**k**) CAN.G587.8. Real-time RT-PCR analysis of (**l**) Can-miR168-5p and its target (**m**) CAN.G637.6. Northern blot analysis was quantified through measuring signal strength using ibright. Actin and U6 were used as internal reference for calibrating the expression of targets and miRNAs, respectively. The error bars represent the mean (±standard error of the mean) of 4 biological replicates. Significant differences between treatments were assessed with Student's *t*-test (* $p < 0.05$; ** $p < 0.01$).

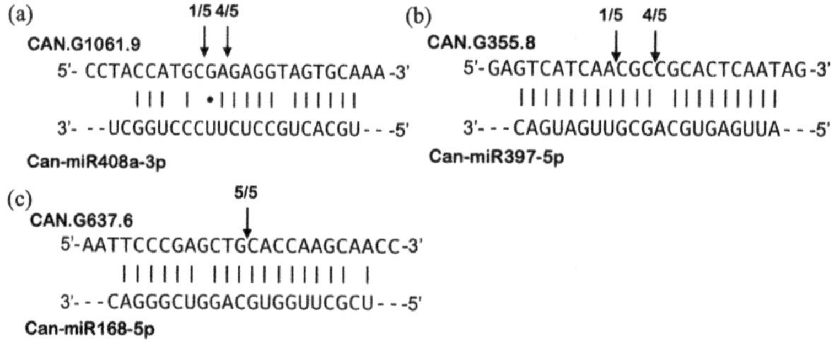

Figure 9. Validation of miRNA targets by 5′ RLM-RACE. The potential cleavage sites of (**a**) Can-miR408a-3p on CAN.G1061.9, (**b**) Can-miR397-5p on CAN.G355.8, (**c**) Can-miR168-5p on CAN.G637.6 in capsicums were mapped. The arrows indicate the cleavage sites and the numbers indicate clone frequencies. The dashes represent the standard base pairing rules and the dots indicate GU wobble base pairing.

4. Discussion

The suppressive effect of elevated temperatures on ETI-mediated resistance or its associated HR response has been demonstrated in both non-viral [20,76,77] and viral pathosystems [20,78]. For example, the *N* gene-mediated HR and resistance to TMV in tobacco are compromised by exposing plants to temperatures above 28 °C [20,78]. In the present study, we found that ETI-mediated resistance to CaCV was overcome at 35 °C/30 °C (day/night) in 8 out of 14 capsicum plants. Interestingly, instead of abolishing HR, distinct local lesions developed in inoculated leaves, and systemic necrotic lesions were also observed in these plants in which resistance is compromised at 35 °C. This result agrees with the previous finding suggesting that HR may fail to restrict virus movement in local lesions at elevated temperatures [26,27,79]. At 32 °C, the *Tsw*-mediated resistance to TSWV in capsicum is compromised, leading to virus movement into uninoculated leaves. Nonetheless, local HR persists in the inoculated leaves of those plants [26,27,79]. Notably, there is limited knowledge regarding the mechanisms that underpin temperature-sensitive ETI-mediated resistance in viral pathosystems [20,29]. In the TMV–tobacco pathosystem, HR is suppressed at elevated temperatures by preventing NLR proteins from translocating into the nucleus to orchestrate plant immune signaling. Tobacco plants mount an effective HR response when the host N protein accumulates in the nucleus after recognition of the TMV coat protein at 22 °C, while plants are unable to induce an HR response at 28 °C due to decreased N protein nuclear localization [20]. In addition, changes in miRNA accumulation that affect the expression of their target genes appear to be linked to the temperature sensitivity of Ny-DG-mediated resistance in the PVY–potato pathosystem. Resistance to PVY^{NTN} was compromised at 28 °C and an enhanced miRNA-mediated downregulation of a specific NLR transcript was observed in PVY-inoculated potato leaves [29].

In previous studies, several miRNAs have been identified in *Capsicum* spp. [41,80–84]. Some of these miRNAs were potentially involved in the interactions between capsicum and plant pathogens [41,84]. In the present study, Illumina sequencing of sRNAs was conducted to investigate the involvement of miRNAs in temperature-sensitive antiviral resistance. By predicting and retrieving miRNA annotations from three sRNA databases—PmiREN, sRNAanno, and PsRNA—105 known and 83 novel miRNAs were identified across four treatment groups: AV, AM, HV, and HM. Given the distinct phenotypic differences observed between CaCV-infected capsicum plants grown at HT and AT, the DE miRNAs revealed by comparing HV and AV treatments allowed us to identify miRNAs potentially involved in temperature-sensitive resistance. Among the DE miRNAs, a significant number expressed higher in the HV than AV were responsive to CaCV infection at HT. In addition, in the enrichment analysis of the upregulated miRNAs in HV compared to AV, targets of Can-miR408a-3p and Can-miR397-5p—which are associated with biological processes, including the metabolism and catabolism of phenylpropanoids and lignin—were identified as the most enriched GO terms.

Lignin deposition plays a crucial role in plant resistance to biotrophic pathogens by confining them to the infection site [85]. In *A. thaliana*, lignification is induced during the ETI response, triggered by infection of avirulent bacterial strains, *Pseudomonas syringae* pv. tomato (*Pst*) DC3000 (AvrRpm1) and *Pst* DC3000 (AvrRpt2). This lignification is associated with localized cell death, restricting the bacterial pathogens to the infection site [85]. Increased lignin accumulation has also been associated with plant resistance to viral pathogens, including TMV, tobacco necrosis virus, southern bean mosaic viruses, and sweetpotato virus disease (SPVD) [86,87]. We, therefore, speculate that the upregulation of Can-miR408a-3p and Can-miR397-5p in HV compared to AV may reduce lignin content by targeting genes in the *LAC* family, which in turn may interfere with restrictions to virus spread from the infection site. Although there is no direct evidence showing the involvement of miR408 and miR397 in lignin-related HR, reduced lignin deposition has been observed in plants overexpressing these two miRNAs (Lu et al. 2013; Song et al. 2018). Moreover, inhibition of miR397 has been shown to increase lignin deposition in the cell wall and reduce the accumulation of SPVD [87].

Based on computational results and their involvement in plant responses to various stresses [88], Can-miR408a-3p, Can-miR397-5p, Can-miR168-5p, and Can-miR164b/c-5p were selected for further analysis of sRNA and target genes expression profiles. Real-time RT-PCR revealed that Can-miR408a-3p, Can-miR397-5p, and Can-miR168-5p were upregulated, whereas Can-miR164b/c-5p was downregulated in HV compared to AV. Additionally, Can-miR408a-3p and Can-miR397-5p were upregulated in HV compared to HM, consistent with our sRNA sequencing data. However, discrepancies were observed between the results of sRNA sequencing and real-time RT-PCR in some pairwise comparisons. For instance, in AV and AM comparisons, real-time RT-PCR indicated an increased level of Can-miR408a-3p and an unchanged level of Can-miR397-5p. Conversely, sRNA sequencing showed decreased levels for both. This inconsistency between sequencing and RT-PCR data has also been documented in prior studies [89,90]. Several factors may contribute to discrepancies between computational sequence analysis and Northern blot or real-time RT-PCR. For instance, biases during sRNA library preparation and amplification, such as adapter ligation bias and varying GC content among miRNAs, can introduce distortions [91–93].

Genes that encode copper-containing proteins, including proteins in the phytocyanin family—cupredoxin, plantacyanin, and uclacyanin—or proteins involved in lignin polymerization such as LACs, have been validated as targets of miR408 in *A. thaliana* [94,95]. For miR397, genes in the *LAC* family are its major target genes in various plant species [96–100]. In the present study, GO enrichment analysis suggests that *LAC* family genes, predicted to be regulated by Can-miR408a-3p and Can-miR397-5p, may be crucial for resistance breakdown at HT. Unexpectedly, real-time RT-PCR analysis did not reveal significant differences in the expression levels of the selected *LAC* genes, including CAN.G671.1, CAN.G394.7, CAN.G351.2, and CAN.G355.8, between plants treated with HV and AV. Their expression was not inversely correlated with the increased pattern of Can-miR408a-3p and Can-miR397-5p. However, negative regulation of CAN.G355.8 by Can-miR397-5p was observed in HV compared to HM conditions, wherein the CAN.G355.8 level was reduced, while the Can-miR397-5p level was increased in HV-treated plants. Furthermore, transcript levels of selected Can-miR397-5p targets, including CAN.G394.7, CAN.G351.2, and CAN.G355.8, were increased by HT in mock-inoculated plants. Given that reduced lignin content is detrimental to *Medicago truncatula* growth at HT [101,102], an induction of *LAC*s at HT may be crucial for heat tolerance in capsicum plants. Taken together, we speculate that the suppression of CAN.G355.8 expression by CaCV at HT is likely to reduce plant tolerance to stresses, which may lead to CaCV resistance breakdown at HT.

In addition to the genes in the *LAC* family, the expression of other targets predicted for Can-miR408a-3p was investigated. CAN.G1061.9 (leucine-rich receptor-like kinase *CLV1-related*) was shown to be cleaved by Can-miR408a-3p through 5′-RACE in the present study. The transcript level of CAN.G1061.9 was lower in CaCV-infected plants grown at HT than those grown at AT, which showed an opposite tendency to Can-miR408a-3p in the HV vs. AV comparison. CAN.G1061.9 encodes a CLV1-related receptor kinase-like protein, which appears to be similar to BARELY ANY MERISTEM 1 (BAM2) in the Plaza 4.0 database [57]. Interestingly, BAM2, which functions redundantly with BAM1, participates in the regulation of cell-to-cell RNAi movement [103]. Therefore, a downregulation of CAN.G1061.9 may affect systemic spread of antiviral RNAi, resulting in reduced resistance of capsicum plants to CaCV at HT. Unlike CAN.G1061.9, a negative correlation between miRNA and its target did not occur for Can-miR408a-3p and CAN.G1305.35 (*Plantacyanin*). The transcript level of CAN.G1305.35 increased during Can-miR408a-3p induction in CaCV-infected plants either grown at HT or AT. This agrees with the prior finding that the levels of miR408 and its targets, *Plantacyanin* and *Uclacyanin*, increased simultaneously in late stages of natural senescence in *A. thaliana* [104]. Emerging evidence suggests that a negative correlation between miRNAs and their targets is not strictly required for target validation [105]. Such exceptions may occur when miRNAs and their targets are expressed partially overlapped or in a cell-type-specific manner [105–107]. For example, miR395 and its target, *SULTR2;1*, are mainly expressed in phloem companion cells and xylem parenchyma cells in roots, respectively [107]. This, therefore, prevents miR395 from downregulating SULTR2;1, leading to an upregulation of both SULTR2;1 and miR395 in roots during sulfur starvation [107].

In plants, ROS function as central regulators in complex signaling networks [108,109]. The ROS burst can trigger HR and induce plant resistance to viruses [25]. However, excessive ROS accumulation occurs when the homeostasis of ROS is imbalanced, leading to irreversible oxidative damage and accelerated senescence [109]. In our study, miRNA profiles were investigated in systemic, uninoculated leaves that displayed necrotic spots. This necrotic spot phenotype may be associated with an increased accumulation of ROS in systemic leaves at elevated temperature [110]. Interestingly, not only an upregulation of miR408 but also a downregulation of miR164, which have been shown to be associated with accelerated leaf senescence [88,111], were observed in HV compared to AV. Repressing miR164 expression was found to induce the expression of its target gene, *ORE1*, leading to early-senescence phenotypes [111,112]. In addition, *NAC* transcription factors targeted by miR164, such as *ANAC021/22* (*NAC1*), *ANAC079/80* (*NAC4*), and *ANAC100* (*NAC5*), were upregulated during senescence [113]. MiRNA168 is crucial in miRNA and siRNA pathways by targeting AGO1, the core component of the RNA-induced silencing complex [114]. A well-known auto-regulatory feedback loop of miR168 and AGO1 is involved in maintaining their homeostasis [114]. During several viral infections in *A. thaliana* and *N. benthamiana*, increased miR168 levels disrupt this homeostasis, resulting in an inhibition of AGO1-mediated antiviral RNA silencing [71,72]. In our study, the transcript levels of two selected targets, CAN.G943.28 (*NAC1*) and CAN.G587.8 (*NAC5*), were upregulated in HV compared to AV, which was negatively correlated with the downregulated Can-miR164b/c-5p. However, no significant expression differences of CAN.G637.6 (AGO1b) were observed in HV compared to AV. Overall, our findings suggest that the upregulation of Can-miR408a-3p and Can-miR397-5p, as well as the downregulation of Can-miR164b-5p and Can-miR164c-5p, may underpin the temperature-sensitive resistance-breaking phenotype of CaCV-infected capsicum at elevated temperature. The functions of these miRNAs, including regulating lignin deposition and leaf senescence, may affect important cellular mechanisms in CaCV-infected capsicum plants, which may lower plant fitness under stress conditions and further impact the capacity of plant resistance to viruses.

5. Conclusions

In this study, we investigated the effect of elevated temperature on a CaCV-resistant capsicum breeding line. Our results showed that ETI-mediated resistance to CaCV is compromised at elevated temperatures (35 °C/30 °C) in some capsicum plants. Through sRNA HTS and computational analysis, we identified 105 known and 83 novel miRNAs across different treatment groups. Of particular interest was the upregulation of Can-miR408a-3p and Can-miR397-5p in HV plants compared to AV among all DE miRNAs. GO enrichment analysis of the target genes (*LACs*) predicted from these miRNAs suggested the potential involvement of phenylpropanoid (GO:0009698) and lignin (GO:0009808) metabolic processes in the capsicum response to CaCV under different temperatures. MiRNAs, including Can-miR408a-3p, Can-miR397-5p, Can-miR164b/c-5p, and Can-miR168-5p, may underlie the temperature-sensitivity of ETI-mediated resistance in CaCV-infected capsicum plants.

Supplementary Materials: The following supporting information can be downloaded at: https://www.mdpi.com/article/10.3390/pathogens13090745/s1, Figure S1: Venn diagram of (a) known and (b) novel miRNAs identified in 12 capsicum libraries; Figure S2: Venn diagrams of four pairwise comparisons showing (a) upregulated miRNAs and (c) their corresponding targets, and (b) downregulated miRNAs and (d) their corresponding targets; Figure S3: Acyclic graph showing hierarchical relationship of the enriched gene ontology (GO) terms in the biological process category; Figure S4: Acyclic graph showing hierarchical relationship of the enriched gene ontology (GO) terms in the molecular function category; Table S1: Sequences of oligonucleotide primers used in this study; Table S2: Summary of small RNA Illumina sequencing results; Table S3: Sequences of identified miRNAs present in libraries across more than one treatment; Table S4: Gene ontology (GO) enrichment analysis of targets that were predicted from upregulated miRNAs in CaCV-infected plants grown at high temperature compared to CaCV-infected plants grown at ambient temperature; Figure S5: Northern hybridizations showing the abundance of miR164 in mock or CaCV-infected capsicum plants grown at ambient temperature (AM and AV) or high temperature (HM and HV); Table S5: GO enrichment analysis of targets that were predicted from downregulated miRNAs in CaCV-infected plants grown at high temperature compared to CaCV-infected plants grown at ambient temperature; Table S6: GO enrichment analysis of targets that were predicted from upregulated miRNAs in CaCV-infected plants grown at high temperature compared to mock-inoculated plants grown at high temperature; Table S7: GO enrichment analysis of targets that were predicted from downregulated miRNAs in CaCV-infected plants grown at high temperature compared to mock-inoculated plants grown at high temperature; Table S8: GO enrichment analysis of targets that were predicted from downregulated miRNAs in CaCV-infected plants grown at ambient temperature compared to mock-inoculated plants grown at ambient temperature; Table S9: GO enrichment analysis of targets that were predicted from upregulated miRNAs in mock-infected plants grown at high temperature compared to mock-inoculated plants grown at ambient temperature; Table S10: Degradome-based miRNAs-targets analysis.

Author Contributions: Conceptualization, W.-A.T., N.M. and R.G.D.; methodology, W.-A.T., N.M., C.A.B. and R.G.D.; validation, W.-A.T., N.M., C.A.B. and R.G.D.; investigation, W.-A.T.; resources, N.M. and R.G.D.; data curation, W.-A.T.; writing—original draft preparation, W.-A.T.; writing—review and editing, N.M., C.A.B. and R.G.D.; visualization, W.-A.T.; supervision, N.M., C.A.B. and R.G.D.; project administration, W.-A.T. and R.G.D. All authors have read and agreed to the published version of the manuscript.

Funding: This research received no external funding.

Institutional Review Board Statement: Not applicable.

Informed Consent Statement: Not applicable.

Data Availability Statement: All data can be accessed on the NCBI database under Bioproject PRJNA1128345.

Acknowledgments: The Authors acknowledge Ritesh Jain, Fernanda Borges Naito, and Stephen Fletcher for scientific discussion; Shirani Widana Gamage for data availability of transcriptome of capsicum plants; Galaxy Australia and UQ High-performance computing for data analysis.

Conflicts of Interest: The authors declare no conflicts of interest. The funders had no role in the design of the study; in the collection, analyses, or interpretation of data; in the writing of the manuscript; or in the decision to publish the results.

References

1. Tripodi, P.; Kumar, S. The Capsicum Crop: An Introduction. In *The Capsicum Genome*; Ramchiary, N., Kole, C., Eds.; Springer International Publishing: Cham, Switzerland, 2019; pp. 1–8.
2. Zou, X.; Zhu, F. Origin, Evolution and Cultivation History of the Pepper. *Acta Hortic. Sin.* **2022**, *49*, 1371–1381. [CrossRef]
3. Khaitov, B.; Umurzokov, M.; Cho, K.-M.; Lee, Y.-J.; Park, K. Importance and production of chilli pepper; heat tolerance and efficient nutrient use under climate change conditions. *Korean J. Agric. Sci.* **2019**, *46*, 769–779. [CrossRef]
4. Penella, C.; Calatayud, A. Pepper Crop under Climate Change: Grafting as an Environmental Friendly Strategy. In *Climate Resilient Agriculture: Strategies and Perspectives*; IntechOpen: London, UK, 2018; pp. 128–155.
5. Lu, J.; Guo, M.; Zhai, Y.; Gong, Z.; Lu, M. Differential Responses to the Combined Stress of Heat and *Phytophthora capsici* Infection Between Resistant and Susceptible Germplasms of Pepper (*Capsicum annuum* L.). *J. Plant Growth Regul.* **2017**, *36*, 161–173. [CrossRef]
6. Parisi, M.; Alioto, D.; Tripodi, P. Overview of Biotic Stresses in Pepper (*Capsicum* spp.): Sources of Genetic Resistance, Molecular Breeding and Genomics. *Int. J. Mol. Sci.* **2020**, *21*, 2587. [CrossRef] [PubMed]
7. Mandal, B.; Jain, R.K.; Krishnareddy, M.; Krishna Kumar, N.K.; Ravi, K.S.; Pappu, H.R. Emerging Problems of Tospoviruses (*Bunyaviridae*) and their Management in the Indian Subcontinent. *Plant Dis.* **2012**, *96*, 468–479. [CrossRef]
8. Sharman, M.; Thomas, J.E.; Tree, D.; Persley, D.M. Natural host range and thrips transmission of capsicum chlorosis virus in Australia. *Australas. Plant Pathol.* **2020**, *49*, 45–51. [CrossRef]
9. Persley, D.M.; Thomas, J.E.; Sharman, M. Tospoviruses—An Australian perspective. *Australas. Plant Pathol.* **2006**, *35*, 161–180. [CrossRef]
10. Health, E.; Panel, o.P.; Bragard, C.; Baptista, P.; Chatzivassiliou, E.; Gonthier, P.; Jaques Miret, J.A.; Justesen, A.F.; MacLeod, A.; Magnusson, C.S.; et al. Pest categorisation of Capsicum chlorosis virus. *Eur. Food Saf. Auth. J.* **2022**, *20*, e07337. [CrossRef]
11. Chen, K.; Xu, Z.; Yan, L.; Wang, G. Characterization of a New Strain of Capsicum chlorosis virus from Peanut (*Arachis hypogaea* L.) in China. *J. Phytopathol.* **2007**, *155*, 178–181. [CrossRef]
12. Krishnareddy, M.; Rani, R.U.; Kumar, K.S.A.; Reddy, K.M.; Pappu, H.R. Capsicum chlorosis virus (Genus *Tospovirus*) Infecting Chili Pepper (*Capsicum annuum*) in India. *Plant. Dis.* **2008**, *92*, 1469. [CrossRef]
13. Widana Gamage, S.M.K.; McGrath, D.J.; Persley, D.M.; Dietzgen, R.G. Transcriptome Analysis of Capsicum Chlorosis Virus-Induced Hypersensitive Resistance Response in Bell Capsicum. *PLoS ONE* **2016**, *11*, e0159085. [CrossRef] [PubMed]
14. McGarth, D. *Capsicum Breeding for Tospovirus Resistance*; Horticultural Australia Ltd.: Sydney, NSW, Australia, 2006.
15. Abudurexiti, A.; Adkins, S.; Alioto, D.; Alkhovsky, S.V.; Avšič-Županc, T.; Ballinger, M.J.; Bente, D.A.; Beer, M.; Bergeron, É.; Blair, C.D.; et al. Taxonomy of the order *Bunyavirales*: Update 2019. *Arch. Virol.* **2019**, *164*, 1949–1965. [CrossRef] [PubMed]
16. Whitfield, A.E.; Ullman, D.E.; German, T.L. Tospovirus-thrips interactions. *Annu. Rev. Phytopathol.* **2005**, *43*, 459–489. [CrossRef]
17. Widana Gamage, S.; Persley, D.M.; Higgins, C.M.; Dietzgen, R.G. First complete genome sequence of a capsicum chlorosis tospovirus isolate from Australia with an unusually large S RNA intergenic region. *Arch. Virol.* **2015**, *160*, 869–872. [CrossRef]
18. Honjo, M.N.; Emura, N.; Kawagoe, T.; Sugisaka, J.; Kamitani, M.; Nagano, A.J.; Kudoh, H. Seasonality of interactions between a plant virus and its host during persistent infection in a natural environment. *Int. Soc. Microb. Ecol. J.* **2020**, *14*, 506–518. [CrossRef] [PubMed]
19. Makarova, S.; Makhotenko, A.; Spechenkova, N.; Love, A.J.; Kalinina, N.O.; Taliansky, M. Interactive responses of potato (*Solanum tuberosum* L.) plants to heat stress and infection with potato virus Y. *Front. Microbiol.* **2018**, *9*, 2582. [CrossRef]
20. Wang, Y.; Bao, Z.; Zhu, Y.; Hua, J. Analysis of temperature modulation of plant defense against biotrophic microbes. *Mol. Plant Microbe Interact.* **2009**, *22*, 498–506. [CrossRef]
21. Anfoka, G.; Moshe, A.; Fridman, L.; Amrani, L.; Rotem, O.; Kolot, M.; Zeidan, M.; Czosnek, H.; Gorovits, R. Tomato yellow leaf curl virus infection mitigates the heat stress response of plants grown at high temperatures. *Sci. Rep.* **2016**, *6*, 19715. [CrossRef]
22. Canto, T.; Aranda, M.A.; Fereres, A. Climate change effects on physiology and population processes of hosts and vectors that influence the spread of hemipteran-borne plant viruses. *Glob. Chang. Biol.* **2009**, *15*, 1884–1894. [CrossRef]
23. Velasquez, A.C.; Castroverde, C.D.M.; He, S.Y. Plant-Pathogen Warfare under Changing Climate Conditions. *Curr. Biol.* **2018**, *28*, R619–R634. [CrossRef]
24. Tsai, W.-A.; Brosnan, C.A.; Mitter, N.; Dietzgen, R.G. Perspectives on plant virus diseases in a climate change scenario of elevated temperatures. *Stress Biol.* **2022**, *2*, 37. [CrossRef] [PubMed]
25. Gouveia, B.C.; Calil, I.P.; Machado, J.P.B.; Santos, A.A.; Fontes, E.P.B. Immune Receptors and Co-receptors in Antiviral Innate Immunity in Plants. *Front. Microbiol* **2017**, *7*, 2139. [CrossRef]
26. de Ronde, D.; Lohuis, D.; Kormelink, R. Identification and characterization of a new class of Tomato spotted wilt virus isolates that break *Tsw*-based resistance in a temperature-dependent manner. *Plant Pathol.* **2019**, *68*, 60–71. [CrossRef]
27. Moury, B.; Selassie, K.G.; Marchoux, G.; Daubèze, A.-M.; Palloix, A. High temperature effects on hypersensitive resistance to Tomato Spotted Wilt Tospovirus (TSWV) in pepper (*Capsicum chinense* Jacq.). *Eur. J. Plant Pathol.* **1998**, *104*, 489–498. [CrossRef]

28. Soler, S.; Diez, M.J.; Nuez, F. Effect of Temperature Regime and Growth Stage Interaction on Pattern of Virus Presence in TSWV-Resistant Accessions of *Capsicum chinense*. *Plant Dis.* **1998**, *82*, 1199–1204. [CrossRef]
29. Szajko, K.; Yin, Z.; Marczewski, W. Accumulation of miRNA and mRNA Targets in Potato Leaves Displaying Temperature-Dependent Responses to Potato Virus Y. *Potato Res.* **2019**, *62*, 379–392. [CrossRef]
30. Dexheimer, P.J.; Cochella, L. MicroRNAs: From Mechanism to Organism. *Front. Cell Dev. Biol.* **2020**, *8*, 409. [CrossRef]
31. Liu, S.-R.; Zhou, J.-J.; Hu, C.-G.; Wei, C.-L.; Zhang, J.-Z. MicroRNA-Mediated Gene Silencing in Plant Defense and Viral Counter-Defense. *Front. Microbiol.* **2017**, *8*, 1801. [CrossRef]
32. D'Ario, M.; Griffiths-Jones, S.; Kim, M. Small RNAs: Big Impact on Plant Development. *Trends. Plant Sci.* **2017**, *22*, 1056–1068. [CrossRef]
33. Borges, F.; Martienssen, R.A. The expanding world of small RNAs in plants. *Nat. Rev. Mol. Cell Biol.* **2015**, *16*, 727–741. [CrossRef]
34. Zhang, B.; Li, W.; Zhang, J.; Wang, L.; Wu, J. Roles of Small RNAs in Virus-Plant Interactions. *Viruses* **2019**, *11*, 827. [CrossRef]
35. Križnik, M.; Baebler, Š.; Gruden, K. Roles of small RNAs in the establishment of tolerant interaction between plants and viruses. *Curr. Opin. Virol.* **2020**, *42*, 25–31. [CrossRef] [PubMed]
36. Shivaprasad, P.V.; Chen, H.-M.; Patel, K.; Bond, D.M.; Santos, B.A.C.M.; Baulcombe, D.C. A MicroRNA Superfamily Regulates Nucleotide Binding Site–Leucine-Rich Repeats and Other mRNAs. *Plant Cell* **2012**, *24*, 859–874. [CrossRef]
37. Li, F.; Pignatta, D.; Bendix, C.; Brunkard, J.O.; Cohn, M.M.; Tung, J.; Sun, H.; Kumar, P.; Baker, B. MicroRNA regulation of plant innate immune receptors. *Proc. Natl. Acad. Sci. USA* **2012**, *109*, 1790–1795. [CrossRef] [PubMed]
38. Sharma, N.; Sahu, P.P.; Prasad, A.; Muthamilarasan, M.; Waseem, M.; Khan, Y.; Thakur, J.K.; Chakraborty, S.; Prasad, M. The Sw5a gene confers resistance to ToLCNDV and triggers an HR response after direct AC4 effector recognition. *Proc. Natl. Acad. Sci. USA* **2021**, *118*, e2101833118. [CrossRef] [PubMed]
39. Zhang, C.; Ding, Z.; Wu, K.; Yang, L.; Li, Y.; Yang, Z.; Shi, S.; Liu, X.; Zhao, S.; Yang, Z.; et al. Suppression of Jasmonic Acid-Mediated Defense by Viral-Inducible MicroRNA319 Facilitates Virus Infection in Rice. *Mol. Plant* **2016**, *9*, 1302–1314. [CrossRef]
40. Yang, J.; Zhang, F.; Li, J.; Chen, J.P.; Zhang, H.M. Integrative Analysis of the microRNAome and Transcriptome Illuminates the Response of Susceptible Rice Plants to Rice Stripe Virus. *PLoS ONE* **2016**, *11*, e0146946. [CrossRef]
41. Tao, H.; Jia, Z.; Gao, X.; Gui, M.; Li, Y.; Liu, Y. Analysis of the miRNA expression profile involved in the tomato spotted wilt orthotospovirus–pepper interaction. *Virus. Res.* **2022**, *312*, 198710. [CrossRef]
42. Zhang, T.; Zhai, J.; Zhang, X.; Ling, L.; Li, M.; Xie, S.; Song, M.; Ma, C. Interactive Web-based Annotation of Plant MicroRNAs with iwa-miRNA. *Genom. Proteom. Bioinform.* **2022**, *20*, 557–567. [CrossRef]
43. Langmead, B.; Trapnell, C.; Pop, M.; Salzberg, S.L. Ultrafast and memory-efficient alignment of short DNA sequences to the human genome. *Genome. Biol.* **2009**, *10*, R25. [CrossRef]
44. Kuang, Z.; Wang, Y.; Li, L.; Yang, X. miRDeep-P2: Accurate and fast analysis of the microRNA transcriptome in plants. *Bioinformatics* **2019**, *35*, 2521–2522. [CrossRef]
45. Guo, Z.; Kuang, Z.; Wang, Y.; Zhao, Y.; Tao, Y.; Cheng, C.; Yang, J.; Lu, X.; Hao, C.; Wang, T.; et al. PmiREN: A comprehensive encyclopedia of plant miRNAs. *Nucleic. Acids. Res.* **2019**, *48*, D1114–D1121. [CrossRef] [PubMed]
46. Chen, C.; Li, J.; Feng, J.; Liu, B.; Feng, L.; Yu, X.; Li, G.; Zhai, J.; Meyers, B.C.; Xia, R. sRNAanno—A database repository of uniformly annotated small RNAs in plants. *Hortic. Res.* **2021**, *8*, 45. [CrossRef]
47. Lunardon, A.; Johnson, N.R.; Hagerott, E.; Phifer, T.; Polydore, S.; Coruh, C.; Axtell, M.J. Integrated annotations and analyses of small RNA-producing loci from 47 diverse plants. *Genome Res.* **2020**, *30*, 497–513. [CrossRef]
48. Meng, J.; Liu, D.; Sun, C.; Luan, Y. Prediction of plant pre-microRNAs and their microRNAs in genome-scale sequences using structure-sequence features and support vector machine. *BMC Bioinform.* **2014**, *15*, 423. [CrossRef] [PubMed]
49. Axtell, M.J.; Meyers, B.C. Revisiting Criteria for Plant MicroRNA Annotation in the Era of Big Data. *Plant Cell* **2018**, *30*, 272–284. [CrossRef] [PubMed]
50. Heberle, H.; Meirelles, G.V.; da Silva, F.R.; Telles, G.P.; Minghim, R. InteractiVenn: A web-based tool for the analysis of sets through Venn diagrams. *BMC Bioinform.* **2015**, *16*, 169. [CrossRef]
51. Friedländer, M.R.; Mackowiak, S.D.; Li, N.; Chen, W.; Rajewsky, N. miRDeep2 accurately identifies known and hundreds of novel microRNA genes in seven animal clades. *Nucleic. Acids Res.* **2012**, *40*, 37–52. [CrossRef]
52. Fletcher, S.J.; Boden, M.; Mitter, N.; Carroll, B.J. SCRAM: A pipeline for fast index-free small RNA read alignment and visualization. *Bioinformatics* **2018**, *34*, 2670–2672. [CrossRef]
53. Love, M.I.; Huber, W.; Anders, S. Moderated estimation of fold change and dispersion for RNA-seq data with DESeq2. *Genome Biol.* **2014**, *15*, 550. [CrossRef]
54. Benjamini, Y.; Hochberg, Y. Controlling the False Discovery Rate: A Practical and Powerful Approach to Multiple Testing. *J. R. Stat. Soc. Ser. B* **1995**, *57*, 289–300. [CrossRef]
55. Bedre, R. Reneshbedre/Bioinfokit: Bioinformatics Data Analysis and Visualization Toolkit. *Zenodo* **2020**. [CrossRef]
56. Dai, X.; Zhuang, Z.; Zhao, P.X. psRNATarget: A plant small RNA target analysis server (2017 release). *Nucleic. Acids Res.* **2018**, *46*, W49–W54. [CrossRef]
57. Van Bel, M.; Diels, T.; Vancaester, E.; Kreft, L.; Botzki, A.; Van de Peer, Y.; Coppens, F.; Vandepoele, K. PLAZA 4.0: An integrative resource for functional, evolutionary and comparative plant genomics. *Nucleic. Acids Res.* **2017**, *46*, D1190–D1196. [CrossRef]

58. Tian, T.; Liu, Y.; Yan, H.; You, Q.; Yi, X.; Du, Z.; Xu, W.; Su, Z. agriGO v2.0: A GO analysis toolkit for the agricultural community, 2017 update. *Nucleic. Acids Res.* **2017**, *45*, W122–W129. [CrossRef] [PubMed]
59. Hong, G.; Zhang, W.; Li, H.; Shen, X.; Guo, Z. Separate enrichment analysis of pathways for up- and downregulated genes. *J. R. Soc. Interface* **2014**, *11*, 20130950. [CrossRef]
60. Benjamini, Y.; Yekutieli, D. The Control of the False Discovery Rate in Multiple Testing Under Dependency. *Ann. Stat.* **2001**, *29*, 1165–1188. [CrossRef]
61. Kang, K.; Zhang, X.; Liu, H.; Wang, Z.; Zhong, J.; Huang, Z.; Peng, X.; Zeng, Y.; Wang, Y.; Yang, Y.; et al. A novel real-time PCR assay of microRNAs using S-Poly(T), a specific oligo(dT) reverse transcription primer with excellent sensitivity and specificity. *PLoS ONE* **2012**, *7*, e48536. [CrossRef]
62. Niu, Y.; Zhang, L.; Qiu, H.; Wu, Y.; Wang, Z.; Zai, Y.; Liu, L.; Qu, J.; Kang, K.; Gou, D. An improved method for detecting circulating microRNAs with S-Poly(T) Plus real-time PCR. *Sci. Rep.* **2015**, *5*, 15100. [CrossRef]
63. Xie, S.; Zhu, Q.; Qu, W.; Xu, Z.; Liu, X.; Li, X.; Li, S.; Ma, W.; Miao, Y.; Zhang, L.; et al. sRNAPrimerDB: Comprehensive primer design and search web service for small non-coding RNAs. *Bioinformatics* **2019**, *35*, 1566–1572. [CrossRef]
64. Tang, Z.; Yu, J.; Xie, J.; Lyu, J.; Feng, Z.; Dawuda, M.M.; Liao, W.; Wu, Y.; Hu, L. Physiological and Growth Response of Pepper (*Capsicum annum* L.) Seedlings to Supplementary Red/Blue Light Revealed through Transcriptomic Analysis. *Agronomy* **2019**, *9*, 139. [CrossRef]
65. Tsai, W.A.; Shafiei-Peters, J.R.; Mitter, N.; Dietzgen, R.G. Effects of Elevated Temperature on the Susceptibility of Capsicum Plants to Capsicum Chlorosis Virus Infection. *Pathogens* **2022**, *11*, 200. [CrossRef] [PubMed]
66. Livak, K.J.; Schmittgen, T.D. Analysis of relative gene expression data using real-time quantitative PCR and the 2(-Delta Delta C(T)) Method. *Methods* **2001**, *25*, 402–408. [CrossRef]
67. Mitter, N.; Dietzgen, R.G. Use of hairpin RNA constructs for engineering plant virus resistance. *Methods Mol. Biol.* **2012**, *894*, 191–208. [CrossRef] [PubMed]
68. Kasschau, K.D.; Xie, Z.; Allen, E.; Llave, C.; Chapman, E.J.; Krizan, K.A.; Carrington, J.C. P1/HC-Pro, a viral suppressor of RNA silencing, interferes with *Arabidopsis* development and miRNA function. *Dev. Cell* **2003**, *4*, 205–217. [CrossRef]
69. Song, C.; Wang, C.; Zhang, C.; Korir, N.K.; Yu, H.; Ma, Z.; Fang, J. Deep sequencing discovery of novel and conserved microRNAs in trifoliate orange (*Citrus trifoliata*). *BioMed Cent. Genom.* **2010**, *11*, 431. [CrossRef]
70. Mallory, A.C.; Elmayan, T.; Vaucheret, H. MicroRNA maturation and action--the expanding roles of ARGONAUTEs. *Curr. Opin. Plant Biol.* **2008**, *11*, 560–566. [CrossRef]
71. Varallyay, E.; Valoczi, A.; Agyi, A.; Burgyan, J.; Havelda, Z. Plant virus-mediated induction of miR168 is associated with repression of ARGONAUTE1 accumulation. *Eur. Mol. Biol. Organ. J.* **2010**, *29*, 3507–3519. [CrossRef]
72. Varallyay, E.; Havelda, Z. Unrelated viral suppressors of RNA silencing mediate the control of ARGONAUTE1 level. *Mol. Plant Pathol.* **2013**, *14*, 567–575. [CrossRef]
73. Fang, X.; Qi, Y. RNAi in Plants: An Argonaute-Centered View. *Plant Cell* **2016**, *28*, 272–285. [CrossRef]
74. Fei, Y.; Mao, Y.; Shen, C.; Wang, R.; Zhang, H.; Huang, J. WPMIAS: Whole-degradome-based Plant MicroRNA-Target Interaction Analysis Server. *Bioinformatics* **2019**, *36*, 1937–1939. [CrossRef]
75. Zhang, L.; Qin, C.; Mei, J.; Chen, X.; Wu, Z.; Luo, X.; Cheng, J.; Tang, X.; Hu, K.; Li, S.C. Identification of MicroRNA Targets of Capsicum spp. Using MiRTrans-a Trans-Omics Approach. *Front. Plant Sci.* **2017**, *8*, 495. [CrossRef]
76. Marques de Carvalho, L.; Benda, N.D.; Vaughan, M.M.; Cabrera, A.R.; Hung, K.; Cox, T.; Abdo, Z.; Allen, L.H.; Teal, P.E. Mi-1-Mediated Nematode Resistance in Tomatoes is Broken by Short-Term Heat Stress but Recovers Over Time. *J. Nematol.* **2015**, *47*, 133–140. [PubMed]
77. Romero, A.M.; Kousik, C.S.; Ritchie, D.F. Temperature Sensitivity of the Hypersensitive Response of Bell Pepper to *Xanthomonas axonopodis* pv. *vesicatoria*. *Phytopathology* **2002**, *92*, 197–203. [CrossRef] [PubMed]
78. Whitham, S.; McCormick, S.; Baker, B. The N gene of tobacco confers resistance to tobacco mosaic virus in transgenic tomato. *Proc. Natl. Acad. Sci. USA* **1996**, *93*, 8776–8781. [CrossRef] [PubMed]
79. Chung, B.N.; Lee, J.H.; Kang, B.C.; Koh, S.W.; Joa, J.H.; Choi, K.S.; Ahn, J.J. HR-Mediated Defense Response is Overcome at High Temperatures in Capsicum Species. *Plant Pathol. J.* **2018**, *34*, 71–77. [CrossRef] [PubMed]
80. Hwang, D.G.; Park, J.H.; Lim, J.Y.; Kim, D.; Choi, Y.; Kim, S.; Reeves, G.; Yeom, S.I.; Lee, J.S.; Park, M.; et al. The hot pepper (*Capsicum annuum*) microRNA transcriptome reveals novel and conserved targets: A foundation for understanding MicroRNA functional roles in hot pepper. *PLoS ONE* **2013**, *8*, e64238. [CrossRef]
81. Seo, E.; Kim, T.; Park, J.H.; Yeom, S.-I.; Kim, S.; Seo, M.-K.; Shin, C.; Choi, D. Genome-wide comparative analysis in Solanaceous species reveals evolution of microRNAs targeting defense genes in *Capsicum* spp. *DNA Res.* **2018**, *25*, 561–575. [CrossRef]
82. Xu, X.-W.; Li, T.; Li, Y.; Li, Z.-X. Identification and Analysis of C. annuum microRNAs by High-Throughput Sequencing and Their Association with High Temperature and High Air Humidity Stress. *Int. J. Bioautomat.* **2015**, *19*, 459–472.
83. Yang, S.; Zhang, Z.; Chen, W.; Li, X.; Zhou, S.; Liang, C.; Li, X.; Yang, B.; Zou, X.; Liu, F.; et al. Integration of mRNA and miRNA profiling reveals the heterosis of three hybrid combinations of *Capsicum annuum* varieties. *GM Crops Food* **2021**, *12*, 224–241. [CrossRef]
84. Apostolova, E.; Hadjieva, N.; Ivanova, D.P.; Yahubyan, G.; Baev, V.; Gozmanova, M. MicroRNA expression dynamics reshape the cultivar-specific response of pepper (*Capsicum annuum* L.) to potato spindle tuber viroid (PSTVd) infection. *Sci. Hortic.* **2021**, *278*, 109845. [CrossRef]

85. Lee, M.H.; Jeon, H.S.; Kim, S.H.; Chung, J.H.; Roppolo, D.; Lee, H.J.; Cho, H.J.; Tobimatsu, Y.; Ralph, J.; Park, O.K. Lignin-based barrier restricts pathogens to the infection site and confers resistance in plants. *Eur. Mol. Biol. Organ. J* **2019**, *38*, e101948. [CrossRef] [PubMed]
86. Kimmins, W.C.; Wuddah, D. Hypersensitive Resistance: Determination of Lignin in Leaves with a Localized Virus Infection. *Phytopathology* **1977**, *67*, 1012–1016. [CrossRef]
87. Li, C.; Liu, X.-X.; Abouelnasr, H.; Mohamed Hamed, A.; Kou, M.; Tang, W.; Yan, H.; Wang, X.; Wang, X.-X.; Zhang, Y.-G.; et al. Inhibition of miR397 by STTM technology to increase sweetpotato resistance to SPVD. *J. Integr. Agric.* **2022**, *21*, 2865–2875. [CrossRef]
88. Ma, C.; Burd, S.; Lers, A. miR408 is involved in abiotic stress responses in *Arabidopsis*. *Plant J.* **2015**, *84*, 169–187. [CrossRef]
89. Moxon, S.; Jing, R.; Szittya, G.; Schwach, F.; Rusholme Pilcher, R.L.; Moulton, V.; Dalmay, T. Deep sequencing of tomato short RNAs identifies microRNAs targeting genes involved in fruit ripening. *Genome Res.* **2008**, *18*, 1602–1609. [CrossRef]
90. Mohorianu, I.; Schwach, F.; Jing, R.; Lopez-Gomollon, S.; Moxon, S.; Szittya, G.; Sorefan, K.; Moulton, V.; Dalmay, T. Profiling of short RNAs during fleshy fruit development reveals stage-specific sRNAome expression patterns. *Plant J.* **2011**, *67*, 232–246. [CrossRef]
91. Linsen, S.E.; de Wit, E.; Janssens, G.; Heater, S.; Chapman, L.; Parkin, R.K.; Fritz, B.; Wyman, S.K.; de Bruijn, E.; Voest, E.E.; et al. Limitations and possibilities of small RNA digital gene expression profiling. *Nat. Methods* **2009**, *6*, 474–476. [CrossRef]
92. Fuchs, R.T.; Sun, Z.; Zhuang, F.; Robb, G.B. Bias in ligation-based small RNA sequencing library construction is determined by adaptor and RNA structure. *PLoS ONE* **2015**, *10*, e0126049. [CrossRef] [PubMed]
93. Dabney, J.; Meyer, M. Length and GC-biases during sequencing library amplification: A comparison of various polymerase-buffer systems with ancient and modern DNA sequencing libraries. *BioTechniques* **2012**, *52*, 87–94. [CrossRef]
94. Abdel-Ghany, S.E.; Pilon, M. MicroRNA-mediated systemic down-regulation of copper protein expression in response to low copper availability in Arabidopsis. *J. Biol. Chem.* **2008**, *283*, 15932–15945. [CrossRef]
95. Trindade, I.; Capitão, C.; Dalmay, T.; Fevereiro, M.P.; Santos, D.M. miR398 and miR408 are up-regulated in response to water deficit in *Medicago truncatula*. *Planta* **2010**, *231*, 705–716. [CrossRef]
96. Li, C.; Li, D.; Zhou, H.; Li, J.; Lu, S. Analysis of the laccase gene family and miR397-/miR408-mediated posttranscriptional regulation in *Salvia miltiorrhiza*. *PeerJ* **2019**, *7*, e7605. [CrossRef]
97. Lu, S.; Li, Q.; Wei, H.; Chang, M.J.; Tunlaya-Anukit, S.; Kim, H.; Liu, J.; Song, J.; Sun, Y.H.; Yuan, L.; et al. Ptr-miR397a is a negative regulator of laccase genes affecting lignin content in *Populus trichocarpa*. *Proc. Natl. Acad. Sci. USA* **2013**, *110*, 10848–10853. [CrossRef] [PubMed]
98. Wang, C.Y.; Zhang, S.; Yu, Y.; Luo, Y.C.; Liu, Q.; Ju, C.; Zhang, Y.C.; Qu, L.H.; Lucas, W.J.; Wang, X.; et al. MiR397b regulates both lignin content and seed number in *Arabidopsis* via modulating a laccase involved in lignin biosynthesis. *Plant Biotechnol. J.* **2014**, *12*, 1132–1142. [CrossRef] [PubMed]
99. Swetha, C.; Basu, D.; Pachamuthu, K.; Tirumalai, V.; Nair, A.; Prasad, M.; Shivaprasad, P.V. Major Domestication-Related Phenotypes in Indica Rice Are Due to Loss of miRNA-Mediated Laccase Silencing. *Plant Cell* **2018**, *30*, 2649–2662. [CrossRef] [PubMed]
100. Huang, S.; Zhou, J.; Gao, L.; Tang, Y. Plant miR397 and its functions. *Funct. Plant Biol.* **2021**, *48*, 361–370. [CrossRef] [PubMed]
101. Zhao, Q.; Tobimatsu, Y.; Zhou, R.; Pattathil, S.; Gallego-Giraldo, L.; Fu, C.; Jackson, L.A.; Hahn, M.G.; Kim, H.; Chen, F.; et al. Loss of function of cinnamyl alcohol dehydrogenase 1 leads to unconventional lignin and a temperature-sensitive growth defect in *Medicago truncatula*. *Proc. Natl. Acad. Sci. USA* **2013**, *110*, 13660–13665. [CrossRef]
102. Liu, Q.; Luo, L.; Zheng, L. Lignins: Biosynthesis and Biological Functions in Plants. *Int. J. Mol. Sci.* **2018**, *19*, 335. [CrossRef]
103. Rosas-Diaz, T.; Zhang, D.; Fan, P.; Wang, L.; Ding, X.; Jiang, Y.; Jimenez-Gongora, T.; Medina-Puche, L.; Zhao, X.; Feng, Z.; et al. A virus-targeted plant receptor-like kinase promotes cell-to-cell spread of RNAi. *Proc. Natl. Acad. Sci. USA* **2018**, *115*, 1388–1393. [CrossRef]
104. Swida-Barteczka, A.; Szweykowska-Kulinska, Z. Micromanagement of Developmental and Stress-Induced Senescence: The Emerging Role of MicroRNAs. *Genes* **2019**, *10*, 210. [CrossRef] [PubMed]
105. Lopez-Gomollon, S.; Mohorianu, I.; Szittya, G.; Moulton, V.; Dalmay, T. Diverse correlation patterns between microRNAs and their targets during tomato fruit development indicates different modes of microRNA actions. *Planta* **2012**, *236*, 1875–1887. [CrossRef]
106. Kidner, C.A.; Martienssen, R.A. Spatially restricted microRNA directs leaf polarity through ARGONAUTE1. *Nature* **2004**, *428*, 81–84. [CrossRef] [PubMed]
107. Kawashima, C.G.; Yoshimoto, N.; Maruyama-Nakashita, A.; Tsuchiya, Y.N.; Saito, K.; Takahashi, H.; Dalmay, T. Sulphur starvation induces the expression of microRNA-395 and one of its target genes but in different cell types. *Plant J.* **2009**, *57*, 313–321. [CrossRef] [PubMed]
108. Singh, R.; Singh, S.; Parihar, P.; Mishra, R.K.; Tripathi, D.K.; Singh, V.P.; Chauhan, D.K.; Prasad, S.M. Reactive Oxygen Species (ROS): Beneficial Companions of Plants' Developmental Processes. *Front. Plant Sci.* **2016**, *7*, 1299. [CrossRef] [PubMed]
109. Pérez-Llorca, M.; Munné-Bosch, S. Aging, stress, and senescence in plants: What can biological diversity teach us? *Geroscience* **2021**, *43*, 167–180. [CrossRef]
110. Singh, A.; Permar, V.; Basavaraj, A.; Bhoopal, S.T.; Praveen, S. Effect of Temperature on Symptoms Expression and Viral RNA Accumulation in Groundnut Bud Necrosis Virus Infected Vigna unguiculata. *Iran. J. Biotechnol.* **2018**, *16*, 227–234. [CrossRef]

111. Li, Z.; Peng, J.; Wen, X.; Guo, H. Ethylene-insensitive3 is a senescence-associated gene that accelerates age-dependent leaf senescence by directly repressing miR164 transcription in *Arabidopsis*. *Plant Cell* **2013**, *25*, 3311–3328. [CrossRef]
112. Li, T.; Gonzalez, N.; Inzé, D.; Dubois, M. Emerging Connections between Small RNAs and Phytohormones. *Trends. Plant Sci.* **2020**, *25*, 912–929. [CrossRef]
113. Podzimska-Sroka, D.; O'Shea, C.; Gregersen, P.L.; Skriver, K. NAC Transcription Factors in Senescence: From Molecular Structure to Function in Crops. *Plants* **2015**, *4*, 412–448. [CrossRef]
114. Dalmadi, Á.; Miloro, F.; Bálint, J.; Várallyay, É.; Havelda, Z. Controlled RISC loading efficiency of miR168 defined by miRNA duplex structure adjusts ARGONAUTE1 homeostasis. *Nucleic. Acids. Res.* **2021**, *49*, 12912–12928. [CrossRef] [PubMed]

pathogens

MDPI

Article

Biocontrol Potential of *Trichoderma* spp. Against *Phytophthora ramorum*

Elisa Becker [1,*], Nirusan Rajakulendran [2] and Simon Francis Shamoun [1]

[1] Pacific Forestry Centre, Canadian Forest Service, Natural Resources Canada, 506 West Burnside Road, Victoria, BC V8Z 1M5, Canada; simon.shamoun@nrcan-rncan.gc.ca
[2] Independent Researcher, Toronto, ON M3H, Canada; nirusan.rajakulendran@gmail.com
* Correspondence: elisa.becker@nrcan-rncan.gc.ca

Abstract: *Phytophthora ramorum*, the cause of Sudden Oak Death and related diseases, threatens over 130 tree and shrub species. We evaluated the biocontrol potential of isolates from nine *Trichoderma* species against *P. ramorum* using growth-rate studies, dual-culture assays, and culture-filtrate assays. Results showed significant variation in *Trichoderma* growth rates and biocontrol potential. Some isolates exhibited rapid growth, effective overgrowth, and lethal effects against *P. ramorum* and produced potent antagonistic metabolites. Faster growth rates only partially correlated with biocontrol efficacy, indicating that factors beyond growth, such as metabolite production, play significant roles. Notably, isolates of *T. koningii*, *T. viride*, and the commercial product SoilGard™ (*T. virens*) showed promising efficacy. We calculated a combined biocontrol variable to rank isolates based on vigour and efficacy to aid in identifying promising candidates. Our findings support the use of *Trichoderma* spp. as biocontrol agents against *P. ramorum* and underscore the need for a multifaceted approach to selecting and optimizing isolates. Our evaluation demonstrated the importance of using different assays to assess specific mechanisms of action of biocontrol candidates. Future research should further explore these interactions to enhance the sustainable management of *P. ramorum*.

Keywords: *Phytophthora ramorum*; sudden oak death; biocontrol; biological control agents; *Trichoderma*; mechanisms of action; antagonistic metabolites; in vitro assay; dual culture assay

1. Introduction

Phytophthora ramorum, Werres, de Cock & Man in't Veld [1], a globally significant invasive plant pathogen, causes substantial ecological and economic damage through diseases such as Sudden Oak Death in the U.S. and Sudden Larch Death and ramorum blight in Europe [2]. Its broad host range threatens forest biodiversity and the viability of the nursery industry. This oomycete affects over 130 species of trees and shrubs, including important nursery crops such as rhododendron, viburnum, and lilac, as well as forest species like maple and oak [3,4]. The pathogen produces sporangia on infected leaves and twigs, releasing motile zoospores that encyst and penetrate hosts under suitable conditions, initiating new infections [1]. In addition to causing foliar and stem diseases, *P. ramorum* survives in soil and can infect rhododendron roots, demonstrating its broad pathogenic capabilities [5,6].

The long-distance spread of *P. ramorum* has predominantly occurred through the transport of infected nursery stock. It has been isolated from diseased plants in nurseries across California, Oregon, Washington, and British Columbia (B.C.), with infected plants shipped to 32 U.S. states between 2003 and 2007 [3]. It has established in Florida and

Mississippi, as evidenced by repeated detections in baiting surveys. In Europe, *P. ramorum* is considered an introduced pathogen found in nurseries and woodland gardens. This has led to quarantine regulations aimed at reducing further spread in North America and Europe. Despite detection in hundreds of nurseries and several wildland areas, its spread in natural environments in western North America remains limited to areas of California and southwestern Oregon. In B.C., *P. ramorum* is a quarantined pest posing a moderate risk to native forest ecosystems [7]. Since its detection in several B.C. nurseries in 2003, eradication actions have been mandated at all sites with infected plants [4,8]. While there are no cures for infected plants, several *P. ramorum* control measures are registered in the U.S. and Canada for preventive use on susceptible plants. These measures, alongside quarantine policies, focus on reducing inoculum levels and limiting human-facilitated pathogen spread. Minimizing the risk of infection is the primary defence against the spread of the pathogen to uninfected plants [9].

Trichoderma spp. are fast-growing, opportunistic fungi that function as avirulent plant symbionts. Their antagonistic abilities against plant pathogens have promoted their use as biological control agents [10–13]. They have shown potential as biocontrol agents against certain *Phytophthora* species [14]. For example, apple seedlings co-cultured with *Phytophthora cactorum* and *Trichoderma* spp. showed significant reductions in root damage and increased plant weight compared to those infected with *P. cactorum* alone [15].

Mycoparasitism, a process where one fungus parasitizes another, often involves the secretion of cell wall-degrading enzymes that hydrolyze the cell walls of target fungi or oomycetes [11]. Antibiosis, the production of inhibitory substances, is another major biocontrol mechanism employed by *Trichoderma*, which produces various antimicrobial compounds [16]. Other modes of action of *Trichoderma* strains include promoting plant growth and plant defence mechanisms and competing for nutrients and space [16–19]. To address the need for effective biocontrol agents against *P. ramorum*, this study investigates the potential of *Trichoderma* spp.

The objective of this study was to investigate the antagonistic properties of isolates from nine *Trichoderma* species, including three commercial biocontrol products, against *P. ramorum*. We conducted three in vitro biological assays to screen *Trichoderma* isolates as potential biological control candidates against *P. ramorum*.

- **Analysis of Temperature-Dependent Growth Rate**: We began by performing a study of temperature-dependent growth rate on 51 *Trichoderma* isolates, hypothesizing that the growth rate of *Trichoderma* correlates with its biocontrol efficacy against *P. ramorum*. *Trichoderma* isolates that grow and colonize substrates rapidly could compete more effectively against *P. ramorum*.
- **Dual-Culture Assay for Direct Interaction**: Next, we assessed the direct interaction, or mycoparasitism, of *Trichoderma* with *P. ramorum*. This assay used a dual-culture approach to measure the rates at which *Trichoderma* overgrew and killed an existing *P. ramorum* culture through direct interaction.
- **Antibiosis Microplate Assay for Antagonistic Effects**: Additionally, we investigated the antibiosis of different *Trichoderma* isolates with respect to *P. ramorum* by testing sterile culture filtrates of *Trichoderma* isolates in a novel in vitro microplate assay to assess their antagonistic effects on *P. ramorum* germination and growth.

To conclude, we calculated a combined biocontrol variable to represent the general vigour and overall efficacy of the *Trichoderma* isolates.

2. Materials and Methods

2.1. Analysis of Temperature-Dependent Growth Rate

To measure the temperature-dependent linear growth rates of *Trichoderma* isolates (Table 1), a growth-rate study was performed at two temperatures: 10 °C and 20 °C. Agar plugs were taken from the leading edges of fresh cultures grown on potato dextrose agar (PDA) plates at 20 °C for 2 days. Growth-rate cultures were initiated on Petri plates (100 mm × 15 mm) containing 25 mL of PDA by placing an agar plug 5 mm from the edge of the plate. The plates were incubated at room temperature (~20 °C) for 4–8 h to allow the fungi to grow from the plug onto the agar surface. The plates were then transferred to incubators set at either 10 °C or 20 °C in darkness. Measurements of mycelial growth were taken along predetermined axes at 24, 48, 72, and 96 h after the acclimation period. Growth during the acclimation period was not included in the calculation of the overall growth rate. The linear growth rate was expressed in mm/day.

Table 1. *Trichoderma* isolates screened in this study, with tree host, geographic location of collection, and name of collector or source company.

Accession	Species [1]	Host	Location	Collected by/Source
PFC 5001	*polysporum*	Black spruce	Comox, BC	T. Osono
PFC 5002	*polysporum*	Douglas fir	Comox, BC	T. Osono
PFC 5003	*polysporum*	Black spruce	Comox, BC	T. Osono
PFC 5004	*hamatum*	Black spruce	Comox, BC	T. Osono
PFC 5005	*hamatum*	Black spruce	Comox, BC	T. Osono
PFC 5006	*hamatum*	Douglas fir	Comox, BC	T. Osono
PFC 5007	*viride*	Black spruce	Comox, BC	T. Osono
PFC 5008	*viride*	Black spruce	Comox, BC	T. Osono
PFC 5009	*viride*	Black spruce	Comox, BC	T. Osono
PFC 5010	*viride*	Douglas fir	Berms, SK	T. Osono
PFC 5011	*viride*	Black spruce	Berms, SK	T. Osono
PFC 5012	*viride*	Jack pine	Berms, SK	T. Osono
PFC 5013	*koningii*	Douglas fir	Groundhog, ON	T. Osono
PFC 5014	*koningii*	Black spruce	Groundhog, ON	T. Osono
PFC 5015	*koningii*	Douglas fir	Groundhog, ON	T. Osono
PFC 5016	*polysporum*	Douglas fir	Groundhog, ON	T. Osono
PFC 5017	*polysporum*	Douglas fir	Groundhog, ON	T. Osono
PFC 5018	*polysporum*	Black spruce	Groundhog, ON	T. Osono
PFC 5019	*viride*	Black spruce	Groundhog, ON	T. Osono
PFC 5020	*viride*	Douglas fir	Groundhog, ON	T. Osono
PFC 5021	*viride*	Black spruce	Groundhog, ON	T. Osono
PFC 5022	*koningii*	Black spruce	CPRS, QC	T. Osono
PFC 5023	*koningii*	Black spruce	CPRS, QC	T. Osono
PFC 5024	*koningii*	Douglas fir	CPRS, QC	T. Osono
PFC 5025	*polysporum*	Douglas fir	CPRS, QC	T. Osono
PFC 5026	*polysporum*	Black spruce	CPRS, QC	T. Osono
PFC 5027	*viride*	Black spruce	CPRS, QC	T. Osono
PFC 5028	*viride*	Black spruce	CPRS, QC	T. Osono
PFC 5029	*viride*	Black spruce	CPRS, QC	T. Osono
PFC 5030	*koningii*	Douglas fir	Nashwaak, NB	T. Osono
PFC 5031	*koningii*	Balsam fir	Nashwaak, NB	T. Osono
PFC 5032	*polysporum*	Douglas fir	Nashwaak, NB	T. Osono
PFC 5033	*polysporum*	Douglas fir	Nashwaak, NB	T. Osono
PFC 5034	*polysporum*	Black spruce	Nashwaak, NB	T. Osono
PFC 5035	*viride*	Douglas fir	Nashwaak, NB	T. Osono
PFC 5036	*viride*	Black spruce	Nashwaak, NB	T. Osono
PFC 5037	*viride*	Douglas fir	Nashwaak, NB	T. Osono
PFC 5092	*hamatum*	Douglas fir	Oregon, USA	M. Elliott
PFC 5093	*pseudokoningii*	Douglas fir	Oregon, USA	M. Elliott

Table 1. *Cont.*

Accession	Species [1]	Host	Location	Collected by/Source
PFC 5094	*saturnisporum*	Douglas fir	Oregon, USA	M. Elliott
PFC 5095	*virens*	Douglas fir	Oregon, USA	M. Elliott
PFC 5096	*virens*	Douglas fir	Oregon, USA	M. Elliott
PFC 5097	*harzianum*	Douglas fir	Oregon, USA	M. Elliott
PFC 5098	*virens*	Douglas fir	Oregon, USA	M. Elliott
PFC 5099	*pseudokoningii*	Douglas fir	Oregon, USA	M. Elliott
PFC 5100	*hamatum*	Douglas fir	Oregon, USA	M. Elliott
PFC 5101	*koningii*	Douglas fir	Oregon, USA	M. Elliott
PFC 5102	*virens*	Douglas fir	Oregon, USA	M. Elliott
PlantHelper™	*atroviride*			AmPac Biotech
T382	*hamatum*			Sylvan Bioproducts
SoilGard™	*virens*			Certis Biologicals

[1] All are members of the *Trichoderma* genus.

2.2. Dual Culture Assay for Direct Interaction

A dual-culture assay was used to evaluate direct interactions between the antagonist and pathogen, following the method of Goldfarb et al. [20]. To allow *P. ramorum* (NA1 lineage, accession 5073, isolated from *Rhododendron*) to establish before introduction of the faster-growing *Trichoderma* cultures, placements were timed accordingly (Figure 1). Each plate containing 15% V8 agar amended with 0.15% $CaCO_3$ was bisected by a line drawn on the plate. An agar plug with an active culture of *P. ramorum* was placed on the center line, approximately 5 mm from the edge of the plate. The plate was incubated at 20 °C for 18 days or until the *P. ramorum* culture reached a diameter of 5.5 cm. To mark the farthest extent of *P. ramorum* growth, a line was drawn perpendicular to the center line. An agar plug of *Trichoderma* from the leading edge of a fresh 2-day-old PDA culture was placed on the center line, with its closest edge 5 mm from the marked extent of *P. ramorum* growth. After 4 days of dual culture growth, eight pairs of 6 mm plugs (one from each side of the center line), spanning a distance of 48 mm within the *P. ramorum* growth area, were collected. One plug from each pair was transferred to a plate containing 15% V8 agar amended with 0.15% $CaCO_3$ to test for the presence of viable *Trichoderma*. The corresponding plug was transferred to a plate containing 15% V8 agar amended with 0.15% $CaCO_3$ plus 100 ppm benomyl (to inhibit *Trichoderma* growth) to test for the presence of viable *P. ramorum*. Re-isolation plates were monitored for the presence of *P. ramorum* or *Trichoderma* for up to 10 days (Figure 1).

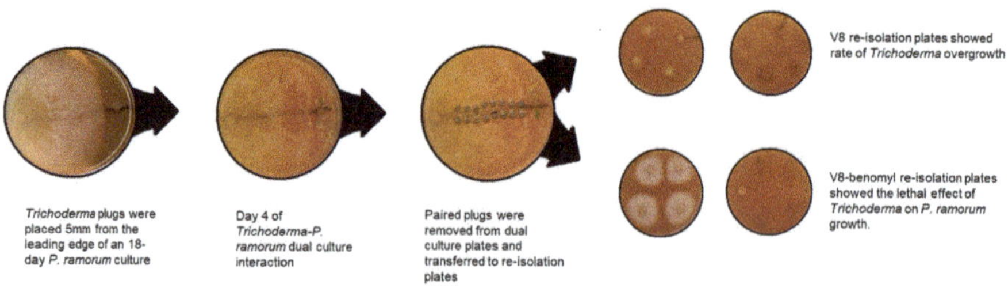

Figure 1. *Trichoderma* isolates in dual culture with 18-day-old *P. ramorum* cultures.

The two parameters measured in this assay were the rate of overgrowth of the pathogen by the candidate *Trichoderma* culture and the rate of lethal effect. The rate of overgrowth was determined by measuring the linear distance on the dual culture plates over which *Trichoderma* overgrew the established *P. ramorum* culture. Growth of both *P. ramorum* and *Trichoderma* on their respective re-isolation plates indicated that *Trichoderma* had overgrown *P. ramorum* but had not killed it. The rate of lethal effect measured how effectively candidate cultures of *Trichoderma* spp. could kill *P. ramorum*. The absence of *P. ramorum* growth on the benomyl-amended re-isolation plates indicated it had been killed by the *Trichoderma* isolate (Figure 1).

2.3. Antibiosis Microplate Assay for Antagonistic Effect

2.3.1. Phytophthora Ramorum Sporangia Production

For sporangia production, fresh agar plugs of *P. ramorum* were transferred into 60 mm plates containing clarified 15% V8 broth amended with 1% $CaCO_3$ and incubated for 3–5 days. After incubation, these liquid cultures were washed with sterile distilled water (sdH_2O) to remove nutrient media, resuspended in sdH_2O, and incubated for 3 days to induce sporangia production. To separate sporangia from mycelium, cultures were filtered through three layers of sterile cheesecloth. To stimulate zoospore release, the sporangia suspension was chilled at 4 °C for 1 h, then incubated at room temperature for 1–2 h, with zoospore release confirmed by microscopy. An aliquot of the zoospore suspension was transferred into a centrifuge tube and vortexed for 30 s to induce encystment. Encysted zoospores were counted using a hemocytometer, and appropriate dilutions were prepared in clarified 15% V8 broth amended with 0.15% $CaCO_3$.

2.3.2. Production of Trichoderma Filtrates

Trichoderma isolates were cultured from agar plugs on PDA plates incubated at 20 °C in the dark for 2–3 days. Plates were subsequently placed under constant light until the cultures started to sporulate. Spore suspensions were prepared by flooding the plates with sdH_2O and mixing with equal volumes of 20% glycerol, and the suspensions were then stored at −20 °C. On the day of inoculation, spores were counted using a hemocytometer and 100 mL of potato dextrose broth (PDB) was inoculated with *Trichoderma* spores to a final concentration of 10,000 spores/mL. Liquid cultures were incubated at room temperature with constant shaking at 130 rpm for 5 days. Mycelium was collected by filtering through three layers of cheesecloth and washed with 25 mL of sdH_2O. Washed mycelium was aseptically transferred into 250 mL flasks containing 100 mL of sdH_2O and incubated at room temperature with constant shaking at 130 rpm for 2 days. This culture filtrate was separated from mycelium by first pouring the liquid through three layers of cheesecloth, then filtering it through a 0.2 μm syringe filter to remove spores and mycelial fragments.

2.3.3. Antibiosis Microplate Assay

To test the activity of sterile filtrates, 100 μL of *P. ramorum* zoospore suspension (1000 zoospores/mL in 15% clarified V8) was mixed with 100 μL of *Trichoderma* culture filtrate in 96-well microplates (Figure 2). To generate the positive controls, 100 μL of sdH_2O was substituted for *Trichoderma* culture filtrate. To generate the negative controls, 100 μL of 15% V8 broth was substituted for the *P. ramorum* zoospore suspension and mixed with 100 μL of *Trichoderma* filtrates. Absorbance at 650 nm was measured at 0, 24, 48, and 72 h. The percent inhibition was calculated using the following formula: $\%\ \text{inhibition} = [(\text{Control} - \text{Treated})/\text{Control}] \times 100$.

2.4. Statistical Analyses

Statistical analyses were performed using R and RStudio with packages including *tidyr, dplyr, psych, ggplot2,* and *RColorBrewer* [21–27]. To create a combined biocontrol variable, values of each relevant variable were standardized (mean of 0 and standard deviation of 1) using the 'scale' function, then combined into a single composite score by averaging the standardized variables.

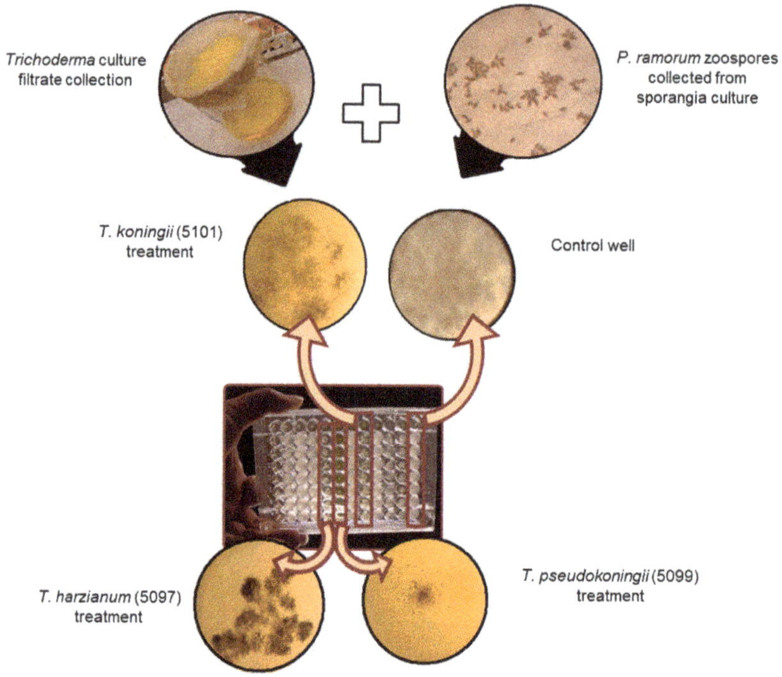

Figure 2. Microplate assay showing the effects of sterile *Trichoderma* culture extracts on *P. ramorum* germination and growth. The control wells contained only *P. ramorum* zoospores and showed uninhibited germination and growth, while the other wells also contained extracts of *Trichoderma* isolates and showed variable inhibition of *P. ramorum* growth relative to the controls. These examples show the least relative inhibition of *P. ramorum* in the top left (isolate 5101) and the most inhibition in the bottom right (isolate 5099).

3. Results

3.1. Analysis of Temperature-Dependent Growth Rate

The mean radial growth of *Trichoderma* isolates at 10 °C after 3 days was 4.76 ± 1.25 cm (Figure 3). At 20 °C, the mean growth after 3 days was 14.72 ± 3.03 cm. The rate of linear growth ranged from 8.0 to 19.4 mm/day among 51 isolates (each with three replicates), with isolate 5035 (*T. viride*) growing the fastest.

3.2. Dual Culture Assay for Direct Interaction-Rate of Overgrowth

Overgrowth was assessed by the ability of *Trichoderma* cultures to overgrow an established *P. ramorum* culture (Figure 4). The mean overgrowth distance after 3 days was 5.95 ± 2.31 cm (Figure 4a). The commercial product SoilGard™ (*T. virens*) exhibited the highest overgrowth rate, at 10.5 mm/day.

3.3. Dual Culture Assay for Direct Interaction-Rate of Lethal Effect

The rate of lethal effect described the ability of *Trichoderma* isolates to mycoparasitize and kill *P. ramorum* through direct interactions (Figure 2). SoilGard (*T. virens*) demonstrated the highest rate of lethal effect at 10.5 mm/day, consistent with its high rate of overgrowth (Figure 4b). The ability of *Trichoderma* spp. to overgrow *P. ramorum* did not necessarily translate to a lethal effect. For example, *T. hamatum* had an average overgrowth rate of 6.0 mm/day; however, none of its isolates were able to mycoparasitize and kill *P. ramorum*. Similarly, *T. polysporum* (5003, 5018, 5025, 5026, 5034), *T. virens* (5095, 5096, 5098), *T. koningii*

(5101), *T. pseudokoningii* (5093, 5094), and *T. harzianum* (5097) coexisted with *P. ramorum* without having a lethal effect on it. In addition, no *Trichoderma* isolate exhibited a lethal effect on *P. ramorum* from a distance. *Trichoderma* isolates had a lethal effect on *P. ramorum* only after overgrowing it.

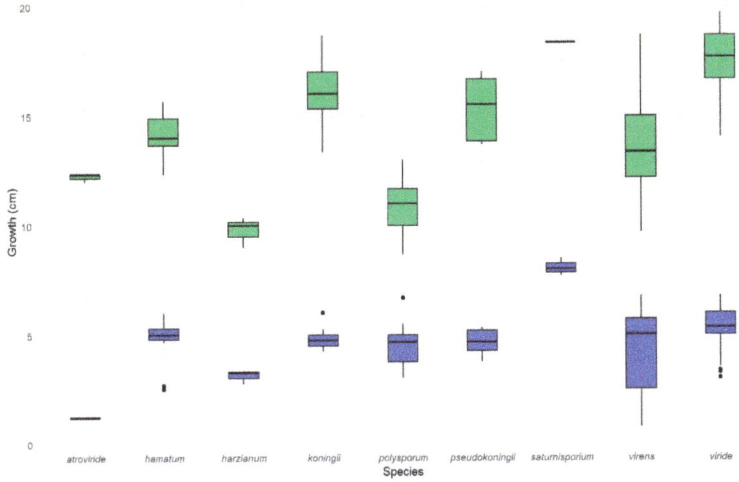

Figure 3. Box plots showing the average radial growth of *Trichoderma* isolates, grouped by species, on PDA at two temperatures (10 °C in blue and 20 °C in green) after 3 days. The numbers of isolates per species were as follows: *atroviride* (1), *hamatum* (6), *harzianum* (1), *koningii* (9), *polysporum* (11), *pseudokoningii* (2), *saturnisporium* (1), *virens* (5), and *viride* (15). Each isolate had 3 replicates. Boxes represent the interquartile range (IQR), with medians indicated by horizontal lines. Whiskers extend to 1.5 times the IQR, and outliers are individual points.

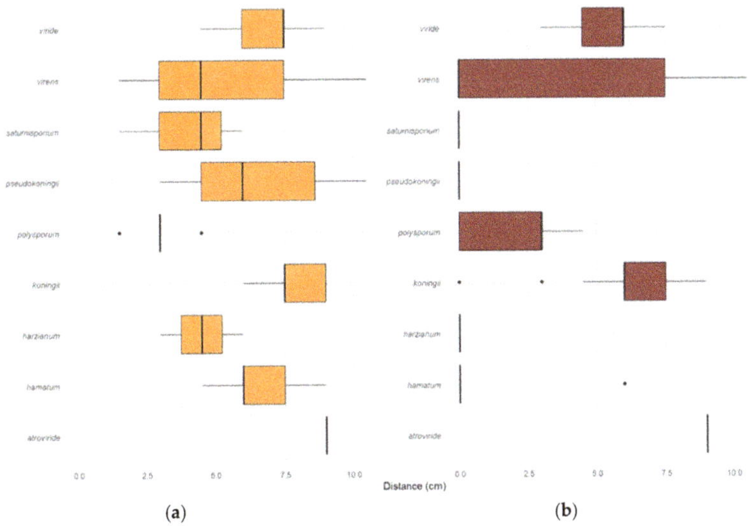

Figure 4. Box plots showing the average overgrowth and lethal effects of *Trichoderma* species in dual culture with 18-day-old *P. ramorum* cultures, as measured in cm after 3 days. Panel (**a**) shows the average radial growth of *Trichoderma* spp. on the surface of media with established *P. ramorum*. Panel (**b**) displays the average distance over which *Trichoderma* species killed *P. ramorum*. Boxes represent the interquartile range (IQR), with medians indicated by horizontal lines. Whiskers extend to 1.5 times the IQR, and outliers are individual points.

3.4. Antibiosis Microplate Assay for Antagonistic Effect

Culture filtrates collected under starvation conditions were tested for their inhibitory effects on *P. ramorum* (Figure 5). The average inhibition of *P. ramorum* growth by sterile culture filtrates of *Trichoderma* isolates was 49 ± 35%. The following isolates produced culture metabolites that inhibited *P. ramorum* growth by more than 75%: *T. virens* (SoilGard), *T. hamatum* (T382), *T. koningii* (5101), *T. pseudokoningii* (5099), *T. viride* (5008), and *T. polysporum* (5017, 5032, 5033, 5016, and 5001).

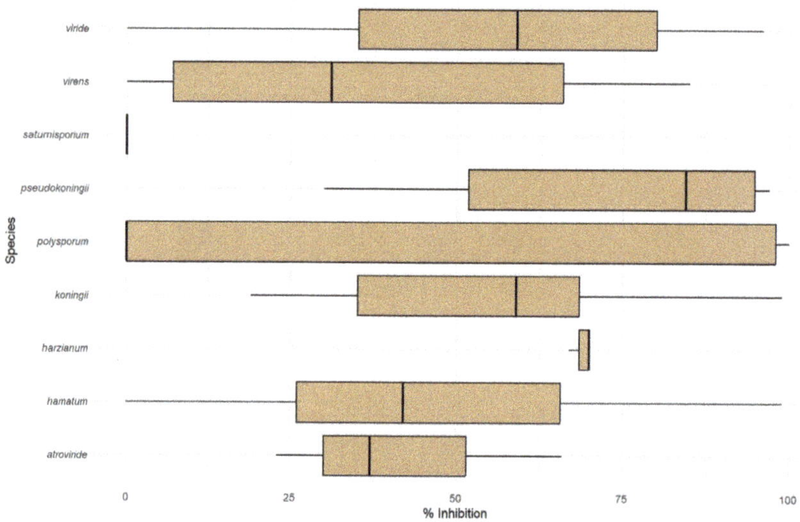

Figure 5. Box plot illustrating the inhibition by *Trichoderma* culture filtrates of *P. ramorum* zoospore germination and growth. The *x*-axis shows percent inhibition of treated *P. ramorum* compared to the control (0–100%). Boxes represent the interquartile range (IQR), with medians indicated by horizontal lines. Whiskers extend to 1.5 times the IQR, and outliers are displayed as individual points.

3.5. Correlation of Assay Variables

A strong positive correlation was observed between the growth of *Trichoderma* isolates at 10 °C and their growth at 20 °C (Figure 6; r = 0.506, $p < 0.001$). Likewise, a strong positive correlation (r = 0.544, $p < 0.001$) suggested that increased growth of *Trichoderma* isolates at 20 °C was associated with increased overgrowth of *P. ramorum*. The strongest positive correlation (r = 0.657, $p < 0.001$) indicated a significant relationship between the overgrowth and the lethal effect on *P. ramorum* exerted by *Trichoderma*. A moderate positive correlation (r = 0.214, $p = 0.008$) suggested that higher values of percent inhibition were associated with higher lethal effect values. A moderate positive correlation (r = 0.167, $p = 0.039$) indicated a relationship between percent inhibition and overgrowth of *P. ramorum* by *Trichoderma*.

3.6. Combined Biocontrol Variable

A combined variable representing the relative general vigour and biocontrol efficacy of the isolates was developed. This was done by standardizing and scaling each of the relevant variables (growth of *Trichoderma* at 10 °C, and at 20 °C, overgrowth of *P. ramorum*, lethal effect on *P. ramorum*, and percent inhibition of *P. ramorum*) and then combining them into a single composite score. The combined biocontrol variable was used to rank each isolate (Figure 7a) and species group (Figure 7b).

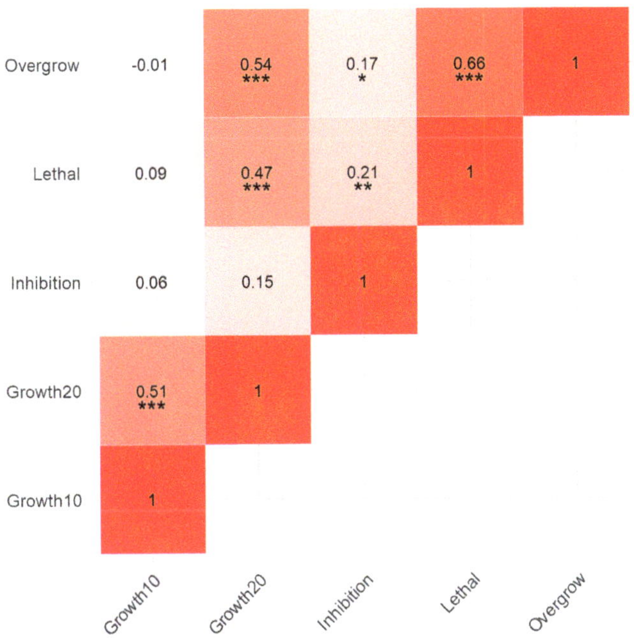

Figure 6. Correlation matrix showing Spearman correlation coefficients between all growth and biocontrol activity variables (growth of *Trichoderma* at 10 °C and 20 °C, overgrowth of *P. ramorum*, lethal effect on *P. ramorum*, and percent inhibition of *P. ramorum*). Darker shades indicate stronger correlations. Significance levels: *** ($p < 0.001$), ** ($p < 0.01$), * ($p < 0.05$).

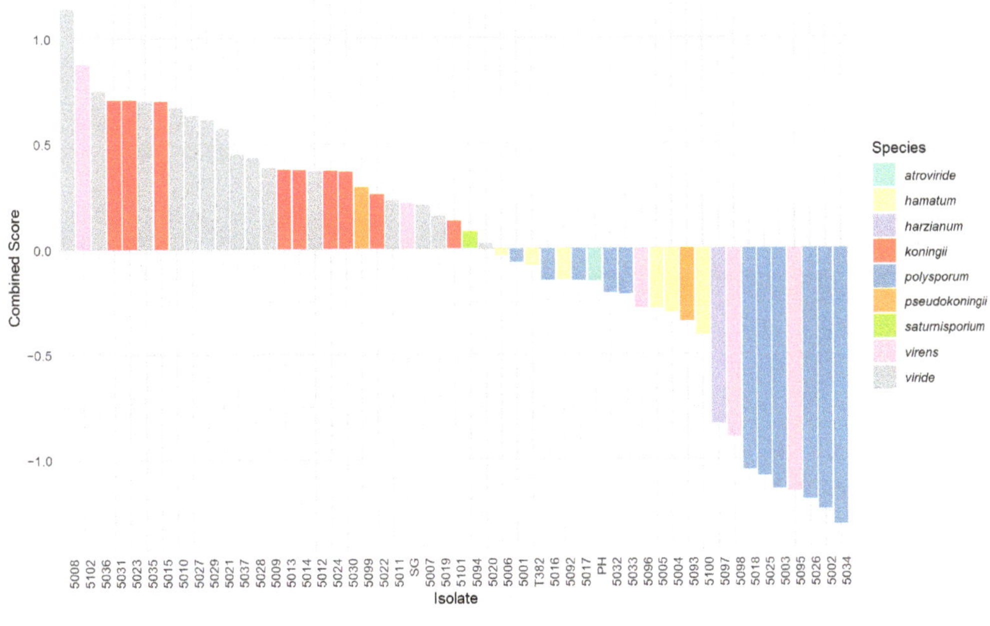

(**a**)

Figure 7. *Cont.*

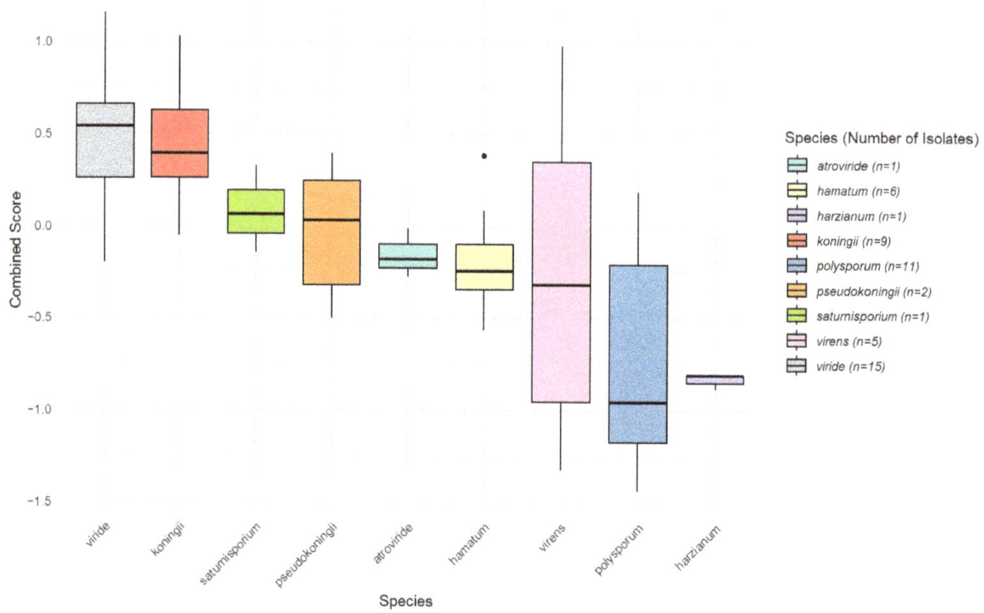

(**b**)

Figure 7. Mean combined scores of *Trichoderma*, as calculated from standardized variables related to biocontrol activity against *P. ramorum* (growth at 10 °C and 20 °C, overgrowth, lethal effect, and percent inhibition). In (**a**), scores of individual isolates are arranged from highest to lowest, with colors indicating species. Each bar represents the average score of an isolate based on three replicates. In (**b**), boxes show the distribution of scores for each species group, with the median as the midline, whiskers representing the interquartile range (IQR), and dots as outliers. Each isolate had 3 replicates. The number of isolates per species is shown in the legend.

4. Discussion

Trichoderma spp. act as potent antagonists against plant pathogens through various direct and indirect modes of action, including mycoparasitism, antibiosis, lytic enzyme secretion, and competition for space and nutrients [10]. Through direct mechanisms such as mycoparasitism, the antagonist secretes lytic enzymes, antibiotics, and other toxic metabolites that synergistically antagonize the pathogen. Ideal biocontrol agents exhibit most of these characteristics, enabling effective control of target pathogens [19,28,29].

Given these diverse biocontrol mechanisms, our study assessed the effectiveness of various *Trichoderma* isolates against *Phytophthora ramorum*, a destructive pathogen causing significant nursery and forestry diseases. We focused on the isolates' growth rates and specific mechanisms of action to identify promising candidates for biocontrol applications against *P. ramorum*. Specifically, we evaluated 51 isolates from nine *Trichoderma* species using criteria reflecting various desired biocontrol abilities or traits.

In our investigation, we hypothesized that the growth rate of *Trichoderma* isolates would be directly correlated with their biocontrol efficacy against *P. ramorum*. Our *Trichoderma* growth rate study indicated a wide variation among the 51 isolates screened, with growth rates at 20 °C ranging from 8.0 to 19.4 mm/day. Notably, 5035 (*T. viride*), the fastest-growing isolate, showed significant biocontrol potential, suggesting that rapid colonization through competition for space is an effective mechanism of action against *P. ramorum*. The correlation between growth rate and biocontrol efficacy was inconsistent, as several fast-growing isolates did not exhibit comparable biocontrol efficacy. While competition for space and nutrients is a classical mechanism of biological control [30], our results suggest that multiple mechanisms, including antibiosis and mycoparasitism, in addition to growth rate, contribute to the biocontrol capabilities of *Trichoderma* against *P. ramorum*.

Mycoparasitism involves direct interactions with the pathogen, such as coiling around the pathogen's hyphae, penetration, and secretion of lytic enzymes [31–33]. These enzymes break down the cell walls of the pathogen, leading to its disintegration [33]. In our study, *T. atroviride*, *T. koningii*, and *T. virens* demonstrated effective mycoparasitism by overgrowing and directly killing *P. ramorum* in dual culture (Figures 1 and 4a,b). This aligns with the results of studies showing that production of lytic enzymes by *Trichoderma* spp. contributes to their efficacy in suppressing root rot caused by *Phytophthora capsici* [18,33], suggesting that a similar mechanism acts against *P. ramorum*.

The dual-culture assays illuminated the complex relationship between *Trichoderma* growth rates and their biocontrol efficacy. While the commercial product SoilGard exhibited the highest rate of overgrowth and greatest lethal effect against *P. ramorum*, it was not statistically different from other isolates with varying growth rates. This underscores that while rapid growth aids in competition for space, it is not the sole determinant of biocontrol efficacy; other mechanisms like antibiosis are also crucial. Moreover, the observation that overgrowth did not always result in a lethal effect suggests that mechanisms beyond mycoparasitism, such as antibiosis, are involved in the biocontrol of *P. ramorum*.

The analysis of culture metabolites produced under starvation conditions provided further insights into the biocontrol mechanisms of *Trichoderma* isolates. *Trichoderma* spp. produce a variety of antibiotic-like substances and secondary metabolites that inhibit pathogen growth. These include compounds like Trichodermin and viridin [33] and volatile organic compounds (VOCs) with antimicrobial properties, contributing to antibiosis [10]. We found that several isolates, notably *T. polysporum*, produced antagonistic metabolites that significantly inhibited *P. ramorum* growth by more than 95%, emphasizing the role of antibiosis.

The results for *T. polysporum* were inconsistent between the two assays. While the microplate assay showed strong antibiosis (Figure 5), the dual-culture assay (Figure 4a,b) suggested limited mycoparasitic activity. This limited mycoparasitic activity may result from the slow linear growth rate of *T. polysporum* (Figure 3), suggesting that rapid growth is essential for effective mycoparasitism. Therefore, antibiosis through metabolite production was found to be a critical biocontrol mechanism of *Trichoderma* isolates against *P. ramorum*, especially for isolates with slower growth rates.

This study evaluated the biocontrol potential of various *Trichoderma* isolates against *P. ramorum* using several methodologies. Growth-rate studies at 20 °C served as a useful screening tool, generally correlating with indicators of biocontrol efficacy like overgrowth, lethal effect, and percentage inhibition. The dual-culture assays assessed direct antagonism by *Trichoderma* isolates through mycoparasitism. They provided insights into specific mechanisms of action, particularly the abilities to overgrow and to kill established cultures of *P. ramorum*. However, the dual-culture assays were complex and time-consuming, limiting their practicality for large-scale screening of mycoparasitic activity.

To efficiently assess the antibiosis mechanism, we developed a new microplate assay evaluating the inhibitory effects of *Trichoderma* culture filtrates on *P. ramorum* zoospore germination and growth. This assay evaluates antibiosis by testing the ability of sterile culture filtrates from *Trichoderma* to inhibit the pathogen. Using measurements of relative light absorbance to quantify *P. ramorum* zoospore germination and growth, this assay allowed us to efficiently screen multiple *Trichoderma* isolates for antibiosis. This efficient method facilitates easy replication, minimizes media usage, and provides rapid results, making it suitable for large-scale screening of biocontrol agents based on antibiosis.

The assays revealed different mechanisms of action. Selecting species and isolates for field studies requires considering the type of application. For instance, in a forest or nursery, a living *Trichoderma* isolate could be used as a foliar spray to prevent *P. ramorum* spread from leaf lesions, or as a soil treatment to minimize splash dispersal or root infection. The microplate assay we developed allows the selection of *Trichoderma* isolates with active metabolites against *P. ramorum* that could be used as a sterile product, eliminating persistence risk after application. These choices may be influenced by the specific *P. ramorum* hosts and the regulatory framework for biocontrol products. Future research should unravel these interactions and mechanisms to enhance the practical application of *Trichoderma* to manage *P. ramorum* infections [12,34]. Understanding the potential mechanisms of action of these isolates will enable us to further screen and tailor candidate agents to meet the specific requirements of diverse applications. Future studies should evaluate field and nursery applications to develop practical biocontrol strategies.

5. Conclusions

The predominantly positive correlations among variables suggest that a combined variable is useful for ranking isolates by biocontrol activity. This combined variable represented the isolates' general vigour and biocontrol efficacy. Integrating these methods provided a comprehensive assessment of *Trichoderma* isolates' biocontrol capabilities and mechanisms. The growth-rate study at 20 °C served as an efficient initial screening method. Although labour-intensive, the dual-culture assays provided detailed insights into direct antagonistic interactions. The new microplate assay enabled rapid, high-throughput evaluation of metabolite-mediated inhibition. Together, these approaches emphasize the importance of using diverse assays to select isolates with multiple mechanisms of action under varied conditions.

Our results partially support the hypothesis that *Trichoderma* isolates' growth rates correlate with their biocontrol efficacy against *P. ramorum*. While rapid growth rates are advantageous and contribute to the competitive suppression of *P. ramorum*, our findings underscore that other factors, including antagonistic metabolites and possibly other mechanisms, play significant roles in the biocontrol process [28,35]. Our results align with those of previous studies demonstrating the importance of rapid colonization and metabolite production in effective biocontrol agents [10,36,37].

Advancements in developing and testing methods for applying promising *Trichoderma* isolates to plants or soil are continually progressing. These findings advance our understanding of biocontrol mechanisms against *P. ramorum* and have broader implications for sustainable disease management in forestry and agriculture [12,13,34,37].

Author Contributions: Conceptualization, E.B. and S.F.S.; methodology, E.B. and N.R.; software, E.B.; validation, E.B. and N.R.; formal analysis, E.B.; investigation, E.B.; resources, S.F.S.; data curation, E.B.; writing—original draft preparation, E.B.; writing—review and editing, E.B., S.F.S. and N.R.; visualization, E.B.; supervision, S.F.S.; project administration, S.F.S.; funding acquisition, S.F.S. All authors have read and agreed to the published version of the manuscript.

Funding: This research was funded by the NRCan CFS-Pest Risk Management Program, the CFIA, and NSERC support to SS. Additional funding was provided by NSERC (Visiting Fellowship to EB) and the YMCA (Student Internship to NR).

Institutional Review Board Statement: Not applicable.

Informed Consent Statement: Not applicable.

Data Availability Statement: The original data presented in the study are openly available in FigShare at https://doi.org/10.6084/m9.figshare.27961578.v1 (accessed on 20 January 2025).

Acknowledgments: The authors thank N. Grünwald for providing the *P. ramorum* isolate used in this study, and AmPac Biotech Corporation, Certis Biologicals, and Sylvan Bioproducts for supplying biocontrol agent samples. We also thank T. Osono and M. Elliott for collecting *Trichoderma* isolates. Special thanks to Robert Kowbel and Grace Sumampong for their excellent technical support.

Conflicts of Interest: The authors declare no conflicts of interest. The funders had no role in the design of the study; in the collection, analyses, or interpretation of data; in the writing of the manuscript; or in the decision to publish the results.

References

1. Werres, S.; Marwitz, R.; Man In't veld, W.A.; De Cock, A.W.A.M.; Bonants, P.J.M.; De Weerdt, M.; Themann, K.; Ilieva, E.; Baayen, R.P. *Phytophthora ramorum* sp. nov., a new pathogen on Rhododendron and Viburnum. *Mycol. Res.* **2001**, *105*, 1155–1165. [CrossRef]
2. Parke, J.L.; Peterson, E.K. Sudden Oak Death, Sudden Larch Death, and Ramorum Blight. *Plant Health Instr.* **2019**, *19*. Available online: https://www.apsnet.org/edcenter/pdlessons/Pages/SuddenOakDeath.aspx (accessed on 20 January 2025).
3. Grünwald, N.J.; Goss, E.M.; Press, C.M. *Phytophthora ramorum*: A Pathogen with a Remarkably Wide Host Range Causing Sudden Oak Death on Oaks and Ramorum Blight on Woody Ornamentals. *Mol. Plant Pathol.* **2008**, *9*, 729–740. [CrossRef]
4. Shamoun, S.F.; Rioux, D.; Callan, B.E.; James, D.; Hamelin, R.; Bilodeau, G.J.; Elliott, M.; Lévesque, C.A.; Becker, E.; McKenney, D.; et al. An Overview of Canadian Research Activities on Diseases Caused by *Phytophthora ramorum*: Results, Progress, and Challenges. *Plant Dis.* **2018**, *102*, 1218–1233. [CrossRef] [PubMed]
5. Rizzo, D.M. *Phytophthora ramorum*: Integrative Research and Management of an Emerging Pathogen in California and Oregon Forests. *Annu. Rev. Phytopathol.* **2005**, *43*, 309–350. [CrossRef] [PubMed]
6. Shishkoff, N. Susceptibility of Some Common Container Weeds to *Phytophthora ramorum*. *Plant Dis.* **2004**, *96*, 1026–1032. [CrossRef] [PubMed]
7. Canadian Food Inspection Agency. *D-01-01: Phytosanitary Requirements to Prevent the Entry and Spread of Phytophthora ramorum (18th Revision)*; CFIA Directive D-01-01; Canadian Food Inspection Agency: Ottawa, ON, Canada, 13 June 2013.
8. Canadian Food Inspection Agency. *PI-010: Regulatory Response Protocol for Nurseries Confirmed with Phytophthora ramorum*; CFIA—PI-010; Canadian Food Inspection Agency: Ottawa, ON, Canada, 29 April 2019.
9. Grünwald, N.J.; Goss, E.M. Evolution and Population Genetics of Exotic and Re-emerging Pathogens: Novel Tools and Approaches. *Annu. Rev. Phytopathol.* **2011**, *49*, 249–267. [CrossRef]
10. Saldana-Mendoza, S.A.; Pacios-Michelena, S.; Palacios-Ponce, A.; Chavez-Gonzalez, M.; Aguilar, C.N. *Trichoderma* as a Biological Control Agent: Mechanisms of Action, Benefits for Crops and Development of Formulation. *World J. Microbiol. Biotechnol.* **2023**, *39*, 269. [CrossRef] [PubMed]
11. Vinale, F.; Sivasithamparam, K.; Ghisalberti, E.L.; Marra, R.; Woo, S.L.; Lorito, M. *Trichoderma*–Plant–Pathogen Interactions. *Soil Biol. Biochem.* **2008**, *40*, 1–10. [CrossRef]
12. Elliott, M.; Shamoun, S.F.; Sumampong, G.; James, D.; Masri, S.; Varga, A. Evaluation of several commercial biocontrol products on European and North American populations of *Phytophthora ramorum*. *Biocontrol Sci. Technol.* **2009**, *19*, 709–721. [CrossRef]
13. Shamoun, S.F.; Elliott, M. *Phytophthora ramorum* Werres, de Cock & Man in't Veld, Sudden Oak Death/Encre des chênes rouges (Peronosporaceae). In *Biological Control Programmes in Canada*; Vankosky, M.A., Martel, V., Eds.; CABI: Wallingford, UK, 2024; pp. 593–601. [CrossRef]
14. Kamoun, S.; Lamour, K. Oomycete Genetics and Genomics: Diversity, Interactions, and Research Tools. In *Phytophthora: Genomics of Plant-Associated Fungi*; Springer: Berlin/Heidelberg, Germany, 2002; pp. 197–213.
15. Smith, V.L.; Wilcox, W.F.; Harman, G.E. Potential for Biological Control of *Phytophthora* Root and Crown Rots of Apple by *Trichoderma* and *Gliocladium* spp. *Phytopathology* **1990**, *80*, 880–885. [CrossRef]
16. Howell, C.R. The Role of Antibiosis in Biocontrol. In *Trichoderma and Gliocladium*; Harman, G.E., Kubicek, C.P., Eds.; Taylor and Francis Ltd.: London, UK, 1998; Volume 2, pp. 173–183.
17. Benítez, T.; Rincón, A.M.; Limón, M.C.; Codón, A.C. Biocontrol Mechanisms of *Trichoderma* Strains. *Int. Microbiol.* **2004**, *7*, 249–260.
18. Gouit, S.; Chair, I.; Belabess, Z.; Legrifi, I.; Goura, K.; Tahiri, A.; Lazraq, A.; Lahlali, R. Harnessing *Trichoderma* spp.: A promising approach to control apple scab disease. *Pathogens* **2024**, *13*, 752. [CrossRef]
19. Ramzan, U.; Abid, K.; Zafar, M.A.; Anwar, A.M.; Nadeem, M.; Tanveer, U.; Fatima, U. *Trichoderma*: Multitalented biocontrol agent. *Int. J. Agric. Biosci.* **2023**, *12*, 77–82. [CrossRef]

20. Goldfarb, B.; Nelson, E.E.; Hansen, E.M. *Trichoderma* spp.: Growth Rates and Antagonism to *Phellinus weirii* in Vitro. *Mycologia* **1989**, *81*, 375–381. [CrossRef]
21. R Core Team. *R: A Language and Environment for Statistical Computing*; Version 4.2.3; R Foundation for Statistical Computing: Vienna, Austria, 2024. Available online: https://www.R-project.org/ (accessed on 20 November 2024).
22. RStudio Team. *RStudio: Integrated Development Environment for R*; Version 2023.03.0; RStudio, PBC: Boston, MA, USA, 2023; Available online: https://www.rstudio.com/ (accessed on 20 November 2024).
23. Wickham, H.; Vaughan, D.; Girlich, M. *tidyr*: Tidy Messy Data. R Package Version 1.1.2. 2024. Available online: https://github.com/tidyverse/tidyr (accessed on 20 November 2024).
24. Wickham, H.; François, R.; Henry, L.; Müller, K. *dplyr: A Grammar of Data Manipulation*, R Package Version 1.1.0; 2024. Available online: https://CRAN.R-project.org/package=dplyr (accessed on 20 November 2024).
25. Revelle, W. *psych: Procedures for Psychological, Psychometric, and Personality Research*; R Package Version 2.3.0; Northwestern University: Evanston, IL, USA, 2024. Available online: https://CRAN.R-project.org/package=psych (accessed on 20 November 2024).
26. Wickham, H. *ggplot2: Elegant Graphics for Data Analysis*; R Package Version 3.4.0; Springer-Verlag New York: New York, NY, USA, 2024; ISBN 978-3-319-24277-4. Available online: https://CRAN.R-project.org/package=ggplot2 (accessed on 20 November 2024).
27. Neuwirth, E. *RColorBrewer: ColorBrewer Palettes*, R Package Version 1.1-3; 2024. Available online: https://CRAN.R-project.org/package=RColorBrewer (accessed on 20 November 2024).
28. Contreras-Cornejo, H.A.; Macías-Rodríguez, L.; Cortés-Penagos, C.; López-Bucio, J. *Trichoderma* spp. as Inducers of Plant Growth and Systemic Resistance. *Microb. Biotechnol.* **2009**, *2*, 509–517. [CrossRef]
29. Tariq, M.; Khan, A.; Asif, M.; Khan, F.; Ansari, T.; Shariq, M. Biological control: A sustainable and practical approach for plant disease management. *Acta Agric. Scand. Sect. B—Soil Plant Sci.* **2020**, *70*, 507–524. [CrossRef]
30. Harman, G.E.; Howell, C.R.; Viterbo, A.; Chet, I.; Lorito, M. *Trichoderma* Species—Opportunistic, Avirulent Plant Symbionts. *Nat. Rev. Microbiol.* **2004**, *2*, 43–56. [CrossRef]
31. Chet, I.; Inbar, J. Biological Control of Fungal Pathogens. *Appl. Biochem. Biotechnol.* **1994**, *48*, 37–43. [CrossRef]
32. Pokhrel, A.; Adhikari, A.; Oli, D.; Paudel, B.; Pandit, S.; GC, B.; Tharu, B.R.R. Biocontrol Potential and Mode of Action of *Trichoderma* Against Fungal Plant Diseases. *Acta Sci. Agric.* **2022**, *6*, 10–21. [CrossRef]
33. Diby, P.; Saju, K.A.; Jisha, P.J.; Sarma, Y.R.; Kumar, A.; Anandaraj, M. *Mycolytic Enzymes Produced by Pseudomonas fluorescens and Trichoderma spp. Against Phytophthora capsici, the Foot Rot Pathogen of Black Pepper (Piper nigrum L.)*; Division of Crop Protection, Indian Institute of Spices Research, Marikkunnu P.O.: Calicut, India, 2005.
34. Woo, S.L.; Ruocco, M.; Vinale, F.; Nigro, M.; Marra, R.; Lombardi, N.; Pascale, A.; Lanzuise, S.; Manganiello, G.; Lorito, M. *Trichoderma*-Based Products and Their Widespread Use in Agriculture. *Open Mycol. J.* **2014**, *8*, 71–126. [CrossRef]
35. Rodrigues, A.O.; May De Mio, L.L.; Soccol, C.R. *Trichoderma* as a Powerful Fungal Disease Control Agent for a More Sustainable and Healthy Agriculture: Recent Studies and Molecular Insights. *Planta* **2023**, *257*, 31. [CrossRef]
36. Hoitink, H.A.J.; Madden, L.V.; Dorrance, A.E. Systemic Resistance Induced by *Trichoderma* spp.: Interactions Between the Host, the Pathogen, the Biocontrol Agent, and Soil Organic Matter Quality. *Phytopathology* **2006**, *96*, 186–189. [CrossRef]
37. Verma, M.; Brar, S.K.; Tyagi, R.D.; Surampalli, R.Y.; Valéro, J. Antagonistic Fungi, *Trichoderma* spp.: Panoply of Biological Control. *Biochem. Eng. J.* **2007**, *37*, 1–20. [CrossRef]

pathogens MDPI

Article

The Tomato lncRNA47258-miR319b-TCP Module in Biocontrol Bacteria Sneb821 Induced Plants Resistance to *Meloidogyne incognita*

Fan Yang [1], Xiaoxiao Wu [2], Lijie Chen [2,*] and Mingfang Qi [1,*]

1 College of Horticulture, Shenyang Agricultural University, Shenyang 110866, China; yangfan20241107@163.com
2 College of Plant Protection, Shenyang Agricultural University, Shenyang 110866, China; wuxiaoxiao1233@163.com
* Correspondence: chenlj2100@126.com (L.C.); qimf2008@163.com (M.Q.)

Abstract: Long non-coding RNAs (lncRNAs) represent a class of non-coding RNAs. In the study of *Pseudomonas putida* Sneb821-induced tomato resistance to *Meloidogyne incognita*, reverse transcription polymerase chain reaction (RT-PCR) was employed to validate 12 lncRNAs in tomato. Among them, the lncRNA47258/miR319b/TCP molecular regulatory module was likely implicated in the process of Sneb821-induced tomato resistance against *M. incognita*. Through the application of tomato hairy root and virus-induced gene silencing (VIGS) technologies for the investigation of lncRNA47258, it was determined that lncRNA47258 could target the *TCP* (*Solyc07g062681.1*) gene and modulate the metabolic pathway of tomato jasmonic acid-related indices, thereby impeding the infection of *M. incognita*. Moreover, the overexpression of the target gene *TCP* (*Solyc07g062681.1*) using tomato hairy root technology demonstrated that it could regulate the jasmonic acid synthesis pathway in tomato, consequently obstructing the infection and suppressing the development of *M. incognita*. Collectively, lncRNA47258/miR319b/*TCP* (*Solyc07g062681.1*) was preliminarily verified to be involved in the Sneb821-induced resistance process against *M. incognita* in tomato.

Keywords: lncRNA; nematode; tomato; biocontrol bacteria

1. Introduction

In agriculture, root-knot nematodes (RKN) are the most widespread and economically damaging pest. These obligate plant-parasitic nematodes, commonly known as RKN, reproduce and feed within plant roots, causing galls or root knots. With a global distribution, RKN infects about 2000 plant species across diverse habitats, contributing to around 5% of global crop losses. The genus RKN includes over ninety species, but *M. incognita* is considered one of the most significant worldwide, and one of the key pests of tomato [1]. The plant hormone jasmonic acid (JA) and its derivatives, like methyl jasmonate (MeJA), play a key role across the plant kingdom. In tomatoes, JA or its derivatives serve as long-distance wound signals. Systemin, a specific inducer of protein inhibitors (PIs), is essential for JA synthesis during insect and nematode attacks and wounding. JA was also found to enhance tomato resistance against root-knot nematodes [2].

Long non-coding RNAs (lncRNAs) represent a class of non-coding RNAs with a length exceeding 200 nucleotides. They can interact with miRNAs and thereby influence gene expression [3]. LncRNAs have been demonstrated to play essential roles in numerous plant biological processes, such as plant pathogen resistance [4]. Functional assays revealed that seedlings with silenced GhlncNAT-ANX2 and GhlncNAT-RLP7 exhibited enhanced resistance to pathogenic fungi, potentially due to the elevated expression of LOX1 and LOX2 [5]. Previous investigations have indicated that lncRNA (DRIR) exerts a positive regulatory effect on drought and salt stress, and Arabidopsis mutants display robust stress tolerance to drought and salt [6]. The overexpression of lncRNA-At5NC056820 in Arabidopsis can enhance the drought tolerance of Arabidopsis to a certain degree [7]. Through lncRNA differential expression and statistical analysis, Li identified four gibberellin-responsive lncRNAs (GARR1, 2, 3, and 4) in wild-type and GA-sensitive dwarf maize, and determined that GARR2 originated from a Gypsy transposon [8]. LncRNA15492 was found to be involved in the interaction between tomato and bacteria [9]. The lncRNA23468/miR482b/NBS-LRR and WRKY1/lncRNA33732/RBOH models could modulate the accumulation of H2O2 in tomato to enhance its disease resistance [10]. Previous studies have verified that the CYP380C6, CYP4CJ1, CYP6DA2, CYP6CY7, and CYP6CY21 genes might participate in the formation of resistance to helix tetramer in cotton aphid. 366492/5, and MSTRG.71880.1 may regulate the expression of CYP6CY21 and CYP380C6 via the "sponge effect" by binding to miRNA [11]. LNC610 is implicated in the regulation of high light-induced anthocyanin production and functions as a positive regulator to promote MdACO1 gene expression and ethylene biosynthesis [12]. LncRNA7, lncRNA2, and their regulatory genes regulate the cotton cell wall defense against Verticillium wilt through auxin-mediated signal transduction, presenting a novel strategic foundation for cotton breeding [13]. The effect of lncRNA354-miR160b on the expression of GhARF17/18 may regulate auxin signaling, consequently affecting plant growth. This study uncovers a lncRNA-related salt stress response mechanism and also reveals that the lncRNA-miRNA-mRNA regulatory module can effectively respond to plant stress resistance [14]. In tomato, a large number of lncRNAs are engaged in the host defense response against potato spindle tuber viroids [15]. Another gene, lncRNA0957, was induced by virus in susceptible tomato plants to augment tomato resistance to etiolated leaf warp virus [16]. Additionally, siRNAs between tomato etiolated leaf virus genes were transferred to tomato plants to target host lncRNAs, thereby modulating disease symptoms [17]. Our previous studies discovered that some miRNAs and lncRNAs might stimulate tomato immunity in response to *P. putida* to resist *M. incognita* infection, and lncRNA47258 may be a key factor in the process of biocontrol bacteria-induced tomato resistance to RKN [18]. These findings offer a novel molecular regulatory model for the mechanism of bacteria-induced plant resistance to *M. incognita* infection in tomato.

2. Materials and Methods

2.1. Species, Plants, and Nematodes Tested

Strain Sneb821 of *P. putida* and the Moneymaker of tomato varieties were deposited by the Northern Institute of Nematodes, Shenyang Agricultural University. *M. incognita* was bred in a greenhouse at the Northern Nematode Institute, Shenyang Agricultural University.

2.2. Materials for Test

pCAMBIA1302, purchased from Beijing Zhuang Meng International Biological Gene Technology Co., LTD.; TRV1 and TRV2 were presented by Shenyang Agricultural University. QuickCut™ Xho I restriction enzyme, purchased from TaKaRa (Dalian, China); A. tumefaciens GV3101, purchased from Shanghai Wedi Biotechnology.

2.3. Total RNA Extraction, Reverse Transcription cDNA, and qPCR Analysis of Tomato

The specific method was referred to in our previous study [18,19].

2.4. Amplification of Target Fragments

The instructions for TaKaRa PrimeSTAR Max DNA Polymerase were followed.

Refer to our previous study [18]. The TRV2 vector was digested with Xho I and BamH I, and the target gene fragment was reversed into the TRV2 vector. pCAMBIA1302 was digested using Xho I.

2.5. E. coli Transformation and Culture of Tomato Hairy Roots

The specific method was referred to in our previous study [18,19].

2.6. Determination of Jasmonic Acid Content

After precooling at −80 °C in the mortar used, the test material stored in an ultra-low temperature refrigerator was removed. Samples were preheated with a small amount of liquid nitrogen, then a small amount of liquid nitrogen was added and ground in an ice bath. Add 4 mL of 80% methanol as the extract solution (the extract solution needs to be precooled at −20 °C), ground into a uniform centrifuge tube in an ice bath, and then transfer 10 mL of the sample to the refrigerator for 4 min. The centrifuge tube adds 4 mL of the extract solution to the remaining precipitate, then stir well, leach at 4 °C for 1h, and centrifuge as above. The supernatants were pooled and the residues discarded. The collected supernatant was tested through a C-18 solid-phase extraction column.

2.7. Nematode Infection and Enzyme Activity Detection

The specific method was referred to in our previous study [18].

2.8. Histochemical Staining and Tomato GUS Staining Were Performed

The specific method was referred to in our previous study [19].

2.9. Statistical Analysis of Data

In this study, SPSS19.0 software was used for data analysis and processing, and the Duncan method was used to analyze the significance of the data in this study, and the data results were expressed as the mean ± standard deviation. Mapping was performed using prism 8.

3. Results

3.1. RT-PCR Validation of Differential lncRNA in Tomato

Specific primers were meticulously designed, and RT-PCR was employed to authenticate the differential lncRNA within the transcriptome (Supplementary Figure S1). Lanes 1–12 corresponded to lncRNA18894, lncRNA21563, lncRNA24059, lncRNA25797, lncRNA35115, lncRNA39939, lncRNA47258, lncRNA44664, lncRNA45969, lncRNA48734, lncRNA51612, and lncRNA7183, all of which were successfully amplified in the transcriptome.

3.2. Expression Analysis of lncRNA47258 in Tomato

In accordance with the ceRNA (competing endogenous RNA) outcomes predicted by the preceding analysis [18], lncRNA47258 is capable of functioning as an endogenous target mimic of miR319b, thereby influencing the expression level of miR319b. The miR319/TCP4 molecular regulatory module is involved in the modulation of the jasmonate signaling pathway within tomato plants. Subsequently, it contributes to the plant's resistance against *M. incognita* [20].

In the present study, the quantitative analysis of lncRNA47258 was conducted. The results demonstrated that the expression of lncRNA47258 was remarkably elevated in tomato at the late stage (6 days post-inoculation and 12 days post-inoculation) following inoculation with nematodes and biocontrol bacteria (RKN + Sneb821). Moreover, the expression of lncRNA47258 was significantly higher in the combined treatment of biocontrol bacteria and nematodes compared to the single treatment (Sneb821-RKN or RKN-Sneb821) (Figure 1). These findings suggested that lncRNA47258 played a crucial role in the resistance of tomato to *M. incognita* infection induced by Sneb821.

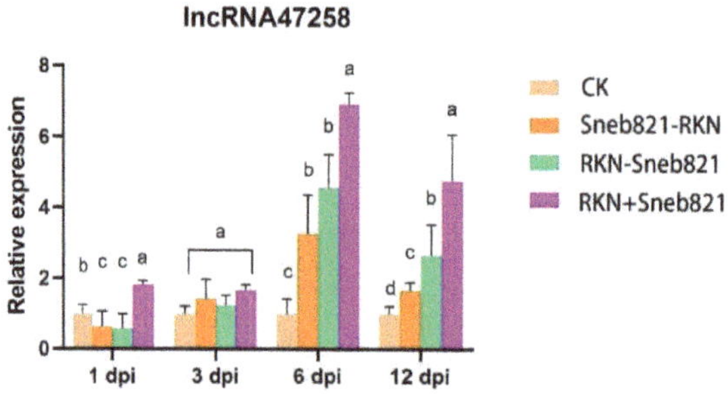

Figure 1. Expression levels of lncRNA47258 in each treatment at each time point. Note: CK (Control Check) represents no treatment, Sneb821-RKN represents inoculation with biocontrol bacteria Sneb821 only, RKN-Sneb821 represents inoculation with *M. incognita* only, and RKN + Sneb821 represents double treatment. The letters above the bars indicate significant difference at a value of $p < 0.05$.

3.3. *Construction of lncRNA47258 Overexpressed Tomato Plants*

lncRNA47258 was found to be overexpressed in the roots of tomato plants. Subsequently, its overexpression vector, pCAMBIA1302-lncRNA47258, was constructed. Based on the fragment sequence of lncRNA47258 within the tomato transcriptome, the genomic location of lncRNA47258 on the tomato genome is presented in Supplementary Figure S2A. Specific primers were designed to amplify the full length of lncRNA. After the PCR reaction, a band approximately 900 bp in size was obtained, which was in accordance with the length of lncRNA47258. Following purification, the amplified products were ligated with the cloning vector pMD20-T and then transformed into Escherichia coli. Positive monoclonal clones were selected for plasmid extraction. The plasmids were identified through enzyme digestion and PCR detection, and bands corresponding to the size of the target gene were obtained (Supplementary Figure S2B). The constructed plasmid pCAMBIA1302-lncRNA47258 was introduced into Agrobacterium rhizogenes MSU440 via the freeze–thaw method and subsequently transformed into tomato plants. GFP fluorescence labeling was employed to verify the transformation of positive tomato plants (Supplementary Figure S2C). Collectively, these results demonstrated that lncRNA47258 was successfully overexpressed in pCAMBIA1302-lncRNA47258 transgenic tomato plants.

3.4. *Overexpression of Tomato lncRNA47258 Inhibited the Infection and Development of M. incognita*

To investigate the potential role of lncRNA47258 in the induction and resistance of Sneb821, the expression of lncRNA was examined in tomato plants overexpressing pCAMBIA1302-lncRNA47258. Notably, lncRNA47258 expression was substantially up-regulated in these overexpressed plants (Figure 2A). Concurrently, the content of jasmonic acid exhibited a significant increase (Figure 2B), suggesting a positive correlation between lncRNA47258 and jasmonic acid synthesis in tomatoes.

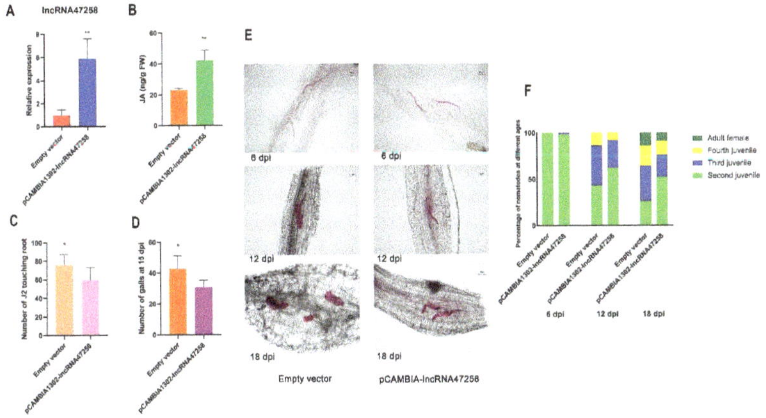

Figure 2. Response of lncRNA47258 to *M. incognita* infection and development. Note: (**A**) Relative expression level of lncRNA47258 in pCAMBIA1302-lncRNA47258 and control group, (**B**) JA content in pCAMBIA1302-lncRNA47258 and control group, (**C**) statistics of J2 in pCAMBIA1302-lncRNA47258 and control group, (**D**) statistics of galls number in pCAMBIA1302-lncRNA47258 and control group, (**E**) development of *M. incognita* in pCAMBIA1302-lncRNA47258 and control group, and (**F**) percentage of different ages of *M. incognita* in pCAMBIA1302-lncRNA47258 and control group. * indicate a significant difference at the $p = 0.05$ level, ** indicate a significant difference at the $p = 0.01$ level.

The constructed pCAMBIA1302-lncRNA47258 overexpression tomato plants were inoculated with second-stage juveniles (J2) of *M. incognita*. Remarkably, the J2 infection amount in the tomato plants was considerably reduced, with approximately 15 fewer nematodes per plant (Figure 2C). At 15 days post-inoculation (dpi), the number of root knots in pCAMBIA1302-lncRNA47258 roots was significantly diminished (Figure 2D).

The development of *M. incognita* in lncRNA47258 overexpression plants and control plants was evaluated at 6 dpi, 12 dpi, and 18 dpi (Figure 2E). By calculating the percentage of nematodes at each instar at each inoculation time point, it was determined that the development of *M. incognita* in tomato plants overexpressing lncRNA47258 was delayed (Figure 2F).

Collectively, these results demonstrate that the overexpression of lncRNA47258 in tomato plants can enhance the jasmonate content in tomato roots, significantly decrease J2 infection in tomato roots, delay nematode development, and inhibit root-knot formation. These findings imply that Sneb821 can augment the jasmonate content in tomato roots by activating the expression of lncRNA47258, thereby conferring resistance against *M. incognita* infection and impeding the development of *M. incognita*.

3.5. Construction of lncRNA47258 VIGS Vector in Tomato

In accordance with the established sequence, the transcriptome lncRNA47258 was screened using the SGN VIGS Tool (https://vigs.solgenomics.net/, accessed on 1 May 2022) to identify the lncRNA47258 silence target fragment. Specific primers were then designed, and following PCR amplification, a band approximately 300 bp in size was obtained from the tomato variety MoneyMaker, which corresponded to the fragment length within the target gene lncRNA47258. After purification, the PCR amplification product was ligated into the cloning vector pMD20-T, yielding the product pMD20-T-lncRNA47258. This was subsequently transformed into Escherichia coli. A positive monoclonal clone was selected, and the plasmid was extracted. The plasmid was then verified through enzyme digestion and PCR analysis, with a band corresponding to the target fragment being observed. Subsequently, the lncRNA47258 fragment was ligated into the expression vector of TRV2 via homologous recombination. The results of the double enzyme digestion are presented in Supplementary Figure S3.

3.6. Silencing lncRNA47258 Reduces the Resistance of Tomato to M. incognita

To investigate the function of lncRNA47258 in tomato plants, a virus-induced gene silencing (VIGS) system was established to suppress its expression. After 21 days of VIGS treatment, quantitative real-time polymerase chain reaction (qRT-PCR) analysis revealed a significant reduction in lncRNA47258 expression in the silenced tomato plants. The expression level of lncRNA47258 was approximately 50% lower compared to that in TRV2 control plants (Figure 3A). The TRV2-lncRNA47258 silenced tomato plants were then utilized to determine the jasmonic acid content. Notably, the jasmonic acid content in the roots of TRV2-lncRNA47258 plants was substantially decreased compared to TRV2 plants (Figure 3B). Six days post-inoculation with *M. incognita*, the number of J2 nematodes infecting the roots was quantified. The results demonstrated a significant increase in J2 infection in TRV2-lncRNA47258 tomato plants, with approximately 10 more J2 nematodes per plant root (Figure 3C,E). Additionally, at 15 days post-inoculation (dpi), the number of root knots in pBI121-lncRNA48734 roots was remarkably elevated (Figure 3D,F,G). These findings imply that the silencing of lncRNA47258 in tomato plants leads to an enhanced J2 infection in the roots. In conjunction with the expression analysis results of lncRNA48734, it is hypothesized that lncRNA48734 exerts a negative regulatory role during *M. incognita* infection. The silencing of lncRNA47258 might inhibit jasmonic acid expression, consequently influencing the resistance of tomato plants against *M. incognita*.

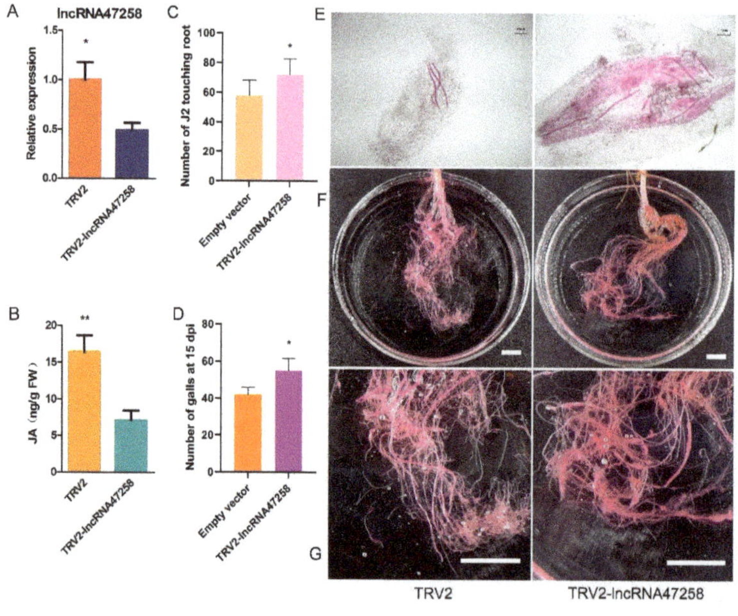

Figure 3. Response of VIGS-lncRNA47258 to *M. incognita* infection. Note: (**A**) Relative expression level of lncRNA47258 in TRV2-lncRNA47258 and control group, (**B**) JA content in TRV2-lncRNA47258 and control group, (**C**) J2 statistics in TRV2-lncRNA47258 and control group, (**D**) statistics of galls number in TRV2-lncRNA47258 and control group, (**E**) infection of J2 in TRV2-lncRNA47258 and the control group, and (**F**,**G**) galls situation in TRV2-lncRNA47258 and control group (E scale bars = 100 μm, F and G scale bars = 1 cm). * indicate a significant difference at the $p = 0.05$ level, ** indicate a significant difference at the $p = 0.01$ level.

3.7. Verification of Target Sites of Tomato lncRNA47258 and Target Gene TCP (Solyc07g062681.1)

Endogenous target mimics (eTMs) of miRNA on lncRNA enable lncRNA to impede the adsorption of miRNA, thereby inhibiting miRNA expression and subsequently influencing the expression of its target genes. The eTM binding site of miR319b was previously predicted within lncRNA47258, and the specific site position was determined to be from 535 nt to 555 nt on lncRNA (Figure 4A). Moreover, compared with TRV2 tomato plants, the expression of miR319b was remarkably up-regulated in TRV2-lncRNA47258 tomato plants (Figure 4B), suggesting that lncRNA47258 functions as an endogenous mimic target of miR319b.

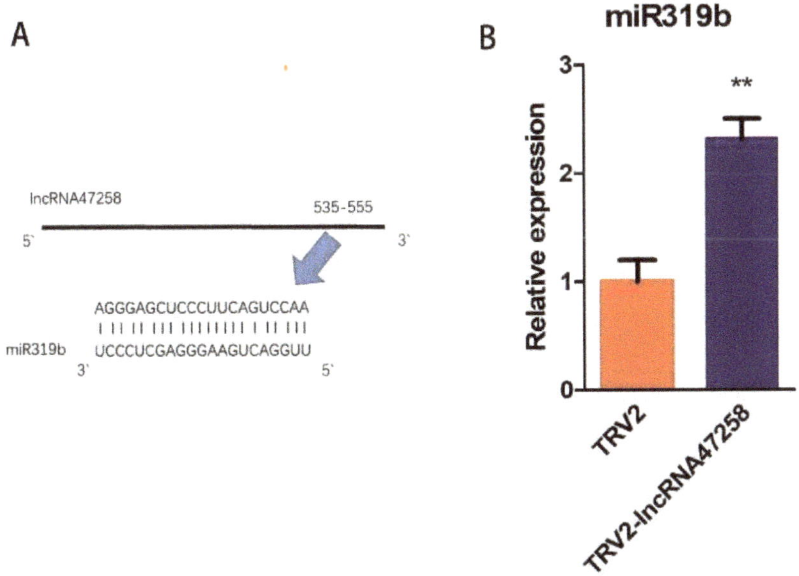

Figure 4. lncRNA47258 contained eTM of miR319b. Note: (**A**) The eTM binding site of lncRNA47258 to miR319b, and (**B**) relative expression level of miR319b in TRV2-lncRNA47258 and control group. ** indicate a significant difference at the $p = 0.01$ level.

The TCP gene serves as a promoting factor for cell proliferation in plants. In rice, the overexpression of the TCP19 gene associated with miR319 can result in the high expression of genes related to multiple signaling pathways, such as the jasmonic acid and auxin pathways [21]. The genes Solyc12g014141.1, Solyc07g062681.1, and Solyc05g012840.1 were investigated. The binding sites of miR319b to these three TCP genes were predicted and subsequently confirmed (Figure 5A). Additionally, the expression of these three TCP genes in TRV2-lncRNA47258 tomato plants was examined, and it was found that the expression of these three genes was significantly down-regulated in comparison with that in TRV2 plants (Figure 5B). These findings suggest that the expression of the tomato TCP gene is regulated by lncRNA47258.

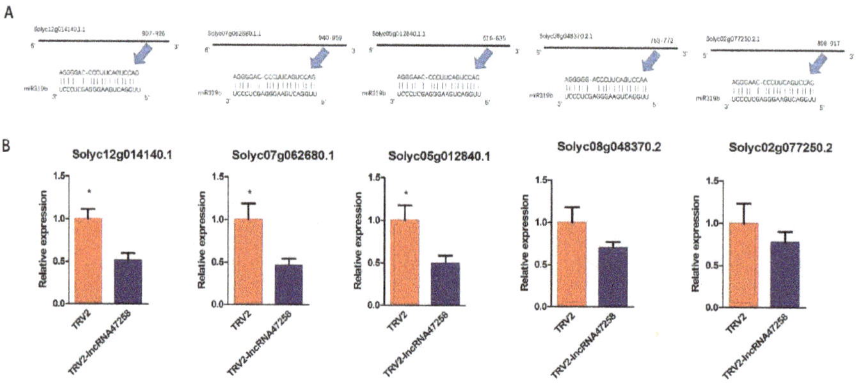

Figure 5. Expression level of TCP gene in tomato plants after silencing the lncRNA47258 gene. Note: (**A**) The eTM binding site of TCP genes to miR319b, and (**B**) relative expression level of TCP genes in TRV2-lncRNA47258 and control group. * indicate a significant difference at the $p = 0.05$ level.

3.8. Construction of Tomato Plants Overexpressing TCP (Solyc07g062681.1), the Target Gene of lncRNA47258

The target gene of lncRNA47258 was identified as TCP (Solyc07g062681.1). Notably, the expression level of TCP was significantly up-regulated in the biocontrol treatment (Supplementary Figure S4A). Subsequently, to further investigate the function of TCP, it was overexpressed in tomato roots. The overexpression vector pCAMBIA1302-Solyc07g062680.1 was constructed. Specific primers were meticulously designed to amplify the full length of the TCP gene. After the PCR reaction, a distinct 1200 bp band was obtained, which precisely corresponded to the expected length of the target gene. Following purification, the amplified products were ligated with the cloning vector pMD20-T and then transformed into Escherichia coli. Positive monoclonal clones were selected for plasmid extraction. The plasmids were verified through enzyme digestion and PCR detection, and bands consistent with the size of the target gene were successfully obtained (Supplementary Figure S4B). The constructed plasmid pCAMBIA1302-Solyc07g062680.1 was introduced into Agrobacterium rhizogenes MSU440 via the freeze–thaw method and subsequently transformed into tomato plants. The transformed tomato plants were validated by GFP fluorescent labeling (Supplementary Figure S4C), thereby confirming the successful overexpression of TCP in tomato plants.

3.9. Overexpression of Tomato TCP (Solyc07g062680.1) Inhibited the Infection and Development of M. incognita

The expression of pCAMBIA1302-Solyc07g062680.1 was found to be remarkably up-regulated in tomato plants with pCAMBIA1302-Solyc07g062680.1 overexpression (Figure 6A). Concomitantly, the content of jasmonic acid was also significantly elevated (Figure 6B), suggesting a positive correlation between TCP and jasmonic acid synthesis in tomato. The constructed pCAMBIA1302-Solyc07g062680.1 overexpressing tomato plants were inoculated with *M. incognita* J2. At 6 days post-inoculation (dpi), the J2 infection in tomato plants was substantially reduced, with approximately 15 fewer J2 per tomato plant (Figure 6C). At 15 dpi, the number of root knots in pCAMBIA1302-Solyc07g062680.1 roots was significantly diminished (Figure 6D). The development of TCP overexpression plants and control *M. incognita* was quantified at 6 dpi, 12 dpi, and 18 dpi (Figure 6E), and the percentage of nematodes at each instar at each time point was calculated (Figure 6F). It was observed that the overexpression of TCP led to a delay in the development of *M. incognita* in tomato roots. Collectively, these results demonstrated that the overexpression of TCP in tomato plants could enhance the jasmonic acid content in tomato roots and significantly mitigate the J2 infection in tomato roots. This indicates that Sneb821 modulates the target gene TCP to augment the jasmonic acid content in tomato roots by activating the expression of lncRNA47258, thereby conferring resistance against infection and inhibiting the development of *M. incognita*.

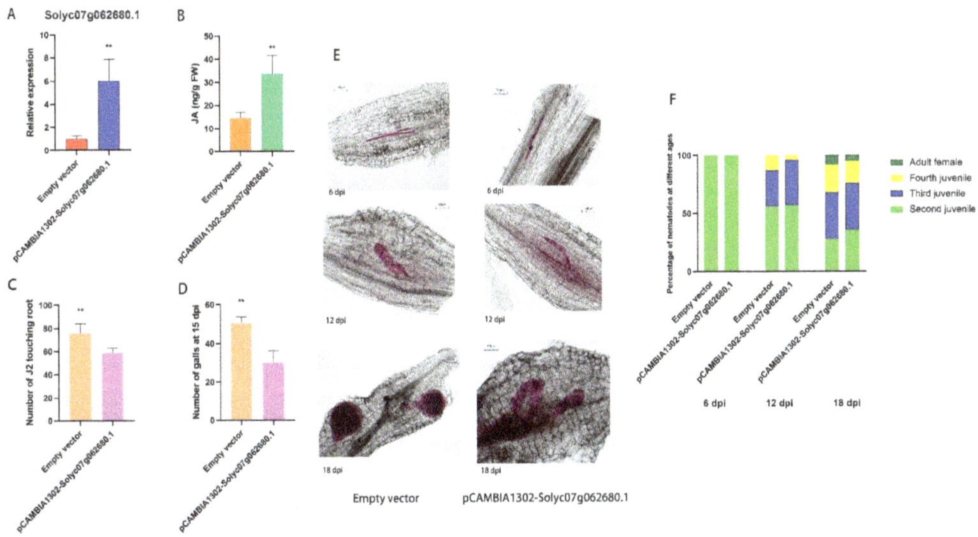

Figure 6. Response of TCP (Solyc07g062680.1) to *M. incognita* infection and development. Note: (**A**) Relative expression level of Solyc07g062680.1 in pCAMBIA1302-Solyc07g062680.1 and control group, (**B**) JA content in pCAMBIA1302-Solyc07g062680.1 and control group, (**C**) statistics of J2 in pCAMBIA1302-Solyc07g062680.1 and control group, (**D**) statistics of galls number in pCAMBIA1302-Solyc07g062680.1 and control group, (**E**) development of *M. incognita* in pCAMBIA1302-Solyc07g062680.1 and control group, and (**F**) percentage of different ages of *M. incognita* in pCAMBIA1302-Solyc07g062680.1 and control group. ** indicate a significant difference at the $p = 0.01$ level.

4. Discussion

Root-knot nematode disease ranks among the most severe soil-borne diseases of crops globally, leading to substantial economic losses in tomato production. Biological control, as an effective control strategy, has garnered increasing attention. Non-coding RNA, which comprises small RNA molecules not encoding proteins, has emerged as a novel research focus in molecular biology. The mechanism underlying ceRNA (endogenous competing RNA) in plant resistance to diseases induced by bio-control bacteria represents an urgent research gap. Previous investigations have demonstrated that Slylnc0195, acting as a "sponge" for miR166, can disrupt the expression of the HD-Zip TF gene in tomato etiolated leaf warp virus disease. The ceRNA network of the Sly-lnc0195-miR166-HD-Zip TF gene can modulate the expression of its target protein HD-Zip TF, thereby influencing tomato's resistance to external diseases [22]. In the study of tomato late blight, a set of ceRNAs, lncRNA23468-miR482b-NBSLRR, has been identified. When lncRNA23468 is overexpressed in tomato, the expression of miR482b is markedly reduced, and the expression of its target gene NBS-LRR is enhanced. These outcomes enhance the resistance of tomato plants to Phytophthora infestans. Conversely, silencing lncRNA23468 results in an elevation in the expression of miR482b and a decline in the expression of NBS-LRR, leading to a reduction in the resistance of tomato plants to the pathogenic phytophthora [23].

In the present study, tomato plants were treated with *P. putida* Sneb821 and inoculated with *M. incognita*. Whole-transcriptome sequencing analysis was conducted, and differentially expressed non-coding RNAs were analyzed and screened. Subsequently, the ceRNA regulatory network module among lncRNA-miRNA-mRNA was constructed. Moreover, gene overexpression and VIGS were employed to validate the mechanism of non-coding RNA and its target genes during tomato resistance to *M. incognita*, aiming to furnish a theoretical foundation for the biological control of root-knot nematodes. The key findings are summarized as follows.

In recent years, it has been revealed that lncRNA, as a novel regulator, can influence miRNA expression via endogenous mimetic target eTM sites within gene sequences [17]. The eTMs containing miR166 and miR399 in tomato slylnc0195 and slylnc1077 affect tomato resistance to etiolated leaf virus by suppressing the expression of miRNA [21]. lncRNA23468, lncRNA13262, and lncRNA01308 possess miR482b eTM. The overexpression of lncRNA23468 in tomato diminishes the expression level of miR482b and augments resistance to late blight. Silencing lncRNA23468 elevates the expression level of miR482b and plant susceptibility to late blight [24].

In this study, lncRNA47258 was predicted to be an eTM for miR319b to adsorb miRNA. The silenced tomato plants exhibited the up-regulated expression of miR319b, a decreased jasmonic acid level, an inhibited expression of the corresponding three TCP target genes, and reduced resistance to *M. incognita*. In contrast, the overexpression of lncRNA47258 increased the jasmonic acid level, up-regulated TCP target genes, and decreased nematode infection and root-knot number. Our results suggest that Sneb821 regulates tomato jasmonic acid synthesis through the lncRNA47258/miR319b/TCP regulatory module, thereby influencing the infection and development of *M. incognita* in roots.

Apart from inhibiting miRNA expression, lncRNAs also impact the expression of related genes. For instance, lncRNA16397 promotes the expression of the glutaredoxin (GRX) gene, thereby modifying tomato resistance to late blight [25]. LncRNA33732 also enhances tomato resistance by inducing the expression of RBOH [10]. Additionally, rice lncRNA ALEX1 plays a regulatory role in plant defense pathways, and its overexpression plants can activate PR genes in the jasmonic acid pathway, thereby strengthening plant resistance to Xanthomonas oryzae [26]. Previous studies have indicated that TCP genes positively regulate jasmonic acid levels in plants and the jasmonic acid pathway plays a vital role in plant resistance to root-knot nematode infection [27]. In this study, lncRNA47258 induced the expression of the TCP gene. In tomato plants with silenced lncRNA47258, the accumulation of Solyc12g0141401.1, Solyc07g062681.1, and Solyc05g012844.1 decreased, and the level of jasmonic acid declined. The overexpression of lncRNA47258 and the target gene TCP verified that lncRNA47258 could regulate the jasmonic acid signaling pathway, increase the content of jasmonic acid in tomato roots, enhance the resistance of tomato to *M. incognita*, retard the development of *M. incognita*, and reduce the formation of root-knot number. The lncRNA47258/miR319b/TCP regulatory module participates in the process of tomato resistance to root-knot nematode induced by biocontrol bacterium Sneb821, and also discloses a novel regulatory mechanism of lncRNAs involved in root-knot nematode biological control.

5. Conclusions

The plant's responses were previously considered as separated versus pathogens, but nematodes are often considered in between [28]. This study's findings are summarized as follows: Sneb821 was demonstrated to interfere with the expression of non-coding RNA genes in tomato plants. It up-regulated the expression of lncRNA while inhibiting the expression of miRNA. Consequently, this regulation led to the up-regulation of the corresponding target genes and an increase in jasmonic acid content, enabling tomato plants to resist the infection of *M. incognita* and suppress the formation of root knots.

The differentially expressed miRNAs within the transcriptome were screened. In combination with previous reports indicating that miRNAs can respond to nematode infection, it was identified that multiple molecular regulatory modules, such as miR156/SPLs, miR319/TCP, miR390/ARFs, miR482/NBS-LRR, and miR396/GRFs, might be implicated in the induction and resistance processes. Thus, considering the in vivo ceRNA mechanism, lncRNAs capable of acting on miR156, miR482, and miR319 were selected for further in-depth investigation. LncRNA47258 was found to affect miR319b to modulate the content of tomato jasmonic acid, thereby inhibiting the impact of *M. incognita* (Figure 7). The results of this study hold significant importance for exploring the biological control mechanisms of plant defense, uncovering novel biomarkers for controlling plant nematode diseases, and devising new strategies for nematode disease control.

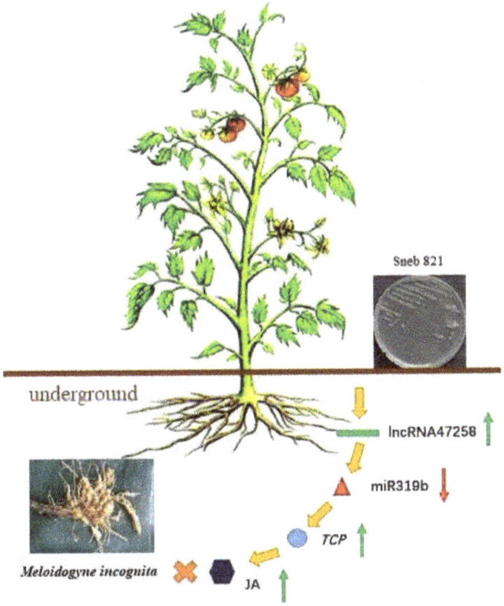

Figure 7. Model of resistance induced by Sneb821 to *M. incognita* in tomato. Note: Arrows indicate positive regulation, and blunt-ended bars indicate inhibition; the red arrows represent down-regulated expression and the green arrows represent up-regulated expression.

Supplementary Materials: The following supporting information can be downloaded at: https://www.mdpi.com/article/10.3390/pathogens14030256/s1, Figure S1: RT-PCR validation of tomato lncRNA. Figure S2: Construction of overexpressed tomato plants of pCAMBIA1302-lncRNA47258. Figure S3: Digestion verification of pMD20-T-lncRNA47258 and TRV2-lncRNA47258. Figure S4: Construction of overexpressed tomato plants of pCAMBIA1302-*Solyc07g062680.1*. Table S1: Primers used in this study.

Author Contributions: Conceptualization, M.Q. and L.C.; methodology, F.Y.; software, F.Y.; validation, F.Y., M.Q. and L.C.; formal analysis, X.W.; investigation, F.Y.; resources, F.Y.; data curation, F.Y.; writing—original draft preparation, F.Y.; writing—review and editing, F.Y.; visualization, M.Q.; supervision, M.Q.; project administration, M.Q.; funding acquisition, L.C. All authors have read and agreed to the published version of the manuscript.

Funding: This work was supported by grants from the National Natural Science Foundation of China (32372481), the National Key Research and Development Program of China (2023YFD1400400), the National Parasitic Resources Center NPRC-2019-194-30, the Project of Shenyang Agricultural University (XJ2023000201).

Institutional Review Board Statement: Not applicable.

Informed Consent Statement: Not applicable.

Data Availability Statement: The original contributions presented in the study are included in the article/Supplementary Material, further inquiries can be directed to the corresponding author.

Conflicts of Interest: The authors declare no conflicts of interest.

References

1. Catani, L.; Manachini, B.; Grassi, E.; Guidi, L.; Semprucci, F. Essential oils as nematicides in plant protection. *Plants* **2023**, *12*, 1418. [CrossRef] [PubMed]
2. Fujimoto, T.; Tomitaka, Y.; Abe, H.; Tsuda, S.; Futai, K.; Mizukubo, T. Expression profile of jasmonic acid-induced genes and the induced resistance against the root-knot nematode in tomato plants after foliar treatment with methyl jasmonate. *J. Plant Physiol.* **2011**, *168*, 1084–1097. [CrossRef] [PubMed]
3. Wierzbicki, A.T. The role of long non-coding RNA in transcriptional gene silencing. *Curr. Opin. Plant Biol.* **2012**, *15*, 517–522. [CrossRef] [PubMed]
4. Zhang, H.; Hu, W.; Hao, J.; Lv, S.; Wang, C.; Tong, W.; Wang, Y.; Wang, Y.; Liu, X.; Ji, W. Genome-wide identification and functional prediction of novel and fungi-responsive lincRNAs in Triticum aestivum. *BMC Genom.* **2016**, *17*, 238–250. [CrossRef]
5. Zhang, L.; Wang, M.; Li, N.; Wang, H.; Qiu, P.; Pei, L.; Xu, Z.; Wang, T.; Gao, E.; Liu, J.; et al. Long noncoding RNAs involve in resistance to *Verticillium dahliae*, a fungal disease in cotton. *Plant Biotechnol. J.* **2018**, *16*, 1172–1185. [CrossRef]
6. Qin, T.; Zhao, H.; Cui, P. A nucleus-localized long non-coding RNA enhances drought and salt stress tolerance. *Plant Physiol.* **2017**, *175*, 1321–1330. [CrossRef]
7. Wu, J.G.; Yang, R.X.; Yang, Z.R. ROS accumulation and antiviral defence control by microRNA528 in rice. *Nat. Plants* **2017**, *3*, 16203. [CrossRef]
8. Li, W.; Chen, Y.; Wang, Y.; Zhao, J.; Wang, Y. Gypsy retrotransposon-derived maize 1 lncRNA GARR2 modulates gibberellin response. *Plant J.* **2022**, *110*, 1433–1446. [CrossRef]
9. Jiang, C.H.; Fan, Z.H.; Li, Z.J.; Niu, D.D.; Li, Y.; Zheng, M.Z.; Wang, Q.; Jin, H.L.; Guo, J.H. Bacillus cereus AR156 triggers induced systemic resistance against *Pseudomonas syringae* pv. tomato DC3000 by suppressing miR472 and activating CNLs-mediated basal immunity in Arabidopsis. *Mol. Plant Pathol.* **2020**, *496*, 1–17. [CrossRef]
10. Cui, J.; Jiang, N.; Meng, J.; Yang, G.; Liu, W.; Zhou, X.; Ma, N.; Hou, X.; Luan, Y. LncRNA33732-respiratory burst oxidase module associated with WRKY1 in tomato-Phytophthora infestans interactions. *Plant J.* **2019**, *97*, 933–946. [CrossRef]
11. Peng, T.; Liu, X.; Tian, F.; Xu, H.; Yang, F.; Chen, X.; Gao, X.; Lv, Y.; Li, J.; Pana, Y.; et al. Functional investigation of lncRNAs and target cytochrome P450 genes related to spirotetramat resistance in *Aphis gossypii* Glover. *Pest Manag. Sci.* **2022**, *78*, 1982–1991. [CrossRef] [PubMed]
12. Yu, J.; Qiu, K.; Sun, W.; Yang, T.; Wu, T.; Song, T.; Zhang, J.; Yao, Y.; Tian, J. A long non-coding RNA functions in high-light-induced anthocyanin accumulation in apple by activating ethylene synthesis. *Plant Physiol.* **2022**, *189*, 66–83. [CrossRef] [PubMed]
13. Zhang, L.; Liu, J.; Cheng, J.; Sun, Q.; Zhang, Y.; Liu, J.; Li, H.; Zhang, Z.; Wang, P.; Cai, C.; et al. lncRNA7 and lncRNA2 modulate cell wall defense genes to regulate cotton resistance to Verticillium wilt. *Plant Physiol.* **2022**, *189*, 264–284. [CrossRef] [PubMed]
14. Zhang, X.; Shen, J.; Xu, Q.; Dong, J.; Song, L.; Wang, W.; Shen, F. Long noncoding RNA lncRNA354 functions as a competing endogenous RNA of miR160b to regulate ARF genes in response to salt stress in upland cotton. *Plant Cell Environ.* **2021**, *44*, 3302–3321. [CrossRef]
15. Zheng, Y.; Wang, Y.; Ding, B.; Fei, Z. Comprehensive transcriptome analyses reveal that potato spindle tuber viroid triggers genome-wide changes in alternative splicing, inducible trans-acting activity of phased secondary small interfering RNAs and immune responses. *J. Virol.* **2017**, *91*, 217–247. [CrossRef]
16. Wang, J.; Yang, Y.; Jin, L.; Ling, X.; Liu, T.; Chen, T.; Ji, Y.; Yu, W.; Zhang, B. Re-analysis of long non-coding RNAs and prediction of circRNAs reveal their novel roles in susceptible tomato following TYLCV infection. *BMC Plant Biol.* **2018**, *18*, 104–106. [CrossRef]
17. Yang, Z.; Yang, C.; Wang, Z.; Yang, Z.; Chen, D.; Wu, Y.; Wu, Y. LncRNA expression profile and ceRNA analysis in tomato during flowering. *PLoS ONE* **2019**, *14*, e0210650. [CrossRef]
18. Yang, F.; Zhao, D.; Fan, H.Y.; Zhu, X.F.; Wang, Y.Y.; Liu, X.Y.; Duan, Y.X.; Xuan, Y.H.; Chen, L.J. Functional analysis of long non-coding RNAs reveal their novel roles in biocontrol of bacteria-induced tomato resistance to *Meloidogyne incognita*. *Int. J. Mol. Sci.* **2020**, *21*, 911. [CrossRef]
19. Yang, F.; Ding, L.; Zhao, D.; Fan, H.Y.; Zhu, X.F.; Wang, Y.Y.; Liu, X.Y.; Duan, Y.X.; Xuan, Y.H.; Chen, L.J. Identification and functional analysis of tomato microRNAs in the biocontrol bacterium Pseudomonas putida induced plant resistance to *Meloidogyne incognita*. *Phytopathology* **2022**, *112*, 2372–2382. [CrossRef]

20. Zhao, W.C.; Li, Z.L.; Fan, J.W.; Hu, C.L.; Yang, R.; Qi, X.; Chen, H.; Zhao, F.K.; Wang, S.H. Identification of jasmonic acid-associated microRNAs and characterization of the regulatory roles of the miR319/TCP4 module under root-knot nematode stress in tomato. *J. Exp. Bot.* **2015**, *66*, 4653–4667. [CrossRef]
21. Wang, S.T.; Sun, X.L.; Hoshino, Y. MiR319 positively regulates cold tolerance by targeting OsPCF6 and OsTCP21 in rice. *PLoS ONE* **2014**, *9*, e91357.
22. Wang, J.; Yu, W.; Yang, Y. Genome-wide analysis of tomato long non-coding RNAs and identification as endogenous target mimic for microRNA in response to TYLCV infection. *Sci. Rep.* **2015**, *5*, 16946. [CrossRef] [PubMed]
23. Jiang, N.; Meng, J.; Cui, J.; Sun, G.; Luan, Y. Function identification of miR482b, a negative regulator during tomato resistance to Phytophthora infestans. *Hortic. Res.* **2018**, *5*, 9–20. [CrossRef] [PubMed]
24. Jiang, N.; Cui, J.; Shi, Y.S.; Yang, G.L.; Zhou, X.X.; Hou, X.X.; Meng, J.; Luan, Y.S. Tomato lncRNA23468 functions as a competing endogenous RNA to modulate NBS-LRR genes by decoying miR482b in the tomato-Phytophthora infestans interaction. *Hortic. Res.* **2019**, *6*, 28–40. [CrossRef]
25. Cui, J.; Luan, Y.; Jiang, N.; Bao, H.; Meng, J. Comparative transcriptome analysis between resistant and susceptible tomato allows the identification of lncRNA16397 conferring resistance to Phytophthora infestans by co-expressing glutaredoxin. *Plant J.* **2017**, *89*, 577–589. [CrossRef]
26. Yu, Y.; Zhou, Y.F.; Feng, Y.Z.; He, H.; Lian, J.P.; Yang, Y.W.; Lei, M.Q.; Zhang, Y.C.; Chen, Y.Q. Transcriptional landscape of pathogen-responsive lncRNAs in rice unveils the role of ALEX1 in jasmonate pathway and disease resistance. *Plant Biotechnol. J.* **2020**, *18*, 679–690. [CrossRef]
27. Fan, C.L.; Hu, L.N.; Zhang, Z. Jasmonic acid mediates tomato's response to root knot nematodes. *J. Plant Growth Regul.* **2015**, *34*, 196–205. [CrossRef]
28. Gupta, R.; Mfarrej MF, B.; Xhemali, B.; Khan, A.; Nadeem, H.; Ahmad, F. Metabolic responses of plants to *Meloidogyne* species parasitism: A review on molecular events and functions. *J. King Saud Univ.-Sci.* **2023**, *36*, 103083. [CrossRef]

MDPI AG
Grosspeteranlage 5
4052 Basel
Switzerland
Tel.: +41 61 683 77 34

Pathogens Editorial Office
E-mail: pathogens@mdpi.com
www.mdpi.com/journal/pathogens

www.ingramcontent.com/pod-product-compliance
Lightning Source LLC
Chambersburg PA
CBHW041949260926
51850CB00019B/2193
* 9 7 8 3 7 2 5 8 5 5 8 9 6 *